Transmission Lines and Networks

McGRAW-HILL ELECTRICAL AND ELECTRONIC ENGINEERING SERIES

Frederick Emmons Terman, *Consulting Editor*
W. W. Harman and J. G. Truxal,
Associate Consulting Editors

PRINCETON UNIVERSITY SERIES

TRANSMISSION LINES
AND NETWORKS

WALTER C. JOHNSON

CHAIRMAN, DEPARTMENT OF ELECTRICAL ENGINEERING
PRINCETON UNIVERSITY

NEW YORK   TORONTO   LONDON

McGRAW-HILL BOOK COMPANY, INC.

1950

TRANSMISSION LINES AND NETWORKS

15 16 17 18 19 20 - MP - 9 8 7 6

32580

PREFACE

THE PURPOSE of this textbook is to present the basic principles of transmission lines and the elementary analysis of passive four-terminal networks
in a form that is appropriate for both power and communication engineers.
The essential prerequisites for the subjects discussed in this book are integral calculus and elementary a-c circuit theory, some acquaintance with
differential equations being helpful but not strictly essential. Additional
mathematical developments are presented where necessary. The material
has been developed over a period of years by using it in the form of mimeo-
graphed class notes, which have been revised several times in the light of
classroom experience.

The basic phenomenon that distinguishes the transmission line from
the lumped circuits is the traveling wave. A clear physical picture of
traveling waves and of their reflections and interference effects is a great
aid in stimulating the interest of the student, and does much to prevent
the subject from becoming a mere succession of impressive but perhaps
unappreciated formulas. Consequently, the book opens with a discussion
of distributed constants and traveling waves, and the physical viewpoint
introduced here is maintained throughout the treatment. The first four
chapters are general in scope and are applicable to power and communications alike. The modern transmission-line chart and its applications to
both lossy and lossless lines are discussed in the fifth chapter. Following
this are several chapters devoted to the special problems of transmission
lines in three frequency ranges: radio-frequency lines, telephone and telegraph lines, and power-transmission lines. From here, the concepts and
notation of transmission-line theory lead naturally into a consideration of
passive four-terminal networks. Applications of the theory to attenuators, impedance-matching networks, and filters are given in some detail.

The more difficult sections, such as those on skin effect and on Foster's
reactance theorem, are preceded by qualitative discussions of these subjects. This allows the more mathematical aspects of these topics to be
omitted when desired, and also permits the student to obtain a physical
understanding of the phenomena before attempting the mathematical
treatment.

No attempt is made in this text to go into the theory of the hollow wave
guide. A highly satisfactory theory of transmission lines can be developed
through conventional circuit theory. On the other hand, only a super-
ficial understanding of hollow wave guides can be obtained on this basis,

for here electromagnetic field theory is the essential basis for analysis. Thus, although there are many basic similarities between the hollow wave guide and the multiple-conductor transmission line operating in its principal mode, the differences between them are just as important as their similarities. The electrical-engineering student is ready for conventional transmission-line theory as soon as he has studied the theory of lumped a-c circuits; hollow wave guides, with their different basis of analysis, have therefore been omitted from this book.

    I am particularly indebted to Professor C. H. Willis, for his encouragement and his many helpful suggestions during the preparation of this material. I am also grateful to Professor H. M. Chandler, Jr. and to Professor W. H. Surber, Jr., for their many valuable criticisms and suggestions. Finally, I wish to express my appreciation of Princeton University's liberal policy in encouraging the publication of this book, and in particular I wish to thank Professor Willis and Dean K. H. Condit for their part in providing the time needed for the preparation of the manuscript.

WALTER C. JOHNSON

PRINCETON, N.J.
June, 1950

CONTENTS

vii

PART II—FOUR-TERMINAL NETWORKS

Part I

TRANSMISSION LINES

CHAPTER 1

DISTRIBUTED CONSTANTS AND TRAVELING WAVES

1.1. Introduction. When electric charges are set into motion, the magnetic field caused by the current of moving charges and the electric field caused by the presence of charges are not established throughout space instantaneously but travel outward at a finite velocity. In air this velocity is approximately that of light in free space, or about 3×10^8 meters/sec.

Imagine two parallel conductors connecting a generator to a load. The voltage impressed by the generator on its end of the line does not reach the load at the same instant but travels down the line at a finite velocity and reaches the load somewhat later. This can be explained by the action of the electric and magnetic fields which are guided by the conductors from the generator to the load, but it is most easily analyzed in terms of the distributed inductance and capacitance of the wires. The velocity of propagation depends upon the medium surrounding the wires, in which the electric and magnetic fields exist. For air-insulated lines the velocity is nearly equal to that of light in free space. It is somewhat lower for solid dielectrics.

When the generator voltage varies sinusoidally with time, the distance that a wave travels in one cycle is equal to the wavelength λ

$$\lambda = \text{velocity} \times \text{period}$$

or

$$\lambda = \frac{v}{f} \tag{1.1}$$

where f is the frequency of the driving source.

Assuming $v = 3 \times 10^8$ meters/sec, we can compute the wavelength for various frequencies: For 60 cycles per second, $\lambda = 3,100$ miles; for 3 megacycles per second, $\lambda = 100$ meters; and for 3,000 megacycles per second, $\lambda = 10$ centimeters.

The time lag between the sending and receiving ends of a transmission line is important whenever the line is so long or the frequency so high that it takes an appreciable portion of a cycle for the wave to travel the full length of the line. This is expressed more conveniently in terms of wavelength: transmission-line theory must be used whenever the length of the line is appreciable compared with a quarter wavelength. When the wires are much shorter than a quarter wavelength, the time lag will be only a small part of a cycle and the system can be analyzed by the more usual a-c circuit theory ("small-circuit" theory).

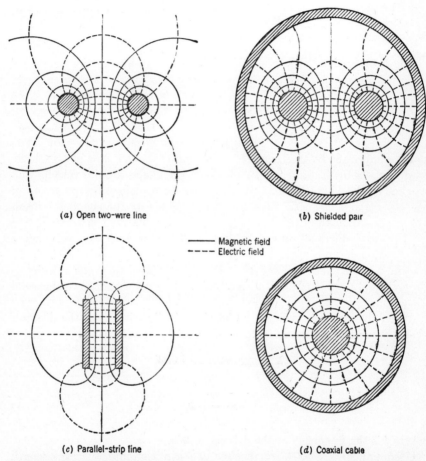

(a) Open two-wire line

(b) Shielded pair

—— Magnetic field
---- Electric field

(c) Parallel-strip line

(d) Coaxial cable

FIG. 1.1. Transverse views of some common types of transmission lines, showing the arrangement of conductors and the configuration of the electric and magnetic fields.

The primary use of transmission lines is to transmit a-c power between points which are separated by distances that are not small compared with a quarter wavelength. At short wavelengths they also find important use as reactive circuit elements, as resonant circuits, as impedance transformers, and in many other ways.

Figure 1.1 shows the arrangement of conductors and the configuration of the electric and magnetic fields for several common types of transmission lines. The open two-wire line is easy to construct, and its characteristics are readily adjusted by changing the spacing of the wires. However, the fields extend far beyond the line, and radiation losses become excessive at the higher radio frequencies. Open-wire lines are not often used at frequencies above a few hundred megacycles. A conducting shield is sometimes placed around the two wires to contain the fields, as shown in Fig. 1.1b. The parallel-strip line shown in Fig. 1.1c is occasionally used to provide a low impedance level.

A coaxial line consists of a hollow tube and a concentric conductor, as shown in Fig. 1.1d. The center conductor may be held in place with dielectric beads or with a continuous solid dielectric which fills the annular space. When a continuous solid insulator is used, the coaxial cable can be made flexible by constructing the outer conductor of a braid of fine wires. A coaxial cable is self-shielded and has no external field except possibly near the terminations. For this reason it is widely used throughout the radio frequency range, and is used effectively at wavelengths as short as 10 cm (3,000 Mc). This is well within the microwave region, which is the name given to the radio spectrum at wavelengths below perhaps a half meter.

Open-wire lines are normally balanced with respect to ground, but a coaxial cable is unsymmetrical and is not balanced with respect to ground.

When the transmission-line problem is analyzed by means of electromagnetic field theory, it is found that the type of transmission studied here is not the only one that can exist on a set of parallel conductors. Our analysis will apply to the so-called *principal mode* in which the electric and magnetic fields are perpendicular to each other and to the direction of the conductors, as shown in Fig. 1.1. This type of traveling wave is often called the transverse electromagnetic, or TEM, wave, and is the only kind that can exist on a transmission line at the lower frequencies. When the frequency becomes so high that the wavelength is comparable with the distance between conductors, other types of waves, of the kind utilized in hollow wave guides, become possible.[1] Except in very special cases, these "higher modes" are considered undesirable on the transmission systems that we are studying; therefore, whenever possible, the spacing between conductors is kept much smaller than a quarter wavelength. Another reason for a small spacing is that, when the distance between the wires of an unshielded line approaches a quarter wavelength, the line acts as an antenna and radiates a considerable portion of the energy that it

[1] For an analysis of higher modes in coaxial cables, see A. B. Bronwell and R. E. Beam, "Theory and Application of Microwaves," Sec. 16.08, McGraw-Hill Book Company, Inc., New York, 1947; and S. Ramo and J. R. Whinnery, "Fields and Waves in Modern Radio," Sec. 9.02, John Wiley & Sons, Inc., New York, 1944.

carries. In our analysis we shall assume a very small spacing and shall neglect radiation losses altogether.

1.2. The Distributed Constants of the Line. Transmission lines are most easily analyzed by an extension of lumped-constant theory. The same theory will apply to all the lines shown in Fig. 1.1.

The most important constants of the line are its distributed inductance and capacitance. When a current flows in the conductors of a transmission line, a magnetic flux is set up around the conductors. Any change in this flux will induce a voltage (the familiar $L\,di/dt$ of lumped-circuit theory). The inductance of the transmission-line conductors is smoothly distributed throughout their length. The distributed inductance, representing the net effect of all the line conductors, will be given the symbol L and will be expressed in henrys per unit length.

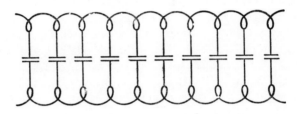

Fɪɢ. 1.2. Schematic representation of the distributed inductance and capacitance of a transmission line.

Between the conductors of the line there exists a uniformly distributed capacitance C. This will be measured in farads per unit length of line. The distributed inductance and capacitance are illustrated in a rough schematic way in Fig. 1.2. When the line is viewed in this way, it is not hard to see that the voltage and current can vary from point to point on the line, and that resonance may exist under certain conditions.

In addition to inductance and capacitance, the conductors also have a resistance R ohms per unit length. This includes the effect of all the conductors. Finally, the insulation of the line may allow some current to leak from one conductor to the other. This is denoted by a conductance G, measured in mhos per unit length of line. The quantity R obviously represents the imperfection of the conductor, while G represents the imperfection of the insulator. The student should not lose sight of the fact that, in the notation of transmission-line theory, G does not denote the reciprocal of R. When solid insulation is used at very high frequencies, the dielectric loss may be considerable. This has the same effect on the line as true ohmic leakage and forms the major contribution to G at these frequencies.

Although the line constants are uniformly distributed along the line, we can gain a rough idea of their effect by imagining the line to be made up of short sections of length Δx, as shown in Fig. 1.3. If L is the inductance per unit length, the inductance of the short section will be $L \cdot \Delta x$ henrys. Similarly, the resistance of the section will be $R \cdot \Delta x$ ohms, the capacitance will be $C \cdot \Delta x$ farads, and the leakage conductance will be $G \cdot \Delta x$ mhos.

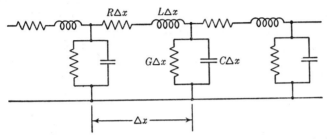

FIG. 1.3.   Approximate representation of a short section of transmission line.

Although the inductance and resistance are shown lumped in one conductor in Fig. 1.3, they actually represent the net effect of both conductors in the short section Δx. As the section lengths Δx are made smaller and smaller, the "lumpy" line of Fig. 1.3 will approach nearer and nearer to the actual smooth line.

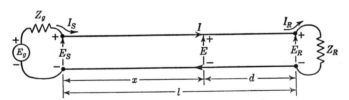

FIG. 1.4.   Schematic diagram of a transmission line.

1.3. Notation and Units.   We shall visualize the basic transmission-line problem in the manner shown by the schematic diagram of Fig. 1.4. The subscripts S and R refer, respectively, to the sending and receiving ends of the line. The line is terminated by a receiving-end impedance (load impedance) of Z_R complex ohms and is fed by a generator which has an open-circuit voltage E_g and an internal impedance Z_g. The unit of length generally used in telephone-line problems is the mile, while for radio-frequency lines the meter is preferred. The line constants L, C, R, and G are expressed in terms of the chosen unit of length; e.g., L is expressed in henrys per mile or in henrys per meter. We shall denote in-

stantaneous voltage and current by e and i, respectively, and shall use
capital letters (E and I) to denote complex a-c quantities.

The conventions of sign are shown in Fig. 1.4. An instantaneous potential difference between lines will be taken to be a positive number of
volts if the upper wire in Fig. 1.4 is positive with respect to the lower one,
and will be taken to be a negative number of volts if the polarity is the
reverse of this. An instantaneous current which flows to the right in the
top wire (and to the left in the bottom one) will be considered positive,
with the reverse directions as negative. A combination either of positive
voltage and positive current, or of negative voltage and negative current,
will correspond to a flow of power to the right.

1.4. The Differential Equations for the Uniform Line. Consider an
infinitesimal section of a line as shown in Fig. 1.5, and consider the instan-

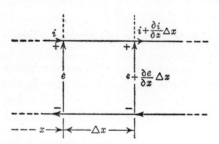

FIG. 1.5. Infinitesimal section of transmission line.

taneous voltage e and the instantaneous current i. The series inductance of the section will be
$L \cdot \Delta x$ henrys, and the series resistance will be $R \cdot \Delta x$ ohms. Similarly, the shunt capacitance will be
$C \cdot \Delta x$ farads and the shunt conductance will be $G \cdot \Delta x$ mhos. Following the conventions of calculus, the
difference between the instantaneous line-to-line voltages at the two
ends of the section will be $(\partial e / \partial x) \Delta x$,
as indicated in Fig. 1.5 (the partial derivative must be used because
there are two independent variables, distance x and time t). The vol-
tage difference $(\partial e / \partial x) \Delta x$ is caused by the current i flowing through the
resistance $R \cdot \Delta x$ and changing at the rate $\partial i / \partial t$ in the inductance $L \cdot \Delta x$.
Thus, we can write

$$-\frac{\partial e}{\partial x} \Delta x = (R \cdot \Delta x)i + (L \cdot \Delta x) \frac{\partial i}{\partial t}$$

The negative sign is used here because positive values of i and of $\partial i / \partial t$
cause e to decrease with increasing x. Upon dividing through by Δx, we
have

$$-\frac{\partial e}{\partial x} = Ri + L \frac{\partial i}{\partial t} \tag{1.2}$$

This is the differential equation that indicates the manner in which the
instantaneous line-to-line voltage e changes along the line.

In a similar manner, the difference in current between the two ends of the section, $(\partial i/\partial x)\Delta x$, will be made up of two parts: (1) the current caused by the voltage e acting on the shunt conductance $G \cdot \Delta x$ and (2) the displacement current through the capacitance $C \cdot \Delta x$ caused by the voltage changing at the rate $\partial e/\partial t$. We can, therefore, write

$$-\frac{\partial i}{\partial x}\, \Delta x = (G \cdot \Delta x)e + (C \cdot \Delta x)\,\frac{\partial e}{\partial t}$$

Dividing by Δx, we obtain a differential equation that indicates the manner in which the current i changes along the line:

$$-\frac{\partial i}{\partial x} = Ge + C\frac{\partial e}{\partial t} \tag{1.3}$$

We now have two partial differential equations with two dependent variables, e and i, and two independent variables, x and t. These equations, together with the boundary conditions relating to the two ends, will in principle yield both the steady-state and the transient solutions. We shall concentrate mainly on the steady-state a-c problem and examine transients only in certain simplified cases.

1.5. Traveling Waves on a Lossless Line. It is illuminating to consider the hypothetical case of a line without loss, for which $R = G = 0$. This approximation is reasonably good when the line losses are much smaller than the energy which travels along the line. Physically short radio-frequency lines can often be solved satisfactorily by this method. Also, the approximation affords a simple and useful, although rather over-idealized, method for calculating the propagation of surges such as those caused by lightning strokes on power lines. The operation of delay lines and of pulse-forming lines is most easily understood by using the lossless theory.

For the lossless condition, Eqs. (1.2) and (1.3) become

$$-\frac{\partial e}{\partial x} = L\frac{\partial i}{\partial t} \tag{1.4}$$

and

$$-\frac{\partial i}{\partial x} = C\frac{\partial e}{\partial t} \tag{1.5}$$

We can eliminate i between the two equations by taking the partial derivative of Eq. (1.4) with respect to x and of Eq. (1.5) with respect to t. The order of differentiation is immaterial, and so $\partial^2 i/\partial x \partial t = \partial^2 i/\partial t \partial x$. Eliminating this quantity between the two equations, we obtain a differential equation for e:

$$\frac{1}{LC}\frac{\partial^2 e}{\partial x^2} = \frac{\partial^2 e}{\partial t^2} \tag{1.6}$$

If e, rather than i, is eliminated between Eqs. (1.4) and (1.5), a relation similar to Eq. (1.6) is obtained for the current:

$$\frac{1}{LC}\frac{\partial^2 i}{\partial x^2} = \frac{\partial^2 i}{\partial t^2} \tag{1.7}$$

Equations (1.6) and (1.7) are one-dimensional forms of the wave equation, the solutions of which are known to consist of waves that can travel in either direction, without change in form or magnitude, at the velocity $1/\sqrt{LC}$. To show this, we shall formulate a mathematical expression for such a traveling wave and then demonstrate that this expression satisfies the differential equation (1.6).

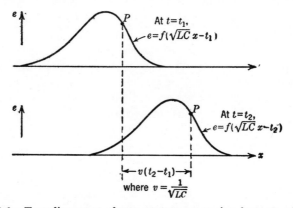

FIG. 1.6. Traveling wave shown at two successive instants of time.

First, we shall show that a wave traveling in the positive x direction with a velocity $1/\sqrt{LC}$ can be expressed mathematically as

$$e = f(\sqrt{LC}x - t) \tag{1.8}$$

where f represents any single-valued function of the argument $\sqrt{LC}x - t$.

The following are specific illustrations of such functions: $\epsilon^{a(\sqrt{LC}x-t)}$, $\sin \omega(\sqrt{LC}x - t)$, and $K(\sqrt{LC}x - t)^2$, where a, ω, and K are constants. Another example, which is intended to be more general and which might be difficult to express analytically, is illustrated in Fig. 1.6. Suppose that an observer travels with the wave shown in Fig. 1.6 in such a way that he stays with a particular point on the wave; for example, the one marked P in the figure. So far as the observer is concerned, the function $f(\sqrt{LC}x - t)$ remains constant in value, which means that he must be moving so that the argument $\sqrt{LC}x - t$ is constant for him. For the point P, then,

$$\sqrt{LC}x - t = \text{a constant}$$

Taking the derivative term by term with respect to time, we obtain an equation containing the velocity dx/dt:

$$\sqrt{LC}\,\frac{dx}{dt} - 1 = 0$$

from which the velocity is found to be

$$\frac{dx}{dt} = \frac{1}{\sqrt{LC}} \qquad \text{length units per second} \qquad (1.9)$$

Similarly, a wave traveling in the negative x direction can be expressed as

$$e = f_2(\sqrt{LC}\,x + t) \qquad (1.10)$$

where f_2 is another single-valued function.

Example 1.   As an example, consider the function $e = K(\sqrt{LC}\,x - t)^2$, where K is a constant. At $t = 0$, this is a parabolic function of x: $e = K(\sqrt{LC}\,x)^2$. At a given value of time t_1 sec later, the function is $e = K(\sqrt{LC}\,x - t_1)^2$, which is again a parabolic function of x but shifted $t_1/\sqrt{LC}$ distance units to the right along the x axis.

Next, we shall show that the traveling wave (1.8) satisfies the differential equation for the lossless line.   To simplify the notation, we write the argument of the function as

$$s = \sqrt{LC}\,x - t \qquad (1.11)$$

and then write the supposed solution as

$$e = f(s) \qquad (1.12)$$

Now we take derivatives of e for substitution into the differential equation. From calculus we can write

$$\frac{\partial e}{\partial x} = \frac{df}{ds}\frac{\partial s}{\partial x}$$

But from Eq. (1.11) we find that $\partial s/\partial x = \sqrt{LC}$, and so

$$\frac{\partial e}{\partial x} = \sqrt{LC}\,\frac{df}{ds}$$

Taking the second derivative with respect to x, we obtain

$$\frac{\partial^2 e}{\partial x^2} = LC\frac{d^2 f}{ds^2} \qquad (1.13)$$

In a similar fashion we can show that

$$\frac{\partial^2 e}{\partial t^2} = \frac{d^2 f}{ds^2} \qquad (1.14)$$

If we now test the solution by substituting Eqs. (1.13) and (1.14) into the differential equation (1.6), we obtain

$$\frac{1}{LC}\left(LC\frac{d^2f}{ds^2}\right) = \frac{d^2f}{ds^2}$$

This is an identity, which verifies that the assumed solution (1.8) satisfies the differential equation.

Example 2. Consider the function $e = K(\sqrt{LC}x - t)^2$, where K is a constant. Taking the partial derivatives with respect to x and t, we obtain

$$\frac{\partial e}{\partial x} = 2\sqrt{LC}K(\sqrt{LC}x - t), \qquad \frac{\partial^2 e}{\partial x^2} = 2LCK$$

$$\frac{\partial e}{\partial t} = -2K(\sqrt{LC}x - t), \qquad \frac{\partial^2 e}{\partial t^2} = 2K$$

Now, we test whether the function is a solution of the differential equation by substituting the foregoing second derivatives into $\partial^2 e/\partial x^2 = LC\partial^2 e/\partial t^2$. The result is the identity $2LCK = 2LCK$, which proves that the function is a solution.

It may at first seem strange that the quantity $1/\sqrt{LC}$ is a velocity, when L is expressed in henrys per unit length and C in farads per unit length. The product *henrys × farads* has the dimensions of seconds squared, as can be seen by recalling the expression for the resonant angular velocity of a simple series circuit:

$$\omega = \frac{1}{\sqrt{\text{henrys} \times \text{farads}}} = \text{dimensionless radians/second}$$

Thus, for the transmission line we have

$$\frac{1}{\sqrt{LC}} = \frac{1}{\sqrt{(\text{henrys/length})(\text{farads/length})}} = \frac{\text{length}}{\text{seconds}}$$

which are the dimensions of velocity. The expression $1/\sqrt{LC}$ gives the velocity in terms of the unit of length used for L and C.

When we calculate in Chap. 3 the inductance and capacitance of a parallel pair of conductors separated by an insulating medium, we shall find that the product LC is independent of the size and separation of the conductors and depends only on the dielectric constant and permeability of the insulating medium. The numerical value of $1/\sqrt{LC}$ for air-insulated conductors is approximately 3×10^8 meters/sec, which checks with experimental determinations of the velocity of light in free space. Solid insulation, with its higher dielectric constant, causes the velocity to be smaller. Also, losses in the line tend to reduce the velocity somewhat.

The differential equation for current is similar to that for voltage, and so we expect its solution to be a corresponding traveling wave.   The solution for i that corresponds to Eq. (1.8) for e is

$$i = \frac{1}{\sqrt{L/C}} f(\sqrt{LC}\, x - t) \tag{1.15}$$

This can be demonstrated by substituting the solutions for e and i into the original differential equations (1.4) and (1.5).   Since the function f represents a voltage, the quantity $\sqrt{L/C}$ must have the dimensions of impedance.   The quantities L and C are characteristics of the line, and the combination $\sqrt{L/C}$ is called the *characteristic impedance* of the lossless line.   We shall denote it by the symbol Z_0.   The characteristic impedance of the lossless line is a real quantity; *i.e.*, it is a resistance and is independent of frequency.   For lines with loss, the characteristic impedance is generally complex and is not independent of frequency.

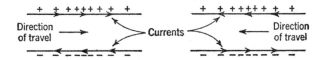

FIG. 1.7.   Voltages and currents caused by traveling waves.

We have now shown that the uniform lossless line can support a wave of voltage which travels at the speed $1/\sqrt{LC}$ in the positive x direction, and that this voltage is accompanied by a similar wave of current, the two being related by the expression $e = \sqrt{L/C}\, i$ at every point.   Since the uniform line looks the same in either direction, we expect that the line can also support a wave traveling in the other direction, as given by

$$e = f_2(\sqrt{LC}\, x + t) \tag{1.16}$$

The truth of this can be verified by substituting Eq. (1.16) into the differential equation (1.6).

The wave of current corresponding to the backward-moving voltage can be shown to be

$$i = -\frac{1}{\sqrt{L/C}} f_2(\sqrt{LC}\, x + t) \tag{1.17}$$

The reason for the negative sign connected with the backward-moving wave of current can be visualized as in Fig. 1.7, which shows a charged region moving along the line (*a*) in the positive direction, and (*b*) in the negative direction.   Using the conventions of sign defined in Fig. 1.4, we see that the voltage is positive in both cases, whereas the current is positive for the wave traveling to the right and negative for the one traveling to the left.   For waves traveling to the right, the voltage and current agree in sign; for waves traveling to the left they are opposite in sign.

The differential equations of the system are linear, and therefore the sum of two separate solutions is also a solution.   More generally, then, we have

$$e = f_1(\sqrt{LC}x - t) + f_2(\sqrt{LC}x + t) \tag{1.18}$$

and

$$i = \frac{1}{Z_0}\left[f_1(\sqrt{LC}x - t) - f_2(\sqrt{LC}x + t)\right] \tag{1.19}$$

where $Z_0 = \sqrt{L/C}$

In a region where two oppositely traveling waves are superimposed, the total voltage and total current no longer have the simple ratio $\sqrt{L/C}$ because of the minus sign in Eq. (1.19).

The relation between the electric and magnetic energies on the line is of interest.   If the instantaneous current at any point is i, the magnetic energy stored in a small length Δx is

$$\mathcal{E}_m = \tfrac{1}{2}(L\,\Delta x)i^2 \tag{1.20}$$

On the other hand, the instantaneous voltage e carries with it an energy stored in the electric field, which, in a length Δx, is

$$\mathcal{E}_e = \tfrac{1}{2}(C\,\Delta x)e^2 \tag{1.21}$$

For a single wave on a lossless line, we have the relation $e = \sqrt{L/C}i$ at every point and at every moment.   Substituting this relation into Eq. (1.21), we obtain

$$\mathcal{E}_e = \frac{1}{2}(C\Delta x)\left(\sqrt{\frac{L}{C}}i\right)^2$$

$$= \tfrac{1}{2}L\,\Delta x\,i^2$$

which is the same as (1.20) for the magnetic energy.   Therefore, on a lossless line carrying a wave in only one direction, the energies associated with the electric and magnetic fields are equal.   These energies travel with the wave of e and i at the speed $1/\sqrt{LC}$.   The energies are not equal in a region where two oppositely traveling waves slide by each other, for, as was mentioned previously, in such a region we no longer have the simple relation $e = \sqrt{L/C}i$.

Example.   Figure 1.8 shows a battery with a steady electromotive force E which is connected at $t = 0$ to one end of an infinitely long line.   After $t = 0$, a rectangular wave of voltage with a magnitude E travels out along the line with a velocity $v = 1/\sqrt{LC}$, accompanied by a current equal to E/Z_0.   At a time t the entire portion of the line between the battery and the point $x = vt$ is charged to the voltage E and carries a direct current E/Z_0, while beyond this point there is no voltage or current whatever.

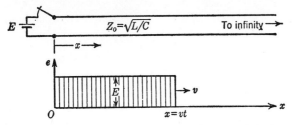

FIG. 1.8.   D-c voltage applied to lossless line of infinite length.

Up to the point $x = vt$, every length Δx of the line carries an energy

$$\mathcal{E} = \mathcal{E}_e + \mathcal{E}_m = \frac{1}{2} C \, \Delta x \, E^2 + \frac{1}{2} L \, \Delta x \left(\frac{E}{Z_0}\right)^2$$

$$= C \, \Delta x \, E^2$$

The total stored energy on the line is, therefore,

$$\mathcal{E}_t = C(vt)E^2$$

and the rate at which this energy is increasing is

$$\frac{\mathcal{E}_t}{t} = CvE^2 = \frac{E^2}{\sqrt{L/C}} = EI$$

This is, of course, equal to the power continuously supplied by the battery.

1.6. Reflections. Figure 1.9 shows one end of a line terminated in an impedance Z_t . Because we do not wish to restrict ourselves to sinusoidal a-c waves, we shall temporarily direct our attention only to pure resistances which, like $\sqrt{L/C}$, are independent of frequency. Imagine an incident wave of voltage, which we shall denote by e^+, traveling to the right, accompanied by a current $i^+ = e^+/Z_0$. At the termination, however, we must have

$$\frac{\text{Total } e}{\text{Total } i} = Z_t \qquad (1.22)$$

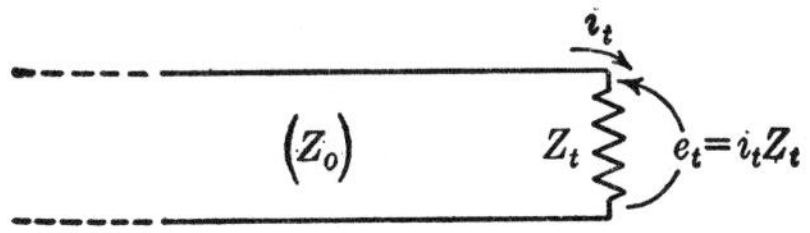

FIG. 1.9.   Transmission line terminated in an impedance Z_t.

Now, unless Z_t is numerically equal to Z_0, this does not satisfy the relation necessary for the line. Part of the incident wave, therefore, will be reflected. Call the reflected voltage and current e^- and i^-, the relation between them being $i^- = -e^-/Z_0$. At the termination, then, Eq. (1.22) can be written as

$$\frac{e_t^+ + e_t^-}{i_t^+ + i_t^-} = Z_t \qquad (1.23)$$

where the subscript t refers to values at the point of termination.

Equation (1.23) can be rewritten in terms of Z_0 as

$$\frac{e_t^+ + e_t^-}{e_t^+/Z_0 - e_t^-/Z_0} = Z_t \qquad (1.24)$$

Solving Eq. (1.24) for the ratio of reflected to incident voltage, we obtain

$$\frac{e_t^-}{e_t^+} = \frac{Z_t - Z_0}{Z_t + Z_0} = k \qquad (1.25)$$

The ratio k is called the reflection coefficient. Observe that k will be zero and there will be no reflection at the termination only when the terminating impedance is equal to the characteristic impedance of the line.

Thus, a terminating impedance different from Z_0 will give rise to a reflected wave which travels away from the termination. The reflection, upon reaching the other end, will itself be reflected if the terminating impedance at that end is different from Z_0.

As an exercise, the student should show that the reflection coefficient for current is the negative of that for voltage.

Example 1. Consider, for example, a d-c generator or a battery with an emf E which is connected at $t = 0$ to one end of two parallel conductors which are terminated at the other end in a resistance R (see Fig. 1.10). Losses in the line will be ignored. For the sake of definiteness, assume that $R = 3Z_0 =$ three times the quantity $\sqrt{L/C}$ of the line.

From $t = 0$ onward, a rectangular wave of voltage with a magnitude E will travel down the line at the velocity $v = 1/\sqrt{LC}$, accompanied by a similar wave of current equal in magnitude to E/Z_0. When the voltage wave reaches the receiving end, it will be reflected with a coefficient which can be obtained from Eq. (1.25):

$$k_R = \frac{3Z_0 - Z_0}{3Z_0 + Z_0} = \frac{1}{2}$$

Therefore, as shown in Fig. 1.10, there will be a reflected wave of voltage with a magnitude $Ek_R = E/2$, accompanied by a current wave equal to $-E/2Z_0$. The first reflected wave will in turn be reflected when it reaches the sending end. The terminating impedance is zero at this end, provided

that the internal resistance of the generator (or battery) is negligible; hence
for the generator end

$$k_g = \frac{-Z_0}{Z_0} = -1$$

The rereflected voltage will therefore be equal to $(E/2)k_g = -E/2$ and,
for this new forward-traveling wave, the accompanying current will be
$-E/2Z_0$. If the successive reflections are followed through, the result

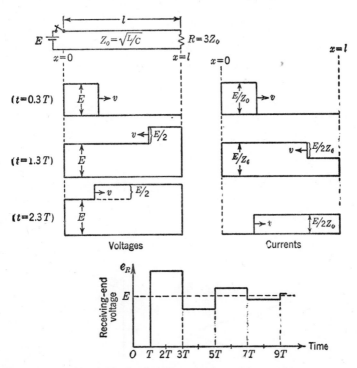

Voltages

Currents

Fig. 1.10.   D-c transients on a lossless line terminated in a resistance equal to $3Z_0$.
The time required for the wave to travel the length of the line is denoted by T, where
$T = l/v$.

shown in Fig. 1.10 will be obtained.   At each moment the ratio of receiving-
end voltage to receiving-end current is equal to the terminal resistance R.
As time goes on, the receiving-end voltage gradually settles down to the
steady-state value E, and the current settles down to the value $E/R = E/3Z_0$.

A space-time diagram, as illustrated in Fig. 1.11, is a convenient means
of keeping track of the various reflections and their sums.   Distance is

plotted horizontally and time is plotted downward.[1] The time required
for a wave to travel the length of the line is denoted by T, where $T = l/v$.

The zigzag lines are traces of the wave fronts of the various reflections.
The numbers attached to the lines indicate the magnitudes of the individual
waves. The magnitude of each reflection is obtained by multiplying the
magnitude of the preceding wave by the reflection coefficient at the point
where reflection takes place. The number shown in each intervening space

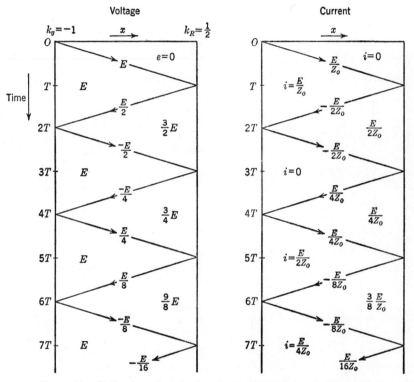

FIG. 1.11. Reflection diagram for the problem shown in Fig. 1.10.

is the sum of the individual waves above that point, and represents the net
current or voltage in that region of the chart. The voltage or current at
any time and position can easily be obtained from the diagram.

Example 2. Figure 1.12 shows an initially uncharged transmission line
which is open-circuited at the far end. At $t = 0$, the switch S is closed,

---

[1] This method can be applied to the calculation of waves of arbitrary shape traveling on lossy lines and is particularly convenient when there are several discontinuities
where reflections can occur. See L. V. Bewley, "Traveling Waves on Transmission
Systems," Chap. IV, John Wiley & Sons, Inc., New York, 1933.

connecting the line to a battery and a series resistance equal to $3Z_0$. The sending end cannot "know" that the line is not infinite until the arrival of the first reflection from the receiving end; therefore, the line will initially

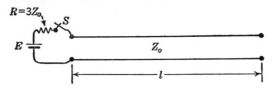

$R=3Z_0$

E

S

Z_0

l

FIG. 1.12.   Open-circuited line charged through a resistance.

look like an impedance Z_0 at the sending end.   Using the voltage-divider principle to calculate the initial sending-end voltage, we find

$$e_s = \frac{Z_0}{Z_0 + R} E = \frac{E}{4} \qquad \text{for } 0 < t < 2l/v$$

A wave of voltage of this value travels to the receiving end, where it is reflected with the coefficient

$$k_R = \lim_{Z_R \to \infty} \left(\frac{Z_R - Z_0}{Z_R + Z_0} \right) = 1$$

The reflection travels back to the generator end, where it is reflected with the coefficient

$$k_g = \frac{3Z_0 - Z_0}{3Z_0 + Z_0} = \frac{1}{2}$$

The successive reflections and rereflections are shown in the diagram of Fig. 1.13, and a graph of sending-end voltage is given in Fig. 1.14.   The build-up of sending-end voltage bears some resemblance to the voltage obtained across a condenser when charged from a battery through a resistance.

Example 3.   Figure 1.15 shows a traveling wave of a shape similar to that often caused on power lines by a lightning stroke.   The wave is assumed to be traveling toward a resistive termination equal to $3Z_0$, and the problem is to find the manner in which the waves of current and voltage will be reflected at the termination.   Although a reflection diagram similar to that of Fig. 1.11 can be used, we shall employ another method which is often useful in simple cases.

The reflection coefficient for voltage is $\frac{1}{2}$, as can be verified by use of Eq. (1.25) with $Z_t = 3Z_0$.   The reflection coefficient for current is the negative of this, or $-\frac{1}{2}$.   We shall calculate the reflection by imagining

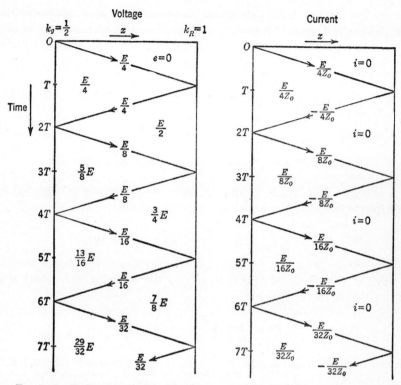

FIG. 1.13.　Reflection diagram for the problem shown in Fig. 1.12.　$T = l/v$.

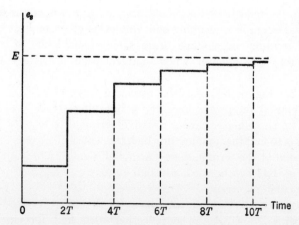

FIG. 1.14.　Graph of sending-end voltage *vs.* time for the problem of Fig. 1.12.　$T = l/v$.

that the line extends beyond its actual termination, as shown in Fig. 1.15, and that this fictitious extension carries the reflections

$$e^- = \tfrac{1}{2}e^+$$

and

$$i^- = -\tfrac{1}{2}i^+$$

The load resistance may be regarded as being replaced by a peculiar sort of mirror set normal to the line, and the fictitious waves to the right of this

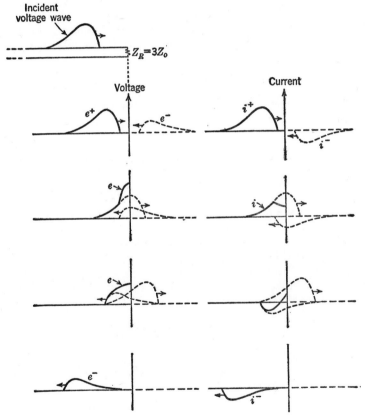

FIG. 1.15. The reflection of a wave from a resistive load equal to $3Z_0$.

may be regarded as the "mirror" reflections of the incident waves. As time goes on, the incident waves disappear into the mirror and the reflected waves emerge, as shown in the successive pictures of Fig. 1.15. The net result is obtained by superposing the two waves. Observe that on the line we always have

$$i^+ = \frac{e^+}{Z_0}$$

and

$$i^- = -\frac{e^-}{Z_0}$$

On the other hand, the superposition of the waves causes the net voltage and current at the load to be in the ratio $3Z_0$, which was the assumed load impedance.

The last picture shows the reflected wave traveling back up the line. Since e^- and i^- are, respectively, half as large as e^+ and i^+, the reflected energy is one-quarter of the incident energy, the remainder having been absorbed by the load resistance.

PROBLEMS

1. *a.* Show by direct substitution that the following expressions for voltage and current satisfy the differential equations (1.4) and (1.5):

$$e = K(\sqrt{LC}x - t), \quad i = K\sqrt{\frac{C}{L}}(\sqrt{LC}x - t)$$

where K is a constant.

b. Repeat part a for $e = K(\sqrt{LC}x + t)$, $i = -K\sqrt{C/L}(\sqrt{LC}x + t)$.

c. Sketch the e and i of part a to scale as functions of x between $x = -10$ meters and $x = 10$ meters, for the following three values of time: $t = 0$, 10^{-8} sec, and 2×10^{-8} sec. Use $K = 2 \times 10^9$ volts/sec, $L = 1.40 \times 10^{-6}$ henry/meter, and $C = 7.94 \times 10^{-12}$ farad/meter.

d. Repeat part c for the e and i of part b.

2. Show by direct substitution that the following expressions for voltage and current satisfy the differential equations (1.4) and (1.5):

$$e = E \cos \omega(t - \sqrt{LC}x)$$

and

$$i = E\sqrt{\frac{C}{L}} \cos \omega(t - \sqrt{LC}x)$$

where ω is a constant.

3. Show that the wave $e = E \cos(\omega t - \beta x)$ travels at the velocity ω/β.

4. Given the wave of voltage $e = E \cos(\omega t - \beta x)$, where $E = 100$ volts, $\omega = 6\pi \times 10^7$ rad/sec, and $\beta = 0.2\pi$ rad/meter.

a. For the three values of time $t = 0$, $\frac{1}{24} \times 10^{-7}$ sec, and $\frac{1}{12} \times 10^{-7}$ sec, sketch e vs. χ for $0 \leq \chi \leq 12.5$ meters.

b. Write the expression for e as a function of t at $\chi = 0$. Repeat for $\chi = 12.5$ meters and compare the two expressions.

5. Show that Eqs. (1.8) and (1.15) satisfy the differential equations (1.4) and (1.5).

6. Eliminate e between Eqs. (1.4) and (1.5) and show that the differential equation for current (1.7) is correct.

7. Show that the reflection coefficient for current is the negative of that for voltage.

8. A lossless transmission line of length l and characteristic impedance Z_0 is terminated in a resistance $Z_R = Z_0/3$. A battery with negligible internal resistance and an emf E is connected across the sending end at $t = 0$.

a. Draw a space-time diagram of the reflections. The time required for a wave to travel the length of the line is T.

b. Sketch the receiving-end voltage and current as functions of time.

c. Sketch the voltage at the center of the line as a function of time.

d. Approximately how long must an air-insulated line be if T is to be 1 μsec?

9. A lossless transmission line of length l and characteristic impedance Z_0 is open-circuited at the far end and is initially uncharged. At $t = 0$, the sending end is connected to a battery with an emf E in series with a resistance R. The time required for a wave to travel the length of the line is T. For the values of R given below, sketch the sending-end voltage and current as functions of time.

a. $R = 4Z_0$.

b. $R = Z_0$.

c. $R = Z_0/4$.

10. A battery with an emf E and a series resistance R are connected at $t = 0$ to the sending end of a lossless transmission line which is short-circuited at the far end.

a. Plot the sending-end voltage and current as functions of time for $R = Z_0/3$. The time required for a wave to travel the length of the line is T.

b. On the graph of part *a*, superimpose another drawn for $R = Z_0/6$. Observe that, as R is decreased, the graph approaches more closely a smooth exponential variation.

11. A battery with an emf E and a series resistance R are connected at $t = 0$ to the sending end of a lossless transmission line which is open-circuited at the far end. The time required for the wave to travel the length of the line is T. Sketch the sending-end voltage as a function of time for $R = 6Z_0$. Compare the result with Fig. 1.14, which was drawn for $R = 3Z_0$. Observe that, as R is increased, the graph approaches more closely a smooth exponential variation.

12. A resistance R is connected from one wire to the other somewhere in the interior of a transmission line. The line extends indefinitely in both directions from the shunting resistor. Imagine that an incident wave of voltage e^+ travels down the line and strikes the junction. Show that the reflected voltage wave is given by

$$-\frac{Z_0}{2R + Z_0} e^+$$

and that the transmitted wave is

$$\frac{2R}{2R + Z_0} e^+$$

13. A lossless transmission line with a characteristic impedance Z_0 branches out into two lines in the form of a Y. Each of the branches has a characteristic impedance equal to that of the main line. A rectangular wave of magnitude E is propagated along the main line and strikes the point where it branches. Find the magnitude of the voltage wave that is reflected from the branch point; also find the magnitude of the voltage waves that run out along the branches. *Hint:* The results of Prob. 12 can be used.

14. For the transmission line of Example 1, Sec. 1.6, sketch the following as functions of x for the three instants of time shown in Fig. 1.10: (*a*) The energy per unit length stored in the electric field, (*b*) the energy per unit length stored in the magnetic field, (*c*) the total stored energy per unit length. The length of the line is 10 meters. The line inductance is 8×10^{-7} henry/meter, the line capacitance is 13.9×10^{-12} farad/meter, and $E = 100$ volts.

15. A rectangular pulse of voltage with a height of 100 volts is propagated down a lossless line that has a characteristic impedance of 400 ohms. The end of the line is open-circuited. Plot the line voltage and the line current *vs.* distance from the open end for (*a*) the instant of time when the leading edge of the pulse has just arrived at the open circuit, (*b*) the instant when half of the pulse has been reflected, and (*c*) the instant when the trailing edge of the pulse is leaving the open circuit.

16. Repeat Prob. 15 for a short circuit, instead of an open circuit, at the end of the line.

17. A voltage wave of the form shown in Fig. 1.15 travels toward the end of a line that is terminated in a resistance $Z_R = Z_0/3$. Sketch the resulting voltage and current for the four instants of time indicated in Fig. 1.15.

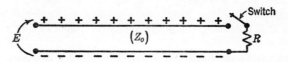

Fig. P18. Line charged to a uniform potential difference.

18. Figure P18 shows a lossless line that is open-circuited on both ends and charged to a voltage E throughout its length. At $t = 0$, the switch is closed, connecting the resistance R to the right end. This starts a wave of voltage, e^-, which originates at the switch and moves leftward along the line. Until the arrival of a reflection, this wave produces a current through R equal to $-e^-/Z_0$ and a voltage across R equal to $E + e^-$. Show that the wave of voltage that is started by closure of the switch is given by

$$e^- = -\frac{Z_0 E}{Z_0 + R}$$

19. A lossless line is open-circuited at both ends and is charged to a voltage E throughout its length (see Fig. P18). At $t = 0$, a resistance $R = Z_0$ is connected across the right-hand end. Draw a space-time reflection diagram and sketch a graph of the voltage across the resistance as a function of time.   Use the symbol T to denote the time required for a wave to travel the length of the line.

20. A lossless line is open-circuited at both ends and is charged to a potential difference of 500 volts throughout its length (see Fig. P18).   The characteristic impedance of the line is 400 ohms and its length is 150 meters.   At $t = 0$, a resistance of 1,500 ohms is connected across the right-hand end.   Using the result of Prob. 18, draw a space-time diagram of the reflections and sketch a graph of the voltage across the resistance as a function of time.

21. Repeat Prob. 20, but use a terminating resistance of 100 ohms.

22. Starting with Eqs. (1.2) and (1.3), eliminate i and show that the differential equation satisfied by the instantaneous voltage on a uniform lossy line is

$$\frac{\partial^2 e}{\partial x^2} = LC \frac{\partial^2 e}{\partial t^2} + (RC + LG) \frac{\partial e}{\partial t} + RGe$$

For historical reasons, this is sometimes called "the telegrapher's equation."   A similar equation is satisfied by the instantaneous current.

23. The so-called *distortionless* line has its constants related so that $RC = GL$. Show by substitution into the telegrapher's equation (Prob. 22) that the following solution is correct for a distortionless line:

$$e = \epsilon^{-xR\sqrt{C/L}} f(\sqrt{LC}x - t)$$

This solution represents a wave that is diminished exponentially in magnitude as it travels along the line.

24. A transmission line with a resistive characteristic impedance Z_0 is terminated in an inductance of L henrys.   An incident voltage wave of constant magnitude, $e^+ = E$, strikes the termination at $t = 0$.

a. Show that the differential equation which is satisfied by the reflected voltage at the receiving end after $t = 0$ is

$$\frac{L}{Z_0} \frac{de_R^-}{dt} + e_R^- = -E$$

b. From the foregoing differential equation and the fact that the inductance behaves like an open circuit at $t = 0$, show that the reflected voltage at the receiving end is

$$e_R^- = -E + 2E\epsilon^{-Z_0 t/L}$$

c. Sketch e_R^- as a function of time, using $E = 100$ volts, $Z_0 = 400$ ohms, and $L = 0.010$ henry.

d. Neglecting losses and using a method similar to that illustrated in Fig. 1.15, sketch the wave shape of the line voltage over a distance of 20 miles from the termination, using the following values of time: 0, 10^{-5}, 2.5×10^{-5}, and 10^{-4} sec. Assume a wave velocity of 180,000 miles/sec.

25. A transmission line with a resistive characteristic impedance Z_0 is terminated in a capacitance of C farads.   An incident voltage wave of constant magnitude, $e^+ = E$, strikes the termination at $t = 0$.

a. Show that the differential equation which is satisfied by the reflected voltage at the receiving end after $t = 0$ is

$$Z_0 C \frac{de_R^-}{dt} + e_R^- = E$$

b. From the foregoing differential equation and the fact that the capacitance behaves like a short circuit at $t = 0$, show that the reflected voltage at the receiving end is

$$e_R^- = E - 2E\epsilon^{-t/Z_0 C}$$

c. Sketch e_R^- as a function of time, using the following data: $E = 100$ volts, $Z_0 = 400$ ohms, and $C = 0.0625 \times 10^{-6}$ farad.

d. Neglecting losses and using the method illustrated in Fig. 1.15, sketch the wave shape of the line voltage over a distance of 20 miles from the termination for the following values of time: 0, 10^{-5}, 2.5×10^{-5}, and 10^{-4} sec. Assume a wave velocity of 180,000 miles/sec.

CHAPTER 2

THE A-C STEADY STATE. LINES WITH NO REFLECTIONS

2.1. The Rotating Vector. The most convenient method of handling sinusoidally varying quantities mathematically is to express them in terms of rotating vectors, and then to represent the vectors by means of complex numbers. In transmission-line theory it is desirable to use a somewhat more abstract notation than is necessary for the simpler a-c circuits, and so we shall devote a small amount of time to examining the mathematical basis for this method of representation.

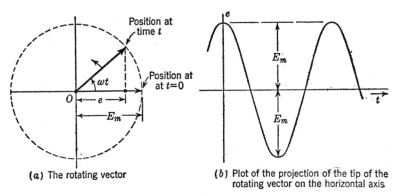

(a) The rotating vector   (b) Plot of the projection of the tip of the rotating vector on the horizontal axis

Fig. 2.1. The generation of a sinusoid by a rotating vector.

In Fig. 2.1 is shown a vector of length E_m whose origin is at O and which rotates counterclockwise with a uniform angular velocity of ω rad/sec. If the vector is horizontal at $t = 0$, then, after a time t has gone by, the vector will make an angle ωt radians with the horizontal axis. It is not hard to see from the geometry of the figure that the projection of the tip of the rotating vector on any stationary straight line is a point that executes a simple harmonic motion with the amplitude E_m. From a geometrical point of view, it is easiest to project on the vertical axis, and this is commonly done in many treatments of a-c circuit theory. However, it will turn out to be more convenient for our present purposes if we project instead on the horizontal axis. From the trigonometry of the figure, it is evident that the projection on this axis is given by

$$e = E_m \cos \omega t \qquad (2.1)$$

The cosine function will go through one complete cycle for each revolution of the generating vector. The period T is the amount of time during which ωt changes by 2π radians; *i.e.*, $T = 2\pi/\omega$ sec. The frequency, or number of cycles per second, is the reciprocal of this, or

$$f = \frac{\omega}{2\pi}$$

The quantity ω is often called the *angular frequency* to distinguish it from f.

Next, we examine the properties of the complex number $\epsilon^{j\theta}$, where θ is a real number and $j = \sqrt{-1}$. Euler's formula states that

$$\epsilon^{j\theta} = \cos \theta + j \sin \theta \qquad (2.2)$$

Therefore, $\epsilon^{j\theta}$ is a complex number with the real part $\cos \theta$ and the imaginary part $\sin \theta$. Now refer to Fig. 2.2, which shows a vector of unit length drawn on the complex plane so that it makes an angle θ with the real axis.

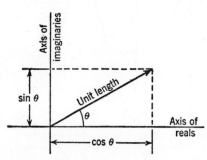

Obviously, the real and imaginary components of this vector are $\cos \theta$ and $\sin \theta$, respectively, which are precisely the components of $\epsilon^{j\theta}$. Therefore, we now have a geometric representation for $\epsilon^{j\theta}$: it is a unit vector on the complex plane and makes an angle θ with the real axis. In the symbolic notation sometimes used,

$$\epsilon^{j\theta} = 1/\underline{\theta}$$

Fig. 2.2. Representation of $\epsilon^{j\theta}$ on the complex plane.

If we now let the angle θ increase uniformly with time, we shall have a unit rotating vector. If the vector starts in the horizontal position at $t = 0$ and rotates with an angular velocity ω (radians per second), its angle at any moment will be $\theta = \omega t$ radians. The unit rotating vector will be expressed algebraically as

$$\epsilon^{j\omega t} = \cos \omega t + j \sin \omega t \qquad (2.3)$$

The projection of the tip of this rotating vector on the horizontal axis (the axis of reals) is the function $\cos \omega t$, as can be seen from the diagram. An equivalent statement is that the real component of $\epsilon^{j\omega t}$ is $\cos \omega t$, and, looking back at Eq. (2.3), we see that this is indeed true. The shorthand notation that is frequently used is the following:

$$\cos \omega t = \Re e[\epsilon^{j\omega t}] \qquad (2.4)$$

where the symbol $\Re e$ is read "the real part of." The imaginary part of $\epsilon^{j\omega t}$ is $\sin \omega t$, and this would do quite as well in our analysis, but, algebraically, it seems more natural to take the real part. This is why we preferred at the beginning of this treatment to project our rotating vectors on the horizontal axis.

Now, we can represent the rotating vector of Fig. 2.1a algebraically in terms of complex numbers as $E_m \epsilon^{j\omega t}$, and its projection on the real axis [compare Eq. (2.1)] as

$$e = \Re e[E_m \epsilon^{j\omega t}] \qquad (2.5)$$

There are several reasons for following this curious procedure. First, it is much more convenient to represent a sinusoidally varying quantity by a vector (or by its associated complex number) than by the trigonometric expression. The advantage becomes more apparent as the number of sinusoidal quantities (perhaps currents and voltages) in a given problem becomes greater. A second reason is the mathematical simplification introduced in taking derivatives of the exponential function which represents the vector, for the exponential does not change its mathematical form when differentiated. For example, the derivative with respect to time of Eq. (2.1) is

$$\frac{de}{dt} = -\omega E_m \sin \omega t \qquad (2.6)$$

On the other hand, the derivative of the exponential form (2.5) is

$$\frac{de}{dt} = \Re e[j\omega E_m \epsilon^{j\omega t}] \qquad (2.7)$$

which represents a vector that starts in the vertical position at $t = 0$. It is not hard to see that the projection of this rotating vector on the horizontal axis is a negative sine function with an amplitude ωE_m, which checks with Eq. (2.6).

To show algebraically that the two expressions are identical, we apply Euler's formula to Eq. (2.7) and write

$$\frac{de}{dt} = \Re e[j\omega E_m(\cos \omega t + j \sin \omega t)]$$

$$= \Re e[\omega E_m(j \cos \omega t - \sin \omega t)]$$

$$= -\omega E_m \sin \omega t$$

It is a common practice to omit the symbol $\Re e$ and to write simply

$$e = E_m \epsilon^{j\omega t} \tag{2.8}$$

and

$$\frac{de}{dt} = j\omega E_m \epsilon^{j\omega t} \tag{2.9}$$

However, when this is done, one must not lose sight of the fact that the instantaneous variation of e is given by the projection of the rotating vector on the axis of reals, a fact that is not indicated explicitly in Eqs. (2.8) and (2.9).

A sinusoidally varying quantity e_2 which leads the foregoing one by an angle α can be expressed as

$$e_2 = E_{2m} \cos(\omega t + \alpha) \tag{2.10}$$

or as

$$e_2 = \Re e[E_{2m}\epsilon^{j(\omega t + \alpha)}] = \Re e[E_{2m}\epsilon^{j\alpha}\epsilon^{j\omega t}] \tag{2.11}$$

or simply as

$$e_2 = E_{2m}\epsilon^{j\alpha}\epsilon^{j\omega t} \tag{2.12}$$

The complex quantity $E_{2m}\epsilon^{j\alpha} = E_{2m}/\underline{\alpha}$ represents the vector in its position at $t = 0$.

To illustrate an elementary application of the foregoing method, consider a circuit containing an inductance L and a resistance R in series. An emf $e = E_m \cos \omega t$ is applied to the circuit, and we wish to find the steady-state alternating current that results. The voltage drop across the inductance is $L\, di/dt$ and that across the resistance is Ri. The sum of the two is equal to the impressed voltage:

$$L\frac{di}{dt} + Ri = E_m \cos \omega t \tag{2.13}$$

Let us write $E_m \cos \omega t$ as $\Re e[E_m\,\epsilon^{j\omega t}]$ and then omit the symbol $\Re e$. Similarly, we shall write the current as $i = \Re e[I_m\epsilon^{j\omega t}]$ where I_m is a complex, or vector, current whose length is equal to the maximum value of the current i. Making these substitutions into Eq. (2.13) and taking the indicated derivative, we obtain the relation

$$(j\omega L + R)I_m\epsilon^{j\omega t} = E_m\epsilon^{j\omega t} \tag{2.14}$$

Canceling $\epsilon^{j\omega t}$ on both sides and solving for the complex current I_m, we obtain

$$I_m = \frac{E_m}{R + j\omega L} \tag{2.15}$$

or

$$I_m = \frac{E_m}{\sqrt{R^2 + \omega^2 L^2}} \,\underline{/-\tan^{-1} \omega L/R} \tag{2.16}$$

The quantity $R + j\omega L$ is, of course, the complex impedance of the circuit. Equation (2.16) gives the current vector in its position at $t = 0$. Usually, the algebra stops at this point, for we have found the essential information, namely, the magnitude of the current and its phase angle with respect to the voltage. But we should not lose sight of the fact that the instantaneous current in the circuit is not given directly by Eq. (2.16). Instead, we must imagine the vector to rotate at the angular speed ω and take its projection. Algebraically this is done by writing

$$i = \Re e[I_m \epsilon^{j\omega t}] \tag{2.17}$$

which, after substituting from Eq. (2.16), expanding, and taking the real part, becomes

$$i = \frac{E_m}{\sqrt{R^2 + \omega^2 L^2}} \cos\left(\omega t - \tan^{-1} \frac{\omega L}{R}\right) \tag{2.18}$$

In all the development so far, we have referred to the maximum value of the sinusoidal wave, E_m. For a sinusoid, the effective value E is related to the maximum value by $E_m = \sqrt{2}E$. If both sides of Eq. (2.16) are divided by $\sqrt{2}$, the result will be expressed in terms of the effective values of current and voltage:

$$I = \frac{E}{R + j\omega L} = \frac{E}{\sqrt{R^2 + \omega^2 L^2}} \,\underline{/- \tan^{-1}\omega L/R} \tag{2.19}$$

This contains the same essential information as the previous expression, namely, the magnitude of the current (now in rms amperes) and its phase angle with respect to the voltage. The equations of circuit theory are usually written in this way because of the convenience of using rms values. Again, however, a word of caution is in order. If we wish to obtain a true picture of the instantaneous current in the circuit, it is now necessary to multiply the expression (2.19) by $\sqrt{2}$ before we rotate the vector and take its projection, or, algebraically speaking, before we multiply by $\epsilon^{j\omega t}$ and take the real part.

For a further discussion of this subject, the reader should refer to the literature.[1]

2.2. A-C Steady-state Solution for the Uniform Line. In the preceding chapter we derived the partial differential equations which apply to the uniform line under both transient and steady-state conditions, and we used these equations to show the properties of traveling waves on a lossless line. The equations that apply under steady-state sinusoidal conditions can be found in either of two ways: they can be derived from the more general partial differential equations (1.2) and (1.3), or they can be found by writing the steady-state relations that must hold across an infinitesimal section of line. We shall first show how the steady state can be derived from the partial differential equations, and shall then show that the second method yields the same result as the first.

From Eqs. (1.2) and (1.3) we have

and
$$\left.\begin{aligned} \frac{\partial e}{\partial x} &= -Ri - L\frac{\partial i}{\partial t} \\[2mm] \frac{\partial i}{\partial x} &= -Ge - C\frac{\partial e}{\partial t} \end{aligned}\right\} \qquad (2.20)$$

As outlined in the preceding section, we shall express the steady-state sinusoidal voltages and currents on the line as the projections of rotating vectors, thus,

and
$$\left.\begin{aligned} e &= \mathfrak{Re}[E_m \epsilon^{j\omega t}] \\[2mm] i &= \mathfrak{Re}[I_m \epsilon^{j\omega t}] \end{aligned}\right\} \qquad (2.21)$$

where E_m and I_m are the complex amplitudes of the voltage and current, respectively, and ω is 2π times the frequency of the driving source. Observe that E_m and I_m will vary along the line and, therefore, will have derivatives with respect to x. Substituting Eqs. (2.21) into the differential equations (2.20), taking the indicated derivatives, and canceling $\epsilon^{j\omega t}$, we obtain

and
$$\left.\begin{aligned} \frac{dE_m}{dx} &= -RI_m - j\omega LI_m \\[2mm] \frac{dI_m}{dx} &= -GE_m - j\omega CE_m \end{aligned}\right\} \qquad (2.22)$$

[1] See, for example, R. H. Frazier, "Elementary Electric-circuit Theory," Chap. IV, McGraw-Hill Book Company, Inc., New York, 1945; E. A. Guillemin, "Communication Networks," Vol. I, Chap. III, John Wiley & Sons, Inc., New York, 1931; W. C. Johnson, "Mathematical and Physical Principles of Engineering Analysis," Chap. VII, McGraw-Hill Book Company, Inc., New York, 1944; MIT Staff, "Electric Circuits," Chap. IV, John Wiley & Sons, Inc., New York, 1940.

Total derivatives are now used because there is only one independent variable, x. The effective values E and I are related to the amplitudes E_m and I_m by $E_m = \sqrt{2}E$ and $I_m = \sqrt{2}I$. Making this substitution, we obtain

and

$$\left.\begin{aligned} \frac{dE}{dx} &= -(R + j\omega L)I \\[2em] \frac{dI}{dx} &= -(G + j\omega C)E \end{aligned}\right\} \qquad (2.23)$$

These relations could also have been deduced from a consideration of the a-c steady-state circuit properties of a small section of line as visualized in Fig. 1.3. The change in voltage across a section of length Δx, which is expressed as $(dE/dx)\Delta x$, is caused by the current I flowing through the series impedance of the section, $R\,\Delta x + j\omega L\,\Delta x$. The minus sign is used because a positive value of I causes E to decrease with increasing x. Similarly, the change in current between the two ends of the section is caused by the voltage E acting on the shunt admittance $G\,\Delta x + j\omega C\,\Delta x$.

It is usual in transmission-line theory to denote the a-c series impedance per unit length of line by the symbol Z and the shunt admittance by the letter Y:

and

$$\left.\begin{aligned} Z &= R + j\omega L \text{ ohms per unit length} \\[1em] Y &= G + j\omega C \text{ mhos per unit length} \end{aligned}\right\} \qquad (2.24)$$

The differential equations (2.23) can then be written more compactly as

$$\frac{dE}{dx} = -ZI \qquad (2.25)$$

and

$$\frac{dI}{dx} = -YE \qquad (2.26)$$

We now have two equations with two unknowns, E and I. To eliminate I, we take the derivative of the first equation with respect to x and obtain

$$\frac{d^2E}{dx^2} = -Z\frac{dI}{dx}$$

Into this we substitute the expression for dI/dx from Eq. (2.26) and obtain

$$\frac{d^2E}{dx^2} = (YZ)E \qquad (2.27)$$

The solution must be a function which, when differentiated twice, yields the original function multiplied by the quantity YZ. One form of this solution is

$$E = A_1 \epsilon^{-\sqrt{YZ}x} + A_2 \epsilon^{\sqrt{YZ}x} \tag{2.28}$$

where the A's are constants with the dimensions of voltage. To find the corresponding expression for I, substitute the above result into Eq. (2.25). Then we obtain

$$I = \frac{1}{\sqrt{Z/Y}} (A_1 \epsilon^{-\sqrt{YZ}x} - A_2 \epsilon^{\sqrt{YZ}x}) \tag{2.29}$$

The quantity $\sqrt{Z/Y}$ is a characteristic of the line and has the dimensions of an impedance: $\sqrt{\dfrac{\text{ohms/unit length}}{\text{mhos/unit length}}} = \text{ohms}$. This is the characteristic impedance of the line. We shall denote it by the symbol Z_0, as in Chap. 1.

$$Z_0 = \sqrt{\frac{Z}{Y}} = \sqrt{\frac{R + j\omega L}{G + j\omega C}} \tag{2.30}$$

For lines with negligible loss, the characteristic impedance reduces to $\sqrt{L/C}$, which is a pure resistance independent of frequency (compare Sec. 1.5).

Observe that the characteristic impedance does not involve the length of the line or the character of the terminating load, but is determined only by the characteristics of the line per unit length. Observe also that it is not an impedance that the line itself possesses. It will be shown later that a load impedance equal to $\sqrt{Z/Y}$ is the only one which will cause no reflection of a received wave, and that, for this termination only, the steady-state a-c sending-end impedance of the line will be equal to $\sqrt{Z/Y}$ regardless of the length of the line. This is the principal significance of the characteristic impedance.

Referring again to Eqs. (2.28) and (2.29), the quantity $\sqrt{YZ}$ is seen to govern the manner in which E and I vary with x; i.e., it governs the way in which the waves are propagated. Therefore, it is given the name *propagation constant*. It will be denoted by the symbol γ. Actually, γ is a function of frequency and might better be called the propagation function. Explicitly,

$$\gamma = \sqrt{YZ} = \sqrt{(R + j\omega L)(G + j\omega C)} \tag{2.31}$$

The propagation constant will, in general, be a complex number. The real part is given the symbol α and is found to determine the way in which the waves die out, or attenuate, as they travel. Hence the real part α is called the *attenuation constant*. The imaginary part of γ is given the symbol β

and is found to determine the variation in phase position of E and I along the line. For this reason β is called the *phase constant*.

The unit of α is called the neper per unit length, and the unit of β is the radian per unit length. The word *neper* is a variation of the spelling of the name Napier.

Example. Consider a typical open-wire telephone line which has $R = 10$ ohms/mile, $L = 0.0037$ henry/mile, $C = 0.0083 \times 10^{-6}$ farad/mile, and $G = 0.4 \times 10^{-6}$ mho/mile. At a frequency of 1,000 cps, we have from Eq. (2.30):

$$Z_0 = \sqrt{\frac{R + j\omega L}{G + j\omega C}} = \sqrt{\frac{10 + j23.2}{(0.4 + j52.1) \times 10^{-6}}}$$

$$= \sqrt{\frac{25.3\underline{/66.8°}}{52.1 \times 10^{-6}\underline{/89.6°}}} = \sqrt{48.5 \times 10^4\underline{/-22.8°}}$$

$$= 697\underline{/-11.40°} = 683 - j\,138 \text{ ohms}$$

From Eq. (2.31) the propagation constant at this frequency is

$$\gamma = \sqrt{(R + j\omega L)(G + j\omega C)} = \sqrt{(25.3\underline{/66.8°})(52.1 \times 10^{-6}\underline{/89.6°})}$$

$$= \sqrt{13.2 \times 10^{-4}\underline{/156.4°}}$$

$$= 0.0363\underline{/78.2°} = 0.0074 + j0.0356 \text{ per mile}$$

Therefore, the attenuation constant is

$$\alpha = 0.0074 \text{ neper/mile}$$

and the phase constant is

$$\beta = 0.0356 \text{ rad/mile}$$

2.3. The Line with No Reflection. The simplest line from a mathematical viewpoint is one on which there is no reflected wave. This condition would be obtained on a hypothetical smooth line of infinite length, for no reflection could ever return from the far end. We shall also show that a similar condition is obtained for a line terminated in its characteristic impedance.

Consider the general solution (2.28) in connection with an infinite line (see Fig. 2.3a). The second term of the solution involves a positive exponent and would tend toward infinity as x increased. This is physically impossible from an energy standpoint; hence, A_2 must be zero. We can evaluate the constant A_1 in terms of the sending-end voltage by substituting $x = 0$ and $E = E_s$ into the solution, thus obtaining (since $A_2 = 0$)

$$E_s = A_1$$

Using this, the solution (2.28) can be written as

$$E = E_s \epsilon^{-\sqrt{YZ}\,z} \tag{2.32}$$

In the preceding section we defined the propagation constant $\gamma = \sqrt{YZ}$ and its real and imaginary parts α and β. In terms of these, we can write

$$E = E_s \epsilon^{-\alpha x} \epsilon^{-j\beta x} = E_s \epsilon^{-\alpha x} \underline{/-\beta x} \tag{2.33}$$

We obtain the corresponding solution for current from Eq. (2.29) by using $A_1 = E_s$ and $A_2 = 0$, which yields

$$I = \frac{E_s}{Z_0} \epsilon^{-\alpha x} \epsilon^{-j\beta x} \tag{2.34}$$

Upon comparison with Eq. (2.33), we see that the current and voltage at every point on the infinite line are related simply by

$$I = \frac{E}{Z_0} \tag{2.35}$$

This expression applies, of course, to the sending end of the line as well as to all other points. Hence, the sending-end, or input, impedance of the infinite line is equal to its characteristic impedance Z_0 .

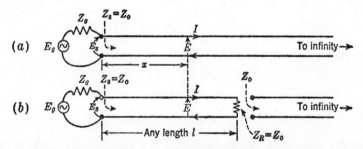

Fig. 2.3. The infinite line and its finite equivalent.

Now suppose that we cut the infinite line at any point $x = l$, as indicated in Fig. 2.3. The line to the right of the cut is still infinite in length and has an input impedance equal to Z_0. So far as the section to the left of the cut is concerned, the infinite portion to the right can be replaced by a lumped receiving-end impedance equal to Z_0. The finite section will still behave as though it were infinite in length, and the solutions (2.33), (2.34) and (2.35) will still apply. The sending-end impedance of the line will be equal to Z_0 regardless of the length l. For low-loss lines this imped- ance is nearly a pure resistance independent of frequency $(\sqrt{L/C})$; hence, moderate changes in frequency will not affect the load voltage or power appreciably.

A line terminated in its characteristic impedance is sometimes called a *correctly terminated* or *nonresonant* line. On a line so terminated, energy flows from the generator, travels down the line in the form of a wave, and, from a circuit standpoint, should all be absorbed by the load without reflection. From the point of view of traveling electric and magnetic fields, it can be seen that a lumped load impedance will not supply a continuation of the fields as did the infinite line. However, the circuit theory is quite accurate except when the conductor spacing is an appreciable part of a quarter wavelength. At short wavelengths, the performance of the line is improved, and the accuracy of the circuit theory is maintained by placing a reflector a quarter wavelength beyond the load. The theory of this is discussed in Chap. 6.

The solutions (2.33) and (2.34) have been given in terms of the sending-end voltage of the line, although, as suggested in Fig. 2.3, the open-circuit voltage of the generator and its internal impedance may be known instead. The sending-end voltage can be found by solving the equivalent sending-end circuit of Fig. 2.4, in which the line is replaced by an equivalent lumped impedance equal to Z_s. When E_s has been computed, Eqs. (2.33) and (2.34) can be used to find the voltage and current at any point on the line.

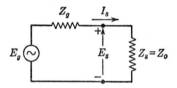

FIG. 2.4. Equivalent sending-end circuit for a line terminated in Z_0.

The power being transmitted down the line at any point can be computed from the relation $P = |E| \cdot |I| \cos\theta$, where θ is the phase angle of the characteristic impedance, or by $|I|^2 R_0$, where R_0 is the resistive component of Z_0. For low-loss lines, Z_0 is very nearly a pure resistance, in which case $\cos\theta \approx 1$ and $R_0 \approx Z_0$.

Example. Consider an open-wire telephone line which is 200 miles in length and is terminated in its characteristic impedance at the receiving end. At the sending end, it is driven by a generator which has a frequency of 1,000 cps, an open-circuit emf of 10 volts rms, and an internal impedance of 500 ohms resistance. At this frequency the line has $Z_0 = 683 - j138$ ohms and $\gamma = 0.0074 + j0.0356$ per mile. Find the sending-end voltage, current, and power, and the receiving-end voltage, current, and power.

Using the equivalent sending-end circuit of Fig. 2.4 with $E_g = 10$ volts, $Z_g = 500 + j0$ ohms, and $Z_s = Z_0 = 683 - j138$ ohms, we can solve for the magnitude of the sending-end current[1]:

[1] In this text, we have been using the italic symbols E, I, and Z to represent the complex, or vector, voltage, current, and impedance. Their magnitudes, or absolute values, will be indicated by placing the symbols between vertical bars, thus: $|E|$, $|I|$, and $|Z|$.

$$|I_s| = \left|\frac{E_g}{Z_g + Z_s}\right| = \frac{10}{|500 + 683 - j138|}$$

$$= \frac{10}{1,190} = 8.40 \times 10^{-3} \text{ amp rms}$$

The magnitude of the sending-end voltage is

$$|E_s| = |I_s Z_s| = 8.40 \times 10^{-3} \sqrt{(683)^2 + (138)^2}$$
$$= 5.85 \text{ volts rms}$$

The average power entering the sending end of the line is

$$P_s = |I_s|^2 R_s = (8.40 \times 10^{-3})^2 \times 683$$
$$= 48.2 \times 10^{-3} \text{ watt}$$

or 48.2 mw.

For the receiving-end voltage we use Eq. (2.33) with x equal to the length of the line. Then, arbitrarily taking E_s to be real, we have

$$E_R = E_s \epsilon^{-\alpha l} \epsilon^{-j\beta l}$$
$$= 5.85 \epsilon^{-0.0074 \times 200} \epsilon^{-j0.0356 \times 200}$$
$$= 1.33 \epsilon^{-j7.12} = 1.33\underline{/-7.12} \text{ radians} = 1.33\underline{/-408°} \text{ volts}$$

The magnitude of the receiving-end voltage is 1.33 volts rms. The line is $7.12/2\pi = 1.134$ wavelengths long, and the receiving-end voltage lags that of the sending end by 1.134 cycles. In so far as the phase position of the receiving-end voltage is concerned, we can substract 2π radians or 360° without changing the result, thus giving

$$E_R = 1.33\underline{/-0.84} \text{ radian} = 1.33\underline{/-48°} \text{ volts}$$

The magnitude of the receiving-end current is

$$|I_R| = \left|\frac{E_R}{Z_R}\right| = \frac{1.33}{697} = 1.91 \times 10^{-3} \text{ amp rms}$$

The average value of the power absorbed by the terminating impedance is

$$P_R = |I_R|^2 R_R = (1.91 \times 10^{-3})^2 \times 683$$
$$= 2.49 \times 10^{-3} \text{ watt}$$

or 2.49 mw.

2.4. The Traveling Wave and Its Characteristics. Equations (2.33) and (2.34) show that, on a line terminated in its characteristic impedance, the voltage and current decrease exponentially in amplitude by the factor $\epsilon^{-\alpha x}$ as the distance from the generator increases. This is caused by the line losses, which absorb energy from the wave as it travels. A second important effect is the progressively increasing lag in phase as x increases, as shown by the factor $\epsilon^{-j\beta x} = \underline{/-\beta x}$. This lag is caused by the finite time required for the wave to travel the distance x.

The traveling wave on the line can be expressed most neatly by using the method of Sec. 2.1 for obtaining instantaneous values: First express Eq. (2.33) in terms of the maximum value, then multiply by the unit rotating vector $\epsilon^{j\omega t}$, and finally take the real part of the resulting expression. If we take E_s to be real, this process gives us

$$e = \Re[\sqrt{2}\,E_s\epsilon^{-\alpha x}\epsilon^{-j\beta x}\epsilon^{j\omega t}]$$
$$= \sqrt{2}\,E_s\epsilon^{-\alpha x}\Re[\epsilon^{j(\omega t-\beta x)}]$$

or

$$e = \sqrt{2}\,E_s\epsilon^{-\alpha x}\cos{(\omega t - \beta x)} \qquad (2.36)$$

Now, we can follow the same reasoning regarding traveling waves that we used in Chap. 1. Imagine that an observer is following the wave $\cos{(\omega t-\beta x)}$ so that he stays with a point of constant phase, for which the angle $\omega t - \beta x = $ a constant. To find the velocity of the point, we take the time derivative of this expression and obtain

$$\omega - \beta\frac{dx}{dt} = 0$$

But dx/dt is the velocity we desire (the *phase velocity*) and this is seen to be

$$v = \frac{\omega}{\beta} \qquad (2.37)$$

It should be observed that this is the velocity at which the steady-state a-c wave and its accompanying electric and magnetic fields are propagated, but it is not the velocity of the electrons in the wire. The electrons may be visualized as executing an oscillatory motion as shown in Fig. 2.5 (this motion is superimposed on their usual random velocity). The phase of the oscillation lags in the direction of motion of the wave, and this gives rise to a sinusoidal distribution of charge which apparently travels along the wire, as suggested in the illustration. The portions marked A are regions of deficiency of charge, and the one marked B is a region of excess charge. The charged regions travel to the right with the phase velocity. An analogous situation is the propagation of sound, which travels at about 1,100 ft/sec in air at sea level. The actual motion of the air molecules, however, is only a sinusoidal one, and the velocity of an individual molecule is not comparable with the velocity of propagation of the wave.

For a line with negligible losses, Eq. (2.31) becomes merely $\gamma = j\omega\sqrt{LC}$; *i.e.*, $\alpha = 0$ and $\beta = \omega\sqrt{LC}$. The velocity ω/β in this case is simply

$$v = \frac{1}{\sqrt{LC}} \qquad (2.38)$$

As mentioned in Sec. 1.5, the product LC for perfect conductors immersed in a lossless dielectric is a constant which depends only on the dielectric constant and the permeability of the dielectric. For air the velocity turns

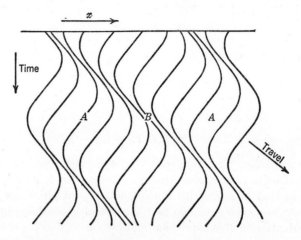

FIG. 2.5. Generation of a traveling wave by individual sinusoidal oscillations.

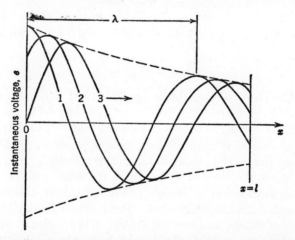

FIG. 2.6. A traveling wave on a lossy line terminated in its characteristic impedance. Drawn for a line of $1\frac{1}{3}$ wavelengths long and a total attenuation of 0.9 neper.

out to be approximately 3×10^8 meters/sec. A solid dielectric will increase C and thus reduce the velocity. This effect is frequently expressed in terms of a velocity constant:

$$\text{Velocity constant} = \frac{\text{actual phase velocity}}{\text{velocity of light in free space}} \tag{2.39}$$

A low-loss coaxial line with a solid dielectric may have a velocity constant of about 0.6 or 0.7. As shown by Eq. (1.1), the wavelength for a given frequency is reduced by the same factor as the phase velocity.

Figure 2.6 shows a traveling wave of voltage on a lossy line at three successive instants of time, as plotted from Eq. (2.36). The amplitude of the wave diminishes exponentially by the factor $\epsilon^{-\alpha x}$ as it travels. The accompanying current wave will be similar in form but will be out of phase with the voltage by an amount equal to the angle of Z_0, since $I = E/Z_0$. The transmitted power will decrease down the line by the factor $\epsilon^{-2\alpha x}$

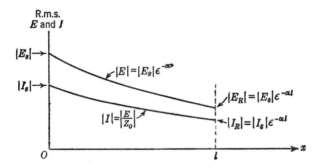

FIG. 2.7. The rms voltage and current along a lossy line terminated in its characteristic impedance. Drawn for a total attenuation of 0.9 neper.

The instantaneous voltage at any point on the line will vary sinusoidally as the traveling wave slides past, and will have an amplitude $\sqrt{2}E_s\epsilon^{-\alpha x}$ and an rms value $E_s\epsilon^{-\alpha x}$. The phase lags progressively down the line; for instance, a quarter wavelength from the generator the voltage lags by a quarter of a cycle, or 90°. The phase lag at any point with respect to the input is βx radians.

The wavelength λ is equal to the distance between successive crests of the wave at any moment; i.e., it is equal to the change in x which makes the angle βx increase by 2π radians. Therefore, $\beta\lambda = 2\pi$ and

$$\lambda = \frac{2\pi}{\beta} \tag{2.40}$$

Comparison with Eq. (2.37) shows that the phase velocity and wavelength are related by

$$\lambda = \frac{2\pi v}{\omega} = \frac{v}{f} \tag{2.41}$$

This relation was first stated in Chap. 1, Eq. (1.1). As an exercise, the student should return to the example given in the preceding section and show that the phase velocity was 176,200 miles/sec and the wavelength 176.2 miles.

The variation of rms voltage and current along a line is sketched in Fig. 2.7. When the over-all losses are small, the factor $\epsilon^{-\alpha l}$ is nearly unity, and if the line is terminated in Z_0, the current and voltage will be practically uniform in magnitude over the whole length. Such a line is said to be "flat."

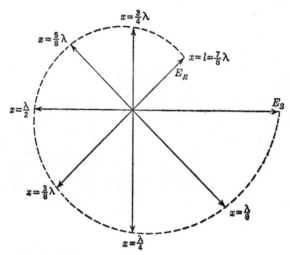

FIG. 2.8. Polar plot of E for a line terminated in Z_0. Drawn for $l = 7\lambda/8$ and an attenuation of 0.8 neper per wavelength.

Figure 2.8 is a polar representation of the vector $E = E_s \epsilon^{-\alpha x} \epsilon^{-j\beta x}$ along the length of a lossy line. The locus of E is a logarithmic spiral, for the magnitude decreases exponentially with increasing angle. At this point the student should review in his mind the process by which any one of these vectors can be translated into the corresponding instantaneous line-to-line voltage.

2.5. Note on Characteristic Impedance. Figure 2.9 shows a symmetrical four-terminal T network

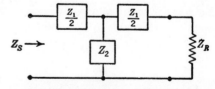

FIG. 2.9. Symmetrical T network.

composed of lumped elements. Network theory shows that there is one particular value of receiving-end impedance which will cause Z_s to be precisely equal to Z_R. This particular impedance is called the characteristic impedance of the network. It is not hard to show that for a symmetrical T network the characteristic impedance is given by

$$Z_0 = \sqrt{Z_1 Z_2 + \frac{Z_1^2}{4}} \qquad (2.42)$$

where Z_1 is the total series impedance of the section and Z_2 is the impedance of the shunt branch.

A simple numerical example which involves only resistances is shown in Fig. 2.10. Here we have $Z_1 = 10$ ohms, $Z_2 = 20$ ohms, and so the characteristic impedance is

$$Z_0 = \sqrt{10 \times 20 + \frac{10^2}{4}} = 15 \text{ ohms}$$

Any number of identical sections can be connected in cascade to form a uniform ladder network, as shown in Fig. 2.10b. If the last section is terminated in Z_0, its input impedance will be Z_0, as the student can show by combining resistances in series and parallel in Fig. 2.10. This will

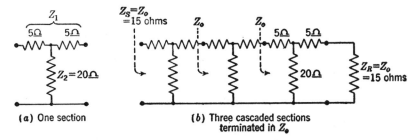

(a) One section   (b) Three cascaded sections terminated in Z_0

FIG. 2.10.   Illustrating the characteristic impedance of a uniform ladder network.

terminate the next-to-last section in Z_0, and so on through the network. Consequently, the sending-end impedance of the whole combination is equal to the characteristic impedance. For this reason the more descriptive name *iterative impedance* is often used in the theory of four-terminal networks.

Suppose that the T network of Fig. 2.9 is to represent a small length of transmission line, in which case

$$Z_1 = (R + j\omega L)\Delta x \quad \text{and} \quad Z_2 = \frac{1}{(G + j\omega C)\Delta x}$$

As Δx is made smaller to approximate a smooth line more closely, the product $Z_1 Z_2$ will remain constant while the quantity Z_1^2 will vanish. In the limit, Eq. (2.42) will yield $Z_0 = \sqrt{(R + j\omega L)/(G + j\omega C)}$, which is the expression previously derived for the smooth line.

2.6. The Decibel and the Neper. We have shown previously that, on a line terminated in its characteristic impedance, the magnitudes of both voltage and current decrease as $\epsilon^{-\alpha x}$, where α is the real part of the propagation constant $\sqrt{YZ}$. The unit of α is the neper per unit length. Consider two points 1 and 2 separated by a distance Δx on a properly terminated line. The ratios of the magnitudes of E and I at the two points are

$$\left|\frac{E_1}{E_2}\right| = \left|\frac{I_1}{I_2}\right| = \epsilon^{\alpha \Delta x} \tag{2.43}$$

The quantity $\alpha \Delta x$ represents the total attenuation, measured in nepers, of the traveling wave between the two points. Taking the logarithm to the base ϵ of both sides of Eq. (2.43), we obtain an expression for the number of nepers of attenuation:

$$\text{Number of nepers} = \log_\epsilon \left|\frac{E_1}{E_2}\right| = \log_\epsilon \left|\frac{I_1}{I_2}\right| \tag{2.44}$$

This definition, expressed in terms of the natural logarithm, explains the association of Napier's name with the unit.

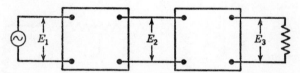

FIG. 2.11. Two four-terminal networks connected in cascade.

A logarithmic scale for the measurement of current or voltage is highly convenient in a transmission system. Figure 2.11 shows two four-terminal networks (perhaps transmission lines) which are connected in cascade. The attenuation in nepers through the separate networks will be denoted by N_1 and N_2, where

$$N_1 = \log_\epsilon \left|\frac{E_1}{E_2}\right|$$

and

$$N_2 = \log_\epsilon \left|\frac{E_2}{E_3}\right|$$

The total attenuation through the combination is obtained by adding the attenuations of the separate networks:

$$\text{Total } N = N_1 + N_2 = \log_\epsilon \left|\frac{E_1}{E_2}\right| + \log_\epsilon \left|\frac{E_2}{E_3}\right|$$

which, by the rule of adding logarithms, becomes

$$N = \log_\epsilon \left|\frac{E_1}{E_3}\right|$$

Thus, the addition of the logarithmic attenuations is equivalent to multiplying the voltage ratios. The convenience of this method becomes greater as more and more networks are added in cascade.

Another reason for using a logarithmic scale is that, within wide limits, the human senses follow a similar law. Thus, the minimum perceptible change in sound intensity that the ear can perceive is roughly proportional to the amount of sound already present.

A second logarithmic unit which is frequently used is the *decibel*, abbreviated db. The decibel is one-tenth the size of the *bel*, which was named in honor of Alexander Graham Bell. The decibel is defined fundamentally in terms of a power ratio:

$$\text{Number of decibels} = 10 \log_{10} \frac{P_1}{P_2} \qquad (2.45)$$

If the powers P_1 and P_2 are associated with equal impedances, as would be the case on a transmission line terminated in its characteristic impedance, the power ratio can be expressed as the square of either the voltage or the current ratio. Under these conditions,

$$\text{Number of decibels} = 20 \log_{10} \left| \frac{E_1}{E_2} \right| = 20 \log_{10} \left| \frac{I_1}{I_2} \right| \qquad (2.46)$$

The decibel is a unit of convenient size. A change in level of 3 db corresponds very nearly to a power ratio of two; and 10 db correspond to a power ratio of 10. A change in sound power of 1 db is roughly the minimum change that is perceptible by the human ear.

The advantages of a logarithmic scale and the convenient size of the decibel have given rise to a tendency to express voltage ratios in terms of decibels even when the impedances are different at the two points. This usage is most common in connection with cascaded voltage amplifiers in which the voltage ratio is of primary interest, and amounts to replacing the definition of Eq. (2.45) by a new one, namely, $20 \log_{10} | E_2/E_1 |$. The two definitions are not equivalent except when the impedance level is the same at both points, and confusion may arise if the usage is not made clear.

The student should show that, in a system having the same impedance level at the two points in question, an attenuation of 1 neper is equivalent to 8.686 db.

2.7. Variation of Z_0, α, and β with Frequency. The characteristic impedance of a uniform line is given by Eq. (2.30) as

$$Z_0 = \sqrt{\frac{R + j\omega L}{G + j\omega C}} \qquad (2.47)$$

At zero frequency, *i.e.*, with *steady-state* direct current, the characteristic impedance reduces to $\sqrt{R/G}$. This may be a somewhat variable quantity on an open-wire line, for G will depend to a large extent on the amount of moisture present on the insulators.

As the frequency increases, G becomes negligible compared with ωC, and R becomes negligible compared with ωL, giving a high-frequency characteristic impedance very nearly equal to $\sqrt{L/C}$.

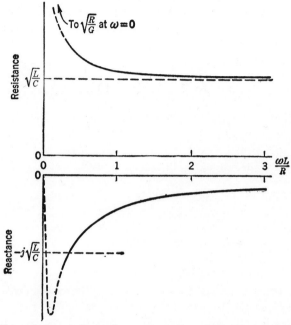

FIG. 2.12.   The resistive and reactive components of the characteristic impedance plotted against the dimensionless frequency variable, $\omega L/R$. The condition $G/C << R/L$ is assumed.

On lines with good insulation the ratio G/C is much smaller than R/L, and so G becomes negligible at a much lower frequency than R.   If $G \ll \omega C$, Eq. (2.47) can be written more simply as

$$Z_0 = \sqrt{\frac{L}{C}}\sqrt{1 - j\frac{R}{\omega L}} \tag{2.48}$$

The resistive and reactive components of this are plotted in Fig. 2.12 as functions of $\omega L/R$. The resistive component approaches the high-frequency value $\sqrt{L/C}$ rather rapidly, while the reactive component tends toward zero more slowly. The dashed portions of the curves indicate roughly the effect of the neglected conductance at the lowest frequencies.

A typical open-wire telephone line, composed of two No. 10 AWG copper wires spaced 12 in. apart, has the following constants: $R = 10.2$ ohms/mile,

$L = 0.00367$ henry/mile, and $C = 0.00821 \times 10^{-6}$ farad/mile. The leak-age conductance of an open-wire line is greatly affected by the amount of moisture present. We shall use $G = 0.3 \times 10^{-6}$ mho/mile, which is an approximate dry-weather value for a line in good condition. On this line the reactance ωL is equal to the resistance R at an angular frequency $\omega = R/L = 2{,}780$ rad/sec, or $f = 442$ cps. This frequency, therefore, will correspond to $\omega L/R = 1$ in Fig. 2.12. The approximate formula (2.48) fails at the lower frequencies where G becomes appreciable compared with ωC. For this line, G is equal to ωC at a frequency of about 6 cps.

The propagation constant for the uniform line is, from Eq. (2.31),

$$\gamma = \alpha + j\beta = \sqrt{(R + j\omega L)(G + j\omega C)} \tag{2.49}$$

The zero-frequency limit of γ is simply $\sqrt{RG}$. This, being real, is an attenuation constant; the phase constant is zero.

As the frequency increases, we finally obtain $\omega L \gg R$ and $\omega C \gg G$. Under these conditions, we can expand Eq. (2.49) by the binomial theorem and retain only the first two terms, i.e., we use $(a + b)^{\frac{1}{2}} \approx a^{\frac{1}{2}} + a^{-\frac{1}{2}}b/2$ for $a \gg b$. Then we obtain

$$\gamma = (j\omega L + R)^{\frac{1}{2}}(j\omega C + G)^{\frac{1}{2}}$$

$$\approx \left[(j\omega L)^{\frac{1}{2}} + \frac{R}{2}(j\omega L)^{-\frac{1}{2}} \right]\left[(j\omega C)^{\frac{1}{2}} + \frac{G}{2}(j\omega C)^{-\frac{1}{2}} \right]$$

Upon expansion and neglect of the small term involving the product RG, the high-frequency propagation constant is found to be

$$\gamma \approx \frac{R}{2}\sqrt{\frac{C}{L}} + \frac{G}{2}\sqrt{\frac{L}{C}} + j\omega\sqrt{LC} \tag{2.50}$$

Since $Z_0 \approx \sqrt{L/C}$ for high frequencies, we now have

$$\alpha \approx \frac{R}{2Z_0} + \frac{GZ_0}{2} \tag{2.51}$$

and, as for the lossless line,

$$\beta \approx \omega\sqrt{LC} \tag{2.52}$$

In Eq. (2.51), the term $R/2Z_0$ is the attenuation caused by energy losses in the conductors, while $GZ_0/2$ is the attenuation caused by energy losses in the insulation.

General expressions for α and β can be found by the following process: Square both sides of Eq. (2.49), separate the result into two equations by equating reals to reals and imaginaries to imaginaries, and then solve for

α and β. This process yields the rather cumbersome expressions

and
$$\left.\begin{array}{l}\alpha^2 = \tfrac{1}{2}[\sqrt{(R^2 + \omega^2 L^2)(G^2 + \omega^2 C^2)} + (RG - \omega^2 LC)] \\[2mm] \beta^2 = \tfrac{1}{2}[\sqrt{(R^2 + \omega^2 L^2)(G^2 + \omega^2 C^2)} - (RG - \omega^2 LC)]\end{array}\right\} \quad (2.53)$$

These equations can be used to study in detail the variation of α and β with frequency. Figures 2.13 and 2.14 show graphs of α and β *vs.* angular frequency as calculated for the typical open-wire telephone line mentioned

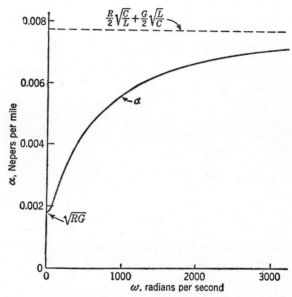

FIG. 2.13. Attenuation function *vs.* angular frequency, calculated for an open-wire line composed of two No. 10 copper conductors spaced 12 in. apart. The high-frequency value, $\dfrac{R}{2}\sqrt{\dfrac{C}{L}} + \dfrac{G}{2}\sqrt{\dfrac{L}{C}}$ is shown for comparison.

previously. Although only the comparatively low-frequency region is shown (0 to 478 cps), both α and β are already seen to be approaching their high-frequency asymptotes.

2.8. The Distortionless Line. The complex signals that are encountered in communications practice can be resolved by the Fourier analysis into sinusoidal components.[1] In linear systems, such as those we are consider-

---

[1] The reader is presumed to have some acquaintance with the Fourier series, by means of which a periodic signal can be resolved into sinusoidal components. The Fourier integral is similarly applicable to the analysis of nonperiodic signals. The Fourier series is discussed in many of the elementary textbooks on circuit theory and in books on engineering mathematics. Introductory treatments of the Fourier

ing, each of these components can be treated separately by the steady-state sinusoidal theory.   When a complex signal is impressed on a transmission system, the received wave will be of the same form as the transmitted one only if all components are attenuated equally and if they all travel at the same velocity.   The first requirement for zero distortion is, therefore, an attenuation constant that is independent of frequency.   Second, to make the phase velocities ω/β equal at all frequencies, the phase constant must be linearly proportional to frequency.   In general, these conditions are not

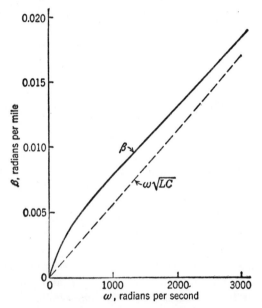

FIG. 2.14.   Phase function *vs.* angular velocity corresponding to Fig. 2.13.   The high-frequency asymptote $\omega\sqrt{LC}$ is plotted for comparison.

satisfied, for α and β are, respectively, the real and imaginary parts of a rather complicated function of frequency: $\sqrt{(R + j\omega L)(G + j\omega C)}$.

A distortionless line is obtained, however, if the constants of the line

---

integral can be found in E. A. Guillemin, "Communication Networks," Vol. II, Chap. XI, John Wiley & Sons, Inc., New York, 1935, and "The Mathematics of Circuit Analysis," Chap. VII, John Wiley & Sons, Inc., New York, 1949; W. C. Johnson, "Mathematical and Physical Principles of Engineering Analysis," Sec. 97, McGraw-Hill Book Company, Inc., New York, 1944; L. A. Pipes, "Applied Mathematics for Engineers and Physicists," Chap. III, Sec. 9, McGraw-Hill Book Company, Inc., New York, 1946; and J. A. Stratton, "Electromagnetic Theory," Sec. 5.7, McGraw-Hill Book Company, Inc.. New York, 1941.

have the following relation:

$$\frac{R}{L} = \frac{G}{C} \tag{2.54}$$

Under these conditions, the propagation constant reduces simply to

$$\gamma = \sqrt{RG} + j\omega\sqrt{LC} \tag{2.55}$$

The line is distortionless, for α is independent of frequency and β varies linearly with frequency. The phase velocity is simply $v = 1/\sqrt{LC}$ for all frequencies. The lossless line is a special case of this, for the condition $R = G = 0$ satisfies Eq. (2.54).

The student should show that a distortionless line has (1) a characteristic impedance equal to $\sqrt{L/C}$, (2) equal electric and magnetic energies associated with a traveling wave, and (3) equal energy losses in the resistance and leakage conductance of the line. The last two conditions hold, of course, only for a single traveling wave and not for the superposition of two oppositely traveling waves.

A condition of low distortion is particularly important for lines in which the ratio of the highest to the lowest frequency is rather large, as it is in telephone lines. On most transmission lines, the energy loss in the resistance of the conductors is greater than the loss in the insulator; that is, R/L is larger than G/C. (Exceptions to this occur when solid dielectrics are used in the ultra-high-frequency region, where the dielectric loss may be quite high.) The representative open-wire telephone line mentioned in the preceding section had $R/L = 2,780$ and $G/C = 36.5$. Even more extreme is the telephone cable, in which the close proximity of the conductors causes the capacitance to be quite high and the inductance low. Typical constants for a cable are $R = 60$ ohms/mile, $L = 0.001$ henry/mile, $C = 0.060 \times 10^{-6}$ farad/mile, and $G = 2 \times 10^{-6}$ mho/mile, giving $R/L = 60,000$ and $G/C = 33$.

It is not ordinarily desirable to increase G artificially, as this will greatly increase the attenuation. On the other hand, a reduction of R to satisfy Eq. (2.54) will require wires of unusually large diameter. To a certain extent, L can be increased and C reduced by a larger spacing between conductors, but there are practical limits to this. The inductance can also be increased by the process known as inductive loading, which is described in the next section. In practice, a certain amount of distortion can be tolerated, and a true distortionless line is rarely attempted.

2.9. Inductive Loading. On some types of transmission lines, the inductance is increased over its normal value either by wrapping the conductors with a high-permeability metal tape or by inserting inductance coils at uniform intervals. This is called inductive loading. Since G/C

is usually smaller than R/L, an increased inductance will make the line approach more nearly a distortionless condition, with the result that the characteristic impedance, attenuation constant, and phase velocity will be more nearly independent of frequency. Usually, however, the most desired effect of inductive loading is the reduction in attenuation that it produces. This will now be investigated, assuming for the moment that the added inductance is uniformly distributed so that the smooth-line formulas will apply.

In Sec. 2.7 it was shown that the high-frequency limit of α is

$$\alpha \approx \frac{R}{2}\sqrt{\frac{C}{L}} + \frac{G}{2}\sqrt{\frac{L}{C}} \tag{2.56}$$

When $R/L > G/C$, the first term of Eq. (2.56) will be larger than the second, indicating that conductor losses contribute the major portion of the attenuation. An increase in L will reduce the first term and increase the second by the same factor, but, up to the point where the two terms are equal, the net effect will be a reduction in the high-frequency limit of α. Physically, this means that an increased L will raise the high-frequency characteristic impedance $\sqrt{L/C}$, and thus a smaller current will be required for a given transmitted power, resulting in lower I^2R losses. A higher voltage will be required, and this will increase the insulation losses, but, so long as these remain the minor portion, the net effect will be a gain.

As an example, consider the line for which Fig. 2.13 was plotted. Without loading, the two terms of Eq. (2.56) are, respectively, 7.63×10^{-3} and 0.10×10^{-3} neper/mile, indicating that the copper loss is 76 times as great as the loss in the insulators. If L were increased by a factor of 4 with no change in the other constants, the first term would be reduced to 3.82×10^{-3} and the second one would be increased to 0.20×10^{-3}, giving a total attenuation constant of 4.02×10^{-3} neper/mile, which is smaller than before. Referring again to Fig. 2.13, the greatest possible improvement would be obtained by reducing the high-frequency asymptote to the same value as the zero-frequency intercept, $\sqrt{RG}$. The line would then be distortionless. If L is increased beyond this point, the second term of Eq. (2.56) dominates and the high-frequency asymptote of α increases.

Inductive loading is much more necessary for telephone cables than for open-wire lines because of the inherently low inductance and high capacitance of the cable. The limitations of space within the cable sheath make desirable a smaller wire than is commonly used on open-wire lines, thus causing a higher resistance. The normal inductance is low because of the proximity of the conductors, and the presence of solid insulation contributes to the high capacitance. As a result, the conductor loss dominates by a large factor. The attenuation and phase velocity vary widely within the audio range, and the high-frequency asymptote of attenuation is quite high. Inductive loading is, therefore, in general use on cable lines.

An audio-frequency transmission line can be smoothly loaded by wrapping the conductors with a high-permeability metal tape. However, this method of construction is quite expensive and is used principally on submarine cables where other methods are impractical. Cable lines on land are "lump-loaded" with inductance coils spaced at uniform intervals. A disadvantage of this method is that the lumped inductances in the line give rise to a low-pass filter effect which distorts the higher frequencies. This effect is reduced as the coils are placed closer together in a better approximation to a smooth line. From physical reasoning, it can be seen that there must be at least several coils per half wavelength at the highest desired frequency if the line is to behave with a reasonable approximation to smoothness.

In the past it was common practice to load open-wire telephone lines to reduce their attenuation. However, the loading had a variable effect which depended on the leakage conductance as determined by moisture on the insulators. Also, the increasing requirements on high-frequency response would have made necessary a rather small spacing between loading coils. Open-wire lines are now used without loading, and vacuum-tube amplifiers are inserted at intervals frequent enough to keep the signals from being attenuated into the noise level from which they could not be recovered.

2.10. Phase and Group Velocities. In Sec. 2.4 we discussed the phase velocity, which is the velocity of a point of constant phase on a sinusoidal traveling wave. As was shown in Eq. (2.37), this is given by the relation

$$v_p = \frac{\omega}{\beta} \qquad (2.57)$$

where we now use the subscript p to distinguish this velocity from another that we shall discuss in a moment.

When a complex signal is impressed on a transmission system which has a phase function β that is not linearly proportional to frequency, the various sinusoidal components of the signal will have different phase velocities and the wave will change shape as it travels along. Under these conditions, the signal cannot be said to travel at any one velocity.

A particularly simple situation with regard to velocities is obtained in the important case of a signal which has all its sinusoidal components grouped closely together in frequency. An example of this kind of wave is an amplitude-modulated signal, in which the amplitude of a sinusoidal "carrier" is varied in proportion to the instantaneous value of a low-frequency signal. One of the components of such a wave has a frequency equal to that of the carrier itself. Each of the frequencies in the original signal produces two "side-band" components, one below and the other above the carrier by the amount of the modulating frequency.[1]   If this set of voltages is impressed on a transmission system, it is found that the envelope of the wave, which has the same shape as the original signal, travels at one velocity, while the actual wave which is enclosed within the envelope travels at another velocity.

Consider, for simplicity, an amplitude-modulated wave which has only two side bands.   The amplitude of the carrier will be denoted by E_1, its angular frequency by ω, and the phase constant of the transmission line at this frequency by β.   The amplitude of each side band will be denoted by E_2, their frequencies by $\omega - \Delta\omega$ and $\omega + \Delta\omega$, respectively, and the corresponding phase constants by $\beta - \Delta\beta$ and $\beta + \Delta\beta$.   The traveling wave consisting of these three components can be written as

$$e = E_2 \cos \left[(\omega - \Delta\omega)t - (\beta - \Delta\beta)x \right] + E_1 \cos (\omega t - \beta x)$$

$$+ E_2 \cos \left[(\omega + \Delta\omega)t - (\beta + \Delta\beta)x \right] \qquad (2.58)$$

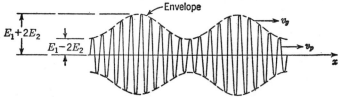

Fig. 2.15.   Illustrating group and phase velocities for a simple amplitude-modulated wave.

If we expand the expressions for the side-band voltages by the trigonometric identity $\cos (a + b) = \cos a \cos b - \sin a \sin b$ and then collect terms, we obtain

$$e = [E_1 + 2E_2 \cos (\Delta\omega\, t - \Delta\beta\, x)] \cos (\omega t - \beta x) \qquad (2.59)$$

We can regard the bracketed expression as the amplitude of a traveling wave $\cos (\omega t - \beta x)$ which moves with the phase velocity ω/β.   The amplitude itself varies sinusoidally between the extreme limits $E_1 + 2E_2$ and $E_1 - 2E_2$, and, furthermore, it moves along the line with a velocity $\Delta\omega/\Delta\beta$. For components grouped closely together, this velocity approaches the derivative $d\omega/d\beta$, which is the reciprocal of the slope of the phase function. This is the so-called *group velocity*:

$$v_g = \frac{d\omega}{d\beta} \qquad (2.60)$$

[1] See F. E. Terman, "Radio Engineering," 3d ed., Secs. 1.5 and 9.1, McGraw-Hill Book Company, Inc., New York, 1947; M.I.T. Staff, "Applied Electronics," pp. 632–638, John Wiley & Sons, Inc., New York, 1943, or any standard textbook on communication theory.

which, in general, is different from the phase velocity ω/β. Note, however, that for the distortionless case, where $\beta = \omega\sqrt{LC}$, the phase and group velocities are both equal to $1/\sqrt{LC}$.

The picture that we now get of the traveling amplitude-modulated wave is illustrated in Fig. 2.15. The envelope, which is the imaginary curve joining the peaks of the actual wave, moves along the line at the group velocity, while the actual wave "slips through" the envelope at the phase velocity.[1]

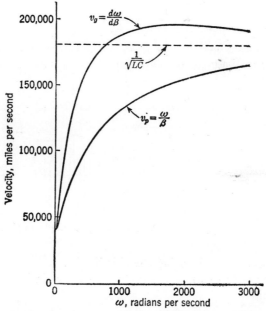

FIG. 2.16. The phase and group velocities corresponding to Fig. 2.14. The distortionless value $1/\sqrt{LC}$ is plotted for comparison.

In Fig. 2.14 was shown a calculated curve of β vs. ω for a representative open-wire telephone line. The corresponding phase and group velocities are plotted in Fig. 2.16.

The concept of group velocity loses its usefulness when the components of a wave are spread so widely in frequency that $\Delta\omega/\Delta\beta$ is not approximately equal to the derivative $d\omega/d\beta$.

It will be observed that, as in the example plotted in Fig. 2.16, the group velocity may exceed the speed of light. However, we are dealing here with a steady-state wave which, in theory, has no beginning and no end. Therefore, we cannot "tag" any of the energy as it enters the transmission system

[1] See H. H. Skilling, "Fundamentals of Electric Waves," 2d ed., pp. 228–231, John Wiley & Sons, Inc., New York, 1948.

and then identify it as it leaves, and so we cannot draw any final conclusion regarding the velocity with which the actual energy is propagated. For this reason a group velocity greater than that of light does not violate the relativity postulate. We can identify the position of the energy in a signal of short duration, but the Fourier analysis of such a wave shows that the components are widely spaced in frequency. The concept of group velocity then loses its usefulness. Further analysis shows that the wave front of a suddenly applied signal cannot travel faster than light, and the relativity postulate is not violated.[1]

PROBLEMS

1. A certain open-wire telephone line has the following constants at 1,000 cps:

$R = 6.75$ ohms/mile
$L = 3.50 \times 10^{-3}$ henry/mile
$C = 0.00872 \times 10^{-6}$ farad/mile
$G = 0.40 \times 10^{-6}$ mho/mile

Using Eqs. (2.30) and (2.31), compute Z_0, α, and β. Determine the phase velocity and the wavelength at this frequency.

2. A certain telephone cable (without loading coils) has the following constants at 1,000 cps:

$R = 86.0$ ohms/mile
$L = 1.1 \times 10^{-3}$ henry/mile
$C = 0.0620 \times 10^{-6}$ farad/mile
G negligible

Using Eqs. (2.30) and (2.31), compute Z_0, α, and β. Determine the phase velocity and the wavelength at this frequency.

3. With inductive loading coils inserted, the telephone cable of Prob. 2 has the following constants at 1,000 cps:

$R = 95.0$ ohms/mile
$L = 151 \times 10^{-3}$ henry/mile
$C = 0.0620 \times 10^{-6}$ farad/mile
G negligible

Compute Z_0, α, β, the phase velocity, and the wavelength at this frequency. Compare the results with those of Prob. 2.

4. If a telephone cable is not "loaded" with series inductance coils, both the inductance and the leakage can be neglected over a considerable range in frequency $(G/C << \omega << R/L)$.

[1] For a more complete treatment of group velocity, see E. A. Guillemin, "Communication Networks," Vol. II, Chap. III, Sec. 6, John Wiley & Sons, Inc., New York, 1935. For a discussion of group and signal velocities, see J. A. Stratton, "Electromagnetic Theory," Secs. 5.17 and 5.18, McGraw-Hill Book Company, Inc., New York, 1941, and R. I. Sarbacher and W. A. Edson, "Hyper and Ultra-high Frequency Engineering," Sec. 5.8, John Wiley & Sons, Inc., New York, 1943.

a. Starting with Eqs. (2.30) and (2.31), neglect L and G and write the resulting expressions for Z_0, α, and β.

b. Using the data $R = 86.0$ ohms/mile, $C = 0.062 \times 10^{-6}$ farad/mile, plot the following quantities against frequency for the range $200 < f < 3{,}000$ cps: α, the phase velocity, and the resistive and reactive components of Z_0.

c. If $L = 1.1 \times 10^{-3}$ henry/mile and $G = 2.0 \times 10^{-6}$ mho/mile, at what frequency will G be equal to ωC? At what frequency will ωL be equal to R?

5. A certain telephone cable, without inductive loading coils, has the following constants: $R = 42.0$ ohms/mile, $L = 1.1 \times 10^{-3}$ henry/mile, $C = 0.0620 \times 10^{-6}$ farad/mile, G negligible. Loading coils are to be added which will provide an additional inductance of 30.0×10^{-3} henry/mile. Also, they will add an estimated 6 ohms resistance per mile at 3,000 cps. Find the spacing between coils which will make them one-sixth of a wavelength apart at 3,000 cps.

6. An open-wire line which is composed of two No. 10 AWG conductors spaced 3 cm apart has the following constants:

At 1,000 cps, $R = 6.55 \times 10^{-3}$ ohm/meter, $L = 1.36 \times 10^{-6}$ henry/meter, $C = 8.84 \times 10^{-12}$ farad/meter, G negligible.

At 100 Mc, $R = 0.606$ ohm/meter, $L = 1.26 \times 10^{-6}$ henry/meter, $C = 8.84 \times 10^{-12}$ farad/meter, G negligible.

For each of the foregoing two frequencies, find α, Z_0, and the neper loss per wavelength, and compare.

7. A d-c source is impressed on the sending end of a cable which has $R = 86.0$ ohms/mile and $G = 1.50 \times 10^{-6}$ mho/mile. Compute Z_0 and α for this condition of operation. Find the ratio of the receiving-end voltage to the sending-end voltage if the cable is 100 miles long and is terminated in its characteristic impedance.

8. Show that, in a system having the same impedance level at the two points in question, an attenuation of 1 neper is equivalent to 8.686 db.

9. Compute the power ratios and, assuming equal impedances, the voltage ratios, corresponding to 1, 3, 10, and 30 db.

10. A typical open-wire telephone line has an attenuation of 0.0050 neper per mile at 1,000 cps, while a typical loaded telephone cable has an attenuation of 0.050 neper/mile at this frequency. For each of these lines, compute the length which will give an attenuation of 10 db.

11. Unloaded telephone cables are sometimes used for carrier telephony. A certain unloaded cable has the following constants at a frequency of 30,000 cps: $R = 52.5$ ohms/mile, $L = 1.1 \times 10^{-3}$ henry/mile, $C = 0.0620 \times 10^{-6}$ farad/mile, G negligible. Compute α, β, the wavelength and phase velocity, and the length of line which gives an attenuation of 25 db.

12. An open-wire telephone line has the following characteristics at 1,000 cps: $Z_0 = 615 - j87$ ohms, $\alpha = 0.00491$ neper/mile, and $\beta = 0.0351$ rad/mile. The line is 150 miles long and is terminated in its characteristic impedance. A 1,000-cps generator provides a sending-end voltage of 5.0 volts.

a. Compute the phase velocity and the wavelength.

b. Compute the sending-end current and power, and the receiving-end power.

c. Plot a polar diagram of E similar to Fig. 2.8, taking E_s to be horizontal.

13. A flexible coaxial cable with a solid dielectric is to be used at a frequency of 300 Mc. Its characteristics are: $Z_0 = 52.0 + j0$ ohms, $\alpha = 0.0156$ neper/meter, velocity constant $= 0.660$ [see Eq. (2.39)]. The line is 75 meters long and is terminated in its characteristic impedance. At the sending end, it is driven from a 300-Mc generator which has an open-circuit voltage $E_g = 50$ volts rms and an internal impedance $Z_g = 52 + j0$ ohms. Find the magnitude of the sending-end voltage and of the receiving-end voltage. Compute the sending-end power and the receiving-end power. How many wavelengths long is the line?

14. In Chap. 1, the reflection coefficient for a wave striking the generator end of a line was shown to be

$$k_g = \frac{Z_g - Z_0}{Z_g + Z_0}$$

Suppose that a line is operating in steady state with a load impedance equal to its characteristic impedance and is driven by a generator that has an open-circuit voltage E_g and an internal impedance Z_g. Show that

$$E_s = \frac{1 - k_g}{2} E_g$$

and that, therefore, at any point x,

$$E = \frac{E_g}{2}(1 - k_g)\epsilon^{-\gamma x}$$

15. Show that a distortionless line has (a) a characteristic impedance equal to $\sqrt{L/C}$, (b) equal electric and magnetic energies associated with a traveling wave, and (c) equal energy losses in the resistance and in the leakage conductance of the line.

16. As shown by Eq. (2.51), the high-frequency attenuation constant of a line is given by

$$\alpha_{\text{h-f}} = \frac{R}{2}\sqrt{\frac{C}{L}} + \frac{G}{2}\sqrt{\frac{L}{C}}$$

The zero-frequency limit of α is $\sqrt{RG}$. Arrange the foregoing equation to express the ratio $\alpha_{\text{h-f}}/\sqrt{RG}$ as a function of the ratio LG/RC. Plot this function over the range $0 < LG/RC < 4$. Observe the minimum that occurs at $LG/RC = 1$, which corresponds to the distortionless condition $R/L = G/C$.

17. At high frequencies, β is given approximately by $\omega\sqrt{LC}$. Starting with the expression for β^2 given by Eq. (2.53) and using the binomial theorem, show that when G is negligible and $\omega L \gg R$, a better approximation to β is

$$\beta = \omega\sqrt{LC}\left(1 + \frac{R^2}{8\omega^2L^2}\right)$$

18. Use the result of Prob. 17 to derive expressions for the phase and group velocities. These expressions will hold, of course, only when G is negligible and when $\omega L \gg R$. Evaluate the two velocities for the cable of Prob. 11.

CHAPTER 3

THE CONSTANTS OF TWO-CONDUCTOR LINES

3.1. A Qualitative Picture of Skin Effect. When an alternating current flows in a conductor, the alternating magnetic flux within the conductor induces an emf. This emf causes the current density to decrease in the interior of the wire and to increase at the outer surface. The result, which is known as the *skin effect*, increases in prominence as the frequency is raised. When ferromagnetic conductors are used, the effect may be appreciable even at commercial power frequencies. At radio frequencies, the current in a wire of moderate size is concentrated in a thin skin at the surface. Analysis shows that when the cross-sectional dimensions of the conductor are much larger than the effective thickness of the "skin" of current, the current density varies exponentially inward from the surface. The distance in which the current density decreases to $1/\epsilon$ of its surface value is called the *nominal depth of penetration* ($\epsilon = 2.718\ldots$). The name may be somewhat misleading, for there is, of course, an appreciable amount of current below this depth. The nominal depth of penetration is analogous to the time constant of an exponential transient. In Sec. 3.2 it is shown that the nominal depth of penetration is given by the relation

$$\delta = \sqrt{\frac{\rho}{\pi f \mu}} \qquad \text{meters} \qquad (3.1)$$

where ρ is the resistivity of the conductor in ohm-meters (sometimes called ohms per meter cube)

f is the frequency, in cycles per second

and μ is the absolute magnetic permeability of the conductor, in rationalized mks (meter-kilogram-second) units of henrys per meter. (The free-space value of μ in these units is $4\pi \times 10^{-7}$ henry/meter.)

For a copper conductor, Eq. (3.1) reduces to

$$\delta = \frac{6.64}{\sqrt{f}} \qquad \text{cm} \qquad (3.2)$$

Thus, the nominal depth of penetration for a copper conductor is about 0.86 cm at 60 cps, and is only 0.0066 cm at 1 Mc.

Further analysis shows that when the external magnetic flux density is not uniform around the surface of the conductor, the current concentrates most strongly at the positions where the magnetic flux is greatest.   If the conductor is cylindrical and the return wire is many radii away (as in many two-wire lines), the magnetic flux near the conductor will be circular and concentric with the wire.   The flux density will be nearly uniform around the periphery, and a high-frequency current will concentrate uniformly on the surface.   However, if the wire is rectangular, the surface flux density will be greatest at the corners of the conductor, and the current will concentrate most strongly there.

When the two conductors of a transmission line are close together, the magnetic flux density is greatest in the region between the conductors, and

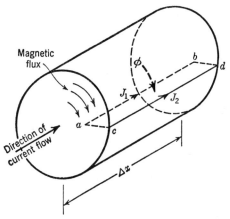

Fig. 3.1.   Two parallel longitudinal paths through a solid cylindrical conductor.

the current will tend to concentrate on the surfaces which face each other. This phenomenon is given the special name *proximity effect*.   It is most noticeable when the line is made of two parallel strips spaced closely together, as in Fig. 1.1c.   On such a line, a high-frequency alternating current will flow mainly on the two facing surfaces of the strips.

In a coaxial line which carries equal and opposite currents in the two conductors, there is no magnetic flux outside the outer conductor.   The flux which is "external" to the conductors, therefore, is in the annular space between them.   An alternating current will tend to concentrate on the surfaces which bound this space: the inner surface of the outer conductor and the outer surface of the inner conductor.

To gain a better qualitative idea of the causes and results of the skin effect, consider a solid cylindrical conductor which carries an alternating current, the return conductor being many radii away so that the magnetic flux within the conductor will run in concentric circles. Figure 3.1 shows a length Δx of such a conductor. Consider the two longitudinal paths ab and cd, ab being in the interior of the conductor and cd at the surface. The current densities along the two paths are denoted by J_1 and J_2, and the magnetic flux linking the rectangle $abdca$ is denoted by ϕ. Around the closed path the algebraic sum of the iR drops plus the induced emf $d\phi/dt$ must be equal to zero. The iR drop along the path ab is equal to $J_1\rho\,\Delta x$ volts, where J_1 is the current density in amperes per square meter, ρ is the resistivity in ohm-meters, and Δx is measured in meters. Similarly, the iR drop along the line cd is $J_2\rho\,\Delta x$. The induced emf around the path $abdca$ is equal to $d\phi/dt$ volts, where ϕ is the flux in webers (1 weber equals 10^8 maxwells). Therefore, around the closed path we can write

$$J_1\,\rho\,\Delta x - J_2\,\rho\,\Delta x + \frac{d\phi}{dt} = 0$$

or

$$J_1\,\rho\,\Delta x + \frac{d\phi}{dt} = J_2\,\rho\,\Delta x \tag{3.3}$$

On direct current the rate of change of flux is zero, and the two current densities J_1 and J_2 are equal. But, on alternating current, with $d\phi/dt$ no longer zero, J_1 must be smaller than J_2 to maintain the equality (3.3). If the line ab were chosen closer to the center of the wire, ϕ would be larger and the current density J_1 would be still smaller; hence, we see that the current density must decrease from the surface of the wire inward. Also, the inequalities in current density grow larger as the time rate of change of flux is raised by an increase in frequency.

Furthermore, one can note from Eq. (3.3) that an increase in ϕ cannot be offset merely by a decrease in J_1 without some sort of shift in phase of the two quantities, for one term involves a derivative while the other does not. This complication cannot be followed through except by a mathematical analysis, but the qualitative picture leads one to expect that the currents at various radii will not be in phase with each other. When we speak of the a-c current carried by a conductor, we mean the integral of the current density over the cross section, taking into account the shift in phase at the various radii.

When the frequency is high enough, the central portions of the conductor will not have to carry any measurable amount of current, for the reason that the $d\phi/dt$ term alone will be sufficient to satisfy Eq. (3.3) for those paths.

When the current is nonuniformly distributed through the cross section of a conductor, the heating of the wire is increased, thus causing a higher effective resistance. The high-frequency resistance of a conductor may be many times its d-c resistance. Along with the increase in a-c resistance there is a reduction in that part of the inductance which is caused by the internal flux, for this flux is diminished as the current moves toward the outside of the conductor. For nonmagnetic conductors which are spaced many radii apart, most of the inductance is caused by the flux which is external to the conductors, and with such lines the reduction in total inductance is not great. In Sec. 3.4 it is shown that the high-frequency resistance of a conductor can be calculated in a very simple way, provided that the flux density is uniform around the periphery. One imagines the actual conductor to be replaced by a fictitious hollow one which has the same surface shape, but which has a wall thickness just equal to the nominal depth of penetration. Then, the d-c resistance of the fictitious hollow conductor is precisely equal to the high-frequency resistance of the actual conductor. At lower frequencies, where the current penetrates deeply into the conductor, a more exact analysis must be used. Such an analysis is given for a cylindrical conductor in Secs. 3.5 and 3.6. The results for the a-c resistance and for the inductance caused by internal flux are presented graphically in Figs. 3.10 and 3.11.

At the higher frequencies the center of a solid conductor is of little use except as a mechanical support. Hollow tubing is frequently used because of its greater rigidity and larger perimeter for a given amount of material.

It is possible to force the current to flow through a greater portion of the cross section of a conductor by making the wire of strands which are woven together so that each strand successively occupies positions at different radii. Such conductors are designated as litzendraht, or litz wire. They are sometimes used at frequencies up to perhaps 1 or 2 Mc, where the depth of penetration becomes about equal to the radius of an individual strand. Beyond this range, solid conductors or hollow tubing is generally used.

3.2. Skin Effect in a Plane Conductor.[1] We shall first analyze a limiting case of skin effect, namely, that where the effective depth of penetration of current into the conductor is much smaller than the cross-sectional dimensions of the conductor. This is generally the case at radio frequencies. In the next section, we shall give a more general analysis of the circularly symmetric case at any frequency.

When the depth of penetration is very much smaller than the radius of curvature of the surface, we can imagine that the surface is perfectly flat over a small region, as indicated in Fig. 3.3. Also, if the thickness of the

[1] Sections 3.2 through 3.6, which present the theory of skin effect in some detail, can be omitted without destroying the continuity of the text. The results given graphically in Figs. 3.10 and 3.11 should, however, be noted.

conductor is many times greater than the depth of penetration, we can assume that the conductor is infinite in depth without appreciable error. We shall express all quantities in the rationalized mks system of units[1] and shall use the following nomenclature:

$\mathcal{E}$ = electric field intensity in volts per meter.

ϕ = magnetic flux in webers (1 weber = 10^8 maxwells).

B = magnetic flux density in webers per square meter.

H = magnetic field intensity in ampere-turns per meter (1 amp-turn/ meter is equivalent to $4\pi/1{,}000$ gilberts/cm, or oersteds).

J = current density in amperes per square meter.

μ = magnetic permeability. (The free-space value in rationalized mks units is $\mu_0 = 4\pi \times 10^{-7}$ henry/meter.)

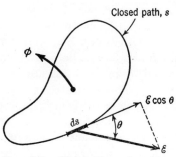

FIG. 3.2. Illustrating the integrand of Eq. (3.4).

ρ = resistivity in ohm-meters.

z = distance from the surface of the conductor, measured perpendicularly inward, in meters.

Two fundamental laws are needed for this analysis. The first of these is Faraday's law of induction, which states that the net emf developed around any closed path is equal to $-d\phi/dt$, where ϕ is the magnetic flux linking the path. The induced emf can be expressed by integrating the tangential component of electric field intensity around the path. The component of the electric field along the direction of the path is $\mathcal{E}\cos\theta$ where θ is the angle between the directions of $\mathcal{E}$ and s (see Fig. 3.2), and so the emf is

$$\text{emf} = \oint \mathcal{E} \cos\theta \, ds = -\frac{d\phi}{dt} \tag{3.4}$$

where ϕ is the total magnetic flux linking the path. The circle on the integral sign indicates that the integral is to be taken around a closed path.

The second law needed is that the magnetomotive force (mmf) around any closed path is equal numerically to the current linking that path (the factor 4π is absent from this expression in the system of units we are using).

---

[1] Explanations of the mks system of units can be found in many books, including A. B. Bronwell and R. E. Beam, "Theory and Application of Microwaves," pp. 459–461, McGraw-Hill Book Company, Inc., New York, 1947, and MIT Staff, "Electric Circuits," Appendix C, John Wiley & Sons, Inc., New York, 1940.

The mmf can be written as the integral of the tangential component of magnetic field intensity around the path, and so

$$\text{mmf} = \oint H \cos \theta \, ds = I \tag{3.5}$$

where θ is now the angle between the directions of H and s.[1]

In Eqs. (3.4) and (3.5), the direction of integration around the path and the positive directions of flux and current, respectively, should be related according to the right-hand rule.

Two other relations will be necessary in our analysis. The first of these is the familiar one relating magnetic flux density to the magnetic field intensity: $B = \mu H$. The other is Ohm's law expressed in terms of current density J and the electric field intensity $\mathcal{E}$, namely, $\mathcal{E} = J\rho$ where ρ is the resistivity of the conductor.

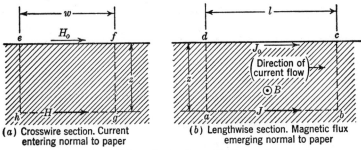

(a) Crosswire section. Current entering normal to paper

(b) Lengthwise section. Magnetic flux emerging normal to paper

FIG. 3.3. A solid conductor bounded by a plane surface.

Now refer to Fig. 3.3 and consider the volume of width w and length l, bounded at the top by the surface of the conductor and at the bottom by a plane lying at a depth z. First, we shall write Eq. (3.4) around the path $abcda$. The electric field intensity along the line ab is, by Ohm's law, equal to $J\rho$, while along the line dc at the surface it is equal to $J_0\rho$. The electric field has the same direction as the current and hence has no component along the lines bc and da. Therefore, the integral in Eq. (3.4) becomes $J\rho l - J_0\rho l$. The total magnetic flux linking the path $abcda$ can be obtained by integrating the magnetic flux density over the area:

$$\phi = l \int_0^z B \, dz$$

[1] The reader who has studied vector analysis will recognize the integrands of Eqs. (3.4) and (3.5) to be the scalar products $\mathcal{E} \cdot ds$ and $H \cdot ds$, respectively. Also, the student will find that in using Eq. (3.5) we shall omit the displacement current. This is permissible in metals even at microwave frequencies because the high conductivity makes the conduction current so much greater than the displacement current.

Therefore, around the path $abcda$, Eq. (3.4) becomes

$$J\rho - J_0\rho = -\frac{d}{dt}\int_0^z B \, dz \qquad (3.6)$$

If we wish to consider only sinusoidally varying quantities, we can use the complex notation introduced in Sec. 2.1 and write

and
$$\left. \begin{array}{l} J = J_m\epsilon^{j\omega t} \\ J_0 = J_{m0}\epsilon^{j\omega t} \\ B = B_m\epsilon^{j\omega t} \end{array} \right\} \qquad (3.7)$$

Substituting these relations into Eq. (3.6), taking the indicated time derivative, and canceling out $\epsilon^{j\omega t}$, we obtain

$$J_m\rho - J_{m0}\rho = -j\omega \int_0^z B_m \, dz \qquad (3.8)$$

The integral sign can be removed by taking the derivative of this expression with respect to z and, since J_{m0} is a constant quantity, we then obtain

$$\rho \frac{dJ_m}{dz} = -j\omega B_m \qquad (3.9)$$

Since $B_m = \mu H_m$, this is equivalent to

$$\rho \frac{dJ_m}{dz} = -j\omega\mu H_m \qquad (3.10)$$

Equation (3.10) has two dependent variables, J_m and H_m; therefore, we need another equation. This can be obtained from the mmf law (3.5). Consider the path $efghe$ in Fig. 3.3. The integral in Eq. (3.5) becomes

$$H_0 w - H w$$

and the right side can be written as

$$w \int_0^z J \, dz$$

Therefore, we have

$$H_0 - H = \int_0^z J \, dz$$

Assuming a sinusoidal variation and using $H = H_m\epsilon^{j\omega t}$, $J = J_m\epsilon^{j\omega t}$, this becomes

$$H_{m0} - H_m = \int_0^z J_m \, dz \qquad (3.11)$$

Taking the derivative with respect to z, we obtain

$$-\frac{dH_m}{dz} = J_m \tag{3.12}$$

To eliminate H_m between (3.10) and (3.12), take the derivative with respect to z of Eq. (3.10). Then, as indicated by Eq. (3.12), substitute J_m in place of $-dH_m/dz$. This gives us the differential equation which indicates the variation of current density with depth:

$$\frac{d^2 J_m}{dz^2} = \frac{j\omega\mu}{\rho} J_m \tag{3.13}$$

The solution to this differential equation must be a function which, when differentiated twice, yields the original function multiplied by $j\omega\mu/\rho$. Such is the case with $C_1\epsilon^{-z\sqrt{j\omega\mu/\rho}}$, or $C_2\epsilon^{z\sqrt{j\omega\mu/\rho}}$, or the sum of these two. But the second function increases with z, and we know physically that this cannot occur in the present problem; hence, the appropriate solution must be simply

$$J_m = C_1\epsilon^{-z\sqrt{j\omega\mu/\rho}}$$

The constant C_1 can be expressed in terms of the current density at the surface by placing $J = J_{m0}$ at $z = 0$, yielding $C_1 = J_{m0}$. Hence, at any depth z, we have

$$\begin{aligned} J_m &= J_{m0}\epsilon^{-z\sqrt{j\omega\mu/\rho}} \\ &= J_{m0}\epsilon^{-z(1+j)\sqrt{\omega\mu/2\rho}} \end{aligned}$$

This is generally expressed as

$$J_m = J_{m0}\epsilon^{-z/\delta}\epsilon^{-jz/\delta} \tag{3.14}$$

where δ is called the *nominal depth of penetration*, or nominal skin depth, and is given by the relation

$$\delta = \sqrt{\frac{2\rho}{\omega\mu}} = \sqrt{\frac{\rho}{\pi f\mu}} \tag{3.15}$$

An expression for the instantaneous current density can be obtained by the method of Sec. 2.1. Equation (3.14) is regarded as giving the position at $t = 0$ of a rotating vector. Instantaneous values are obtained by projecting the rotating vector on the horizontal axis. Multiplying Eq. (3.14) by $\epsilon^{j\omega t}$ and taking the real part, we obtain

$$J = J_{m0}\epsilon^{-z/\delta}\cos(\omega t - z/\delta) \tag{3.16}$$

This relation is similar in form to the one that we obtained for a wave traveling without reflection on a smooth transmission line [compare with

Eq. (2.36)]. The physical picture is that of a wave which originates at the surface of the conductor and which travels perpendicularly inward, attenuating very rapidly as it goes. The wave is diminished to $1/\epsilon$ of its surface value in a distance equal to the nominal depth of penetration, δ. The skin of current has no definite depth, but the foregoing definition of δ provides a convenient measure of relative thickness. At a depth of 5δ, the current density is ϵ^{-5}, or 0.00674, of its value at the surface. A conductor thicker than this will behave almost as though it had infinite depth.

The attenuation and phase constants of the traveling wave are each equal to $1/\delta$, and so the phase of the wave is shifted 1 radian as its magni-

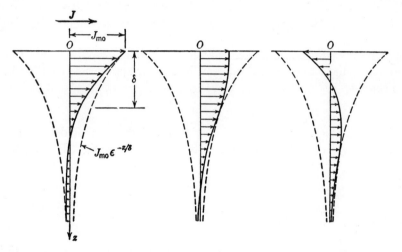

FIG. 3.4. A wave traveling downward in a solid metal body, shown at three successive instants of time.

tude is reduced by the factor $1/\epsilon = 0.368$. The wavelength is $2\pi/\beta = 2\pi\delta$, and the phase velocity is $\omega/\beta = 2\pi\delta f$. A sketch of the traveling wave of current is shown in Fig. 3.4. Because of the high attenuation, the shape of the wave is greatly distorted.

The magnitude of the depth of penetration is particularly interesting. For copper, $\mu \approx \mu_0 = 4\pi \times 10^{-7}$ henry/meter, and $\rho \approx 1.74 \times 10^{-8}$ ohm-meter. Then, using Eq. (3.15),

$$\delta = \sqrt{\frac{1.74 \times 10^{-8}}{\pi f \times 4\pi \times 10^{-7}}} = \frac{0.0664}{\sqrt{f}} \quad \text{meter}$$

$$= \frac{6.64}{\sqrt{f}} \quad \text{cm}$$

Thus, at a frequency of 60 cps, the nominal depth of penetration in copper is 0.857 cm, or about 0.34 in. At 10 kc it is 0.066 cm, and at 10 Mc it is about 0.0021 cm. At radio frequencies, the nominal depth of penetration in copper is very small indeed.

An expression for the total current flowing in the conductor can be obtained by integrating the current density with respect to depth. Considering a width w, the total current is

$$I_m = w \int_0^\infty J_m \, dz \qquad \text{amp}$$

Substituting from Eq. (3.14) and integrating, we obtain

$$I_m = \frac{w\delta}{1+j} J_{m0} = \frac{w\delta}{\sqrt{2}} J_{m0} \underline{/-45^\circ} \tag{3.17}$$

Therefore, the net alternating current that flows through a cross section of width w has a crest value $w\delta J_{m0}/\sqrt{2}$ and lags the current density at the surface by 45°. This net current is the one we refer to in the transmission-line equations.

The equation for the magnetic field intensity, H, within the conductor can be found by substituting (3.14) into (3.10). Taking the derivative with respect to z, this results in

$$H_m = \frac{\rho J_{m0}}{\mu\omega\delta} \frac{1+j}{j} \epsilon^{-(1+j)z/\delta}$$

Using Eq. (3.15) for δ, we can write

$$H_m = \frac{\delta}{2} J_{m0}(1-j)\epsilon^{-(1+j)z/\delta} \tag{3.18}$$

Comparison with Eq. (3.14) shows that, at each point, the magnetic field intensity is related to the current density by

$$H_m = \frac{\delta}{2} J_m(1-j) = \frac{\delta}{\sqrt{2}} J_m \underline{/-45^\circ} \tag{3.19}$$

Therefore, the magnetic field intensity at every point is proportional in magnitude to the current density at that point, but lags by 45°, or one-eighth of a cycle. Also, comparison with Eq. (3.17) shows that the net integrated current per unit width is numerically equal to the magnetic field intensity at the surface, or, for a width w,

$$I_m = wH_{m0} \tag{3.20}$$

This could have been proved directly from the magnetomotive force law (3.5)

3.3. Internal Impedance.   Consider a short section of transmission line, as shown in Fig. 3.5.   We wish to establish the basis for determining the inductance and resistance of the line for use in the fundamental line equations.

Consider the path *abcda* in Fig. 3.5, where *ab* is a longitudinal path through one conductor and *cd* is a longitudinal path through the other. Around this path we can write the emf law given by Eq. (3.4), and from the resulting expression we can separate out the inductive and resistive voltages. The equation is easiest to write if the lines *ab* and *cd* are at the surfaces of

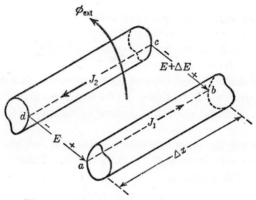

Fig. 3.5.   Section of a two-conductor line.

the conductors, for then the path is linked only by the flux that is external to the conductors.   Around this path, then, we can write

$$J_1\rho_1\,\Delta x + (E + \Delta E) + J_2\rho_2\,\Delta x - E = -\frac{d\phi_{\text{ext}}}{dt}$$

where ϕ_{ext} is the flux external to the conductors.   This equation can be rewritten as

$$-\Delta E = \frac{d\phi_{\text{ext}}}{dt} + (\rho_1 J_1 + \rho_2 J_2)\Delta x \qquad (3.21)$$

The quantity $-\Delta E$ is the decrease in the line-to-line voltage in the distance Δx.   Part of this drop is caused by the voltage $d\phi_{\text{ext}}/dt$ which is induced by the flux external to the conductors.   The rest is associated with the conductors themselves and may be said to be an *internal* impedance drop.   The internal impedance drop per unit length for either conductor is equal to the product $J\rho$ at the surface of the conductor and, by Ohm's law, this is equal to the longitudinal electric field intensity ε in volts per meter along the surface.   Dividing the volts-per-meter drop by the total current carried by the wire, we obtain the internal impedance. Thus, using maximum values,

$$Z_{\text{int}} = \frac{\mathcal{E}_{ms}}{I_m} = \frac{J_{ms}\rho}{I_m} \qquad \text{ohms/meter length} \qquad (3.22)$$

where I_m is the total current carried by the conductor and J_{ms} is the current density at the surface of the conductor.

Since the current density at the surface is not in phase with the total current, the ratio will be complex. The real part of (3.22) will be identified as the effective resistance of the conductor and the imaginary part as the internal reactance, both per unit length.

The quantity $-\Delta E$ in Eq. (3.21) would be unaltered if we chose the paths ab and cd to be inside the conductors. In this case we would have to consider a flux somewhat greater than ϕ_{ext} and the current densities J_1 and J_2 at this depth would be correspondingly smaller, the increase in one being exactly offset by a reduction in the other, as indicated in the discussion of Sec. 3.1.

Formulas for the inductance caused by the external flux will be derived for several common transmission-line configurations in later sections.

3.4. The Internal Impedance of a Plane Conductor. For the internal impedance of a plane conductor, we substitute Eq. (3.17) into (3.22) and obtain for a width w,

$$Z_{\text{int}} = \frac{\rho}{w\delta}(1 + j) \qquad \text{ohms/meter length} \qquad (3.23)$$

The real part of this is the effective resistance:

$$R = \frac{\rho}{w\delta} \qquad \text{ohms/meter length} \qquad (3.24)$$

The imaginary part is the internal reactance. Dividing by ω to obtain the internal inductance, we have

$$L_i = \frac{\rho}{\omega w\delta} \qquad \text{henrys/meter length} \qquad (3.25)$$

It is not difficult to show that the effective resistance as defined above, when multiplied by the square of the net effective current, gives the correct result for the average power dissipated in the conductor.

It is worth while to note that the effective resistance per unit length as given by Eq. (3.24) is equal to the d-c resistance of a rectangular slab of width w and depth δ. This provides an easily remembered method for computing the high-frequency resistance of a conductor in certain simple cases. The nominal depth of penetration must be much smaller than the cross-sectional dimensions of the conductor in order for our analysis to hold. If the surface flux density is uniform, then, according to Eq. (3.20), the current will be distributed in a uniform skin. Under these conditions, the effective resistance of the conductor will be equal to the d-c resistance of a fictitious hollow conductor which has the same surface shape, but which has a thickness equal to the nominal depth of penetration. This simple method does not hold when the shape of the conductor is such that the magnetic flux density at the surface is nonuniform, for the current will then be distributed in a corresponding nonuniform manner, and this increases the losses still further.

Example. Find the high-frequency resistance and the internal inductance of a cylindrical conductor of radius a, assuming that the magnetic field is uniform around the periphery and that $\delta \ll a$. This might represent the central conductor of a coaxial line or one of the wires of a two-wire line when the spacing is much greater than the radius.

Using Eq. (3.24) with w set equal to the perimeter of the wire, we have

$$R = \frac{\rho}{2\pi a \delta} = \frac{1}{2a} \sqrt{\frac{\rho f \mu}{\pi}} \qquad \text{ohms/meter length} \qquad (3.26)$$

where a is the radius in meters, ρ is expressed in ohm-meters, and μ is in henrys per meter.

Under our assumption of a well-developed skin effect, the a-c resistance is seen to increase as the square root of the frequency.

Assuming a copper conductor with no tinning or silver plating on the surface, and taking $\rho = 1.74 \times 10^{-8}$ ohm-meter and $\mu = 4\pi \times 10^{-7}$ henry/meter, we obtain

$$R = 4.17 \times 10^{-8} \frac{\sqrt{f}}{a} \qquad \text{ohms/meter} \qquad (3.27a)$$

if a is measured in meters, or

$$R = 4.17 \times 10^{-6} \frac{\sqrt{f}}{a} \qquad \text{ohms/meter} \qquad (3.27b)$$

where a is measured in *centimeters*.

The internal inductance of the conductor is obtained from Eq. (3.25), using $w = 2\pi a$. Then

$$L_i = \frac{\rho}{2\pi a \omega \delta} = \frac{1}{4\pi a} \sqrt{\frac{\rho \mu}{\pi f}} \qquad \text{henrys/meter length} \qquad (3.28)$$

The internal inductance is seen to decrease as $1/\sqrt{f}$ when the skin effect is well developed.

It should be emphasized again that the foregoing methods do not apply when the nominal depth of penetration is at all comparable with the cross-sectional dimensions of the conductor. A more general analysis for a cylindrical conductor at any frequency is given in the next section.

3.5. Skin Effect in a Cylindrical Conductor. The analysis of this case proceeds in much the same way as for the plane case. We shall assume cylindrical symmetry. This is achieved naturally in the coaxial cable, but, for a two-wire line, it amounts to assuming that the return conductor is many radii away so that the electric and magnetic fields are nearly uniform around the conductors.

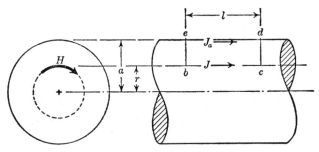

FIG. 3.6.   A solid cylindrical conductor.

Consider Fig. 3.6, which shows a portion of a cylindrical conductor. Using the emf law (Eq. 3.4) on the path *bcdeb*, we have

$$J\rho l - J_a \rho l = -\frac{d\phi}{dt}$$

where ϕ is the flux linking the path and is given by

$$\phi = l \int_r^a B \, dr$$

Therefore, we have

$$J\rho - J_a \rho = -\frac{d}{dt} \int_r^a B \, dr$$

Assuming a sinusoidal variation with time and using $J = J_m \epsilon^{j\omega t}$, $B = B_m \epsilon^{j\omega t}$, we obtain

$$J_m \rho - J_{ma} \rho = -j\omega \int_r^a B_m \, dr$$

Taking the derivative of this expression to remove the integral and noting that the variable r is the lower limit and that this reverses the sign, we obtain

$$\rho \frac{dJ_m}{dr} = j\omega B_m$$

Using μH_m for B_m, this becomes

$$\rho \frac{dJ_m}{dr} = j\omega\mu H_m \qquad (3.29)$$

The second necessary equation is obtained from the mmf law, Eq. (3.5). Writing this around the circular path at the radius r and setting the mmf equal to the enclosed current, we have

$$2\pi r H = \int_0^r J 2\pi r \, dr$$

Using $H = H_m \epsilon^{j\omega t}$ and $J = J_m \epsilon^{j\omega t}$, we obtain

$$2\pi r H_m = \int_0^r J_m 2\pi r \, dr$$

Taking the derivative with respect to r and rearranging, this becomes

$$\frac{dH_m}{dr} + \frac{1}{r} H_m = J_m \qquad (3.30)$$

To eliminate H_m, we solve (3.29) for H_m and substitute into (3.30). This yields the differential equation for the current density J_m:

$$\frac{d^2 J_m}{dr^2} + \frac{1}{r} \frac{dJ_m}{dr} = \frac{j\omega\mu}{\rho} J_m \qquad (3.31)$$

This differs from the corresponding equation for the plane case (3.13) by the new term $(1/r)dJ_m/dr$. Because of the variable coefficient $1/r$ that this involves, the solution is not one of the elementary functions. Equation (3.31) is a special form of Bessel's differential equation which is more usually written in the form

$$\frac{d^2 y}{dr^2} + \frac{1}{r} \frac{dy}{dr} + k^2 y = 0 \qquad (3.32)$$

This equation possesses two independent solutions which are known as Bessel functions of the first and second kinds.[1] The second kind approaches infinity as r approaches zero and hence is not a solution to our problem and

---

[1] Equation (3.32) is the Bessel equation whose solutions are said to be of zero order. The Bessel equation whose solutions are of the nth order is

$$\frac{d^2 y}{dr^2} + \frac{1}{r} \frac{dy}{dr} + \left(k^2 - \frac{n^2}{r^2} \right) y = 0$$

In our present problem we have $n = 0$. For a more general discussion of Bessel functions see N. W. McLachlan, "Bessel Functions for Engineers," Oxford University Press, New York, 1934.

is of no interest to us here. While the k of Eq. (3.32) is real in the most elementary case, the present problem involves a complex k, for we have $k^2 = -j\omega\mu/\rho$.

We can obtain the desired solution by writing it in the form of an infinite series and determining the coefficients by substituting the series into the differential equation. We therefore assume a solution of the form

$$y = a_0 + a_1 r + a_2 r^2 + a_3 r^3 + \cdots + a_n r^n + \cdots \qquad (3.33)$$

Taking the first and second derivatives, and substituting into Eq. (3.32), we obtain

$$2a_2 + 3\cdot 2a_3 r + 4\cdot 3a_4 r^2 + \cdots + n(n-1)a_n r^{n-2} + \cdots$$
$$+ a_1 r^{-1} + 2a_2 + 3a_3 r + \cdots + na_n r^{n-2} + \cdots$$
$$= -k^2(a_0 + a_1 r + a_2 r^2 + \cdots + a_{n-2} r^{n-2} + \cdots)$$

This is an identity; *i.e.*, it must hold for all values of r. In order for this to be possible, each power of r must have the same coefficient on both sides of the equation. Equating coefficients of like powers of r; first for r^{-1}, then for r^0 (the constant terms), next for r, and so on, we obtain the set of relations:

$$a_1 = 0$$
$$2a_2 + 2a_2 = -k^2 a_0$$
$$3\cdot 2a_3 + 3a_3 = -k^2 a_1$$
$$\dots\dots\dots\dots\dots\dots\dots\dots\dots$$
$$n(n-1)a_n + na_n = -k^2 a_{n-2}$$

First, we observe that the odd coefficients are all related to a_1, which is zero; therefore, all the odd coefficients must be zero. The even ones are all related to a_0:

$$a_2 = -\frac{k^2}{2^2} a_0$$

$$a_4 = -\frac{k^2}{2^4} a_2 = \frac{k^4}{2^2 4^2} a_0$$

and so on. In general, the following recursion formula can be used to find any a_n from the preceding a_{n-2}:

$$a_n = -\frac{k^2}{n^2} a_{n-2} \qquad (3.34)$$

The infinite series solution can now be written as

$$y = a_0\left[1 - \frac{(kr)^2}{2^2} + \frac{(kr)^4}{2^2\cdot 4^2} - \frac{(kr)^6}{2^2\cdot 4^2\cdot 6^2} + \cdots\right] \qquad (3.35)$$

The bracketed series is convergent for all finite values of kr, whether real or complex. When k is real, the series is known as the Bessel function of the first kind and zero order and is denoted by the symbol J_0. Its values are tabulated in many reference books. Its graph bears some resemblance to that of a damped cosine function. However, in our present problem we have

$$k^2 = -\frac{j\omega\mu}{\rho}$$

and, recalling from the previous section that the nominal depth of penetration is given by

$$\delta = \sqrt{\frac{2\rho}{\mu\omega}}$$

we can express k^2 as

$$k^2 = -j\frac{2}{\delta^2}$$

The solution for current density as a function of radius can, therefore, be written from Eq. (3.35) as

$$J_m = a_0\left[1 + j\frac{(\sqrt{2}r/\delta)^2}{2^2} - \frac{(\sqrt{2}r/\delta)^4}{2^2\cdot4^2} - j\frac{(\sqrt{2}r/\delta)^6}{2^2\cdot4^2\cdot6^2} + \cdots\right]$$

The terms are alternately real and imaginary. Separating them into two series, we have

$$J_m = a_0\left\{\left[1 - \frac{(\sqrt{2}r/\delta)^4}{2^2\cdot4^2} + \frac{(\sqrt{2}r/\delta)^8}{2^2\cdot4^2\cdot6^2\cdot8^2} - \cdots\right]\right.$$
$$\left. + j\left[\frac{(\sqrt{2}r/\delta)^2}{2^2} - \frac{(\sqrt{2}r/\delta)^6}{2^2\cdot4^2\cdot6^2} + \cdots\right]\right\} \qquad (3.36)$$

The two parts of this particular special form of the Bessel function are given special names, and the values of the bracketed series are tabulated as functions of the argument $\sqrt{2}r/\delta$. The names are derived from the words "Bessel function real" and "Bessel function imaginary"; thus we have

$$\text{ber } u = 1 - \frac{u^4}{2^2\cdot4^2} + \frac{u^8}{2^2\cdot4^2\cdot6^2\cdot8^2} - \cdots \left.\begin{matrix} \\ \\ \\ \\ \\ \end{matrix}\right\}$$

and $\qquad\qquad\qquad\qquad\qquad\qquad\qquad\qquad\qquad\qquad\qquad\qquad (3.37)$

$$\text{bei } u = \frac{u^2}{2^2} - \frac{u^6}{2^2\cdot4^2\cdot6^2} + \frac{u^{10}}{2^2\cdot4^2\cdot6^2\cdot8^2\cdot10^2} - \cdots$$

These functions and their first derivatives are plotted in Figs. 3.7 and 3.8.[1]

---

[1] A particularly useful set of tables will be found in H. B. Dwight, "Mathematical Tables," pp. 214–221, McGraw-Hill Book Company, Inc., New York, 1941.

The current density as a function of radius can now be expressed as

$$J_m = a_0 \left(\text{ber } \frac{\sqrt{2}r}{\delta} + j \text{ bei } \frac{\sqrt{2}r}{\delta} \right) \qquad (3.38)$$

The quantity a_0 is essentially a constant of integration, to be determined by the boundary conditions. If we denote the current at the outer radius by J_{ma}, we have upon substitution,

$$J_{ma} = a_0 \left(\text{ber } \frac{\sqrt{2}a}{\delta} + j \text{ bei } \frac{\sqrt{2}a}{\delta} \right)$$

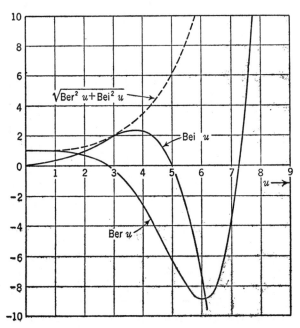

FIG. 3.7. The Bessel functions ber u and bei u.

We can solve this for a_0 and then write the general solution (3.38) as

$$J_m = J_{ma} \frac{\text{ber } \sqrt{2}r/\delta + j \text{ bei } \sqrt{2}r/\delta}{\text{ber } \sqrt{2}a/\delta + j \text{ bei } \sqrt{2}a/\delta} \qquad (3.39)$$

The absolute value of this gives the magnitude of the current density at the radius r; its angle gives the phase of the current. Graphs of the magnitude of current density vs. radius are given in Fig. 3.9 for various values of a/δ. The dashed curves show, for comparison, the exponential distribution as calculated for the plane case. It will be seen that, when the radius is much greater than the nominal depth of penetration, the distribution as calculated from the plane case is quite accurate. The graph for $a/\delta = 2$ corresponds to a frequency of about 10,500 cps for a copper wire 0.102 in. in diameter (No. 10 AWG), and the one for $a/\delta = 5$ corresponds to approximately 66,000 cps for this wire.

3.6. The Internal Impedance of a Cylindrical Conductor. As shown by Eq. (3.22), we can find the internal impedance of the cylindrical conductor (per unit length) by evaluating the longitudinal electric field intensity at the surface and dividing this by the total current carried by the wire.

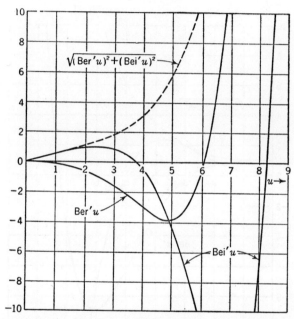

Fig. 3.8. The first derivatives of the ber and bei functions.

The total current can be found by integrating Eq. (3.39) over the cross section of the conductor; however, the easiest method is to observe from the magnetomotive force law (3.5) that the integral of H around the periphery of the wire is equal to the current enclosed. In terms of maximum values,

$$2\pi a H_{ma} = I_m \tag{3.40}$$

The magnetic field intensity at the surface of the wire is, from Eq. (3.29),

$$H_{ma} = \frac{\rho}{j\omega\mu} \left[\frac{dJ_m}{dr} \right]_{r=a} \tag{3.41}$$

The derivative dJ_m/dr can be found from Eq. (3.39). We shall use primes to denote derivatives of ber and bei with respect to their arguments, *i.e.*

$$\frac{d}{dr}\, \text{ber}\, \frac{\sqrt{2}r}{\delta} = \frac{\sqrt{2}}{\delta}\frac{d}{du}\, \text{ber}\, u = \frac{\sqrt{2}}{\delta}\, \text{ber}'\, u$$

and

$$\frac{d}{dr}\, \text{bei}\, \frac{\sqrt{2}r}{\delta} = \frac{\sqrt{2}}{\delta}\frac{d}{du}\, \text{bei}\, u = \frac{\sqrt{2}}{\delta}\, \text{bei}'\, u$$

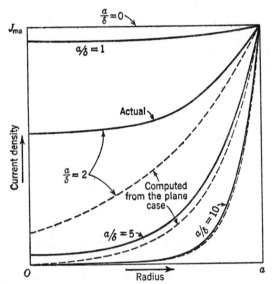

FIG. 3.9. Distribution of alternating current within a solid cylindrical conductor. The dashed exponential curves are computed from the plane case and are plotted for comparison.

Using this notation, we substitute (3.39) into (3.41) and then use (3.40) to find I_m. After simplification, this results in

$$I_m = \sqrt{2}\pi a\delta J_{ma}\left(\frac{\text{bei}'\, q - j\, \text{ber}'\, q}{\text{ber}\, q + j\, \text{bei}\, q}\right) \tag{3.42}$$

where $q = \sqrt{2}a/\delta$.

Now, from Eq. (3.22), the internal impedance is given by

$$Z_{\text{int}} = \frac{J_{ma}\,\rho}{I_m} \qquad \text{ohms/meter length}$$

Substituting for I_m from (3.42), this becomes

$$Z_{\text{int}} = \frac{\rho}{\sqrt{2}\pi a\delta}\left(\frac{\text{ber}\, q + j\, \text{bei}\, q}{\text{bei}'\, q - j\, \text{ber}'\, q}\right) \qquad \text{ohms/meter} \tag{3.43}$$

Rationalizing this expression to find the real and imaginary parts, we obtain for the effective resistance and the internal inductance of the cylindrical conductor:

$$R = \frac{\rho}{\sqrt{2}\pi a\delta} \frac{\text{ber } q \text{ bei}' q - \text{bei } q \text{ ber}' q}{(\text{bei}' q)^2 + (\text{ber}' q)^2} \qquad \text{ohms/meter} \qquad (3.44)$$

and

$$\omega L_i = \frac{\rho}{\sqrt{2}\pi a\delta} \frac{\text{bei } q \text{ bei}' q + \text{ber } q \text{ ber}' q}{(\text{bei}' q)^2 + (\text{ber}' q)^2} \qquad \text{ohms/meter} \qquad (3.45)$$

It can be shown from the series expressions defining the ber and bei functions that, as the frequency approaches zero ($q \to 0$), the above expressions reduce to

$$R_0 = \frac{\rho}{\pi a^2} \qquad \text{ohms/meter}$$

and

$$L_{i0} = \frac{\mu}{8\pi} \qquad \text{henrys/meter}$$

$$(3.46)$$

The above equation for resistance is, of course, the same as would be found from the elementary formula $\rho \times$ length/area. The expression for the low-frequency internal inductance will be derived in a different way in Sec. 3.8. The foregoing formulas were derived for a single conductor and must be multiplied by 2 for a two-wire line.

It is particularly convenient to have the ratios of the resistance and internal inductance at any frequency to the values at zero frequency. From Eqs. (3.44) to (3.46), these ratios can be written as

$$\frac{R}{R_0} = \frac{q}{2} \frac{\text{ber } q \text{ bei}' q - \text{bei } q \text{ ber}' q}{(\text{bei}' q)^2 + (\text{ber}' q)^2} \qquad (3.47)$$

and

$$\frac{L_i}{L_{i0}} = \frac{4}{q} \frac{\text{bei } q \text{ bei}' q + \text{ber } q \text{ ber}' q}{(\text{bei}' q)^2 + (\text{ber}' q)^2} \qquad (3.48)$$

where $q = \sqrt{2}a/\delta$.

The ratios R/R_0 and L_i/L_{i0} are plotted in Figs. 3.10 and 3.11 as functions of the ratio of radius to nominal skin depth, a/δ. When the depth of penetration is much smaller than the radius of the wire, the resistance and internal inductance approach the values that were computed on the basis of the plane case [see Eqs. (3.26) and (3.28)]:

$$R \approx \frac{\rho}{2\pi a\delta} = R_0 \frac{a}{2\delta} \qquad \text{ohms/meter}$$

and

$$L_i \approx \frac{\rho}{2\pi a\omega\delta} = L_{i0} \frac{2\delta}{a} \qquad \text{henrys/meter}$$

(3.49)

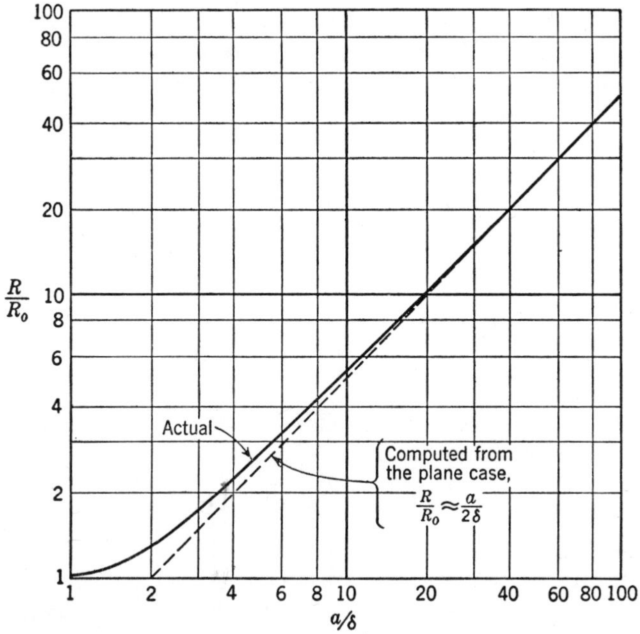

Fig. 3.10. Resistance ratio for a cylindrical conductor, assuming a uniform magnetic field around the periphery. $R_0 = \rho/\pi a^2$ ohms/meter for one conductor, $\delta = \sqrt{\rho/\pi f\mu}$ meter, a = radius of the conductor.

To obtain the inductance per unit length of a two-wire transmission line, the internal inductance L_i should be multiplied by 2 to take account of both wires, and the result added to the inductance caused by the flux that is external to the wires. The externally caused inductance is almost always the major portion of the total. As the frequency is raised, the internal inductance tends toward zero.

3.7. The Field about a Long Cylindrical Conductor. In preparation for the derivation of formulas for the capacitance and the external inductance of parallel-wire and coaxial lines, we shall review briefly the derivation of the magnetic field about a cylindrical current-carrying conductor and the electric field about a charged conductor.

Consider an isolated cylindrical conductor, as shown in Fig. 3.12, and assume that the conductor is so long that end effects can be ignored. The conductor is carrying a current I in the axial direction. From symmetry considerations, the magnetic flux will be circular about the wire.

We shall first determine the flux outside the conductor. The mmf acting around a closed path encircling the conductor is numerically equal to the

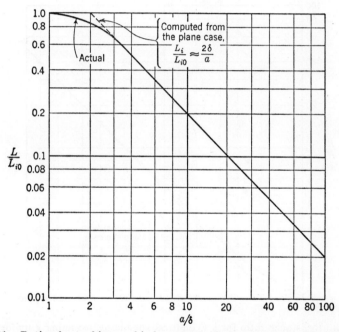

Fig. 3.11. Ratio of actual internal inductance to the low-frequency internal inductance of a cylindrical conductor. $L_{i0} = \mu/8\pi$ henry/meter for one conductor.

current I, regardless of the distribution of current across the cross section (the factor 4π being absent in the mks rationalized system of units). Therefore, around a circle of radius r,

$$H = \frac{\text{mmf}}{2\pi r} = \frac{I}{2\pi r} \tag{3.50}$$

The flux density outside the wire is, therefore,

$$B = \mu H = \frac{\mu I}{2\pi r} \qquad \text{for } r > a \tag{3.51}$$

where a is the radius of the wire and μ is the absolute permeability of the insulating medium.   In general, this permeability will be very nearly equal to that of free space, $\mu_0 = 4\pi \times 10^{-7}$ henry/meter.

Next, consider the flux inside the wire.   We shall assume for the moment that the frequency is low enough so that skin effect is negligible and the current is distributed uniformly.   Then, inside a circle of radius r within the conductor, the enclosed current will be Ir^2/a^2.   Around this circle,

$$H = \frac{Ir^2/a^2}{2\pi r} = \frac{Ir}{2\pi a^2}$$

Then, for $r < a$,

$$B = \frac{\mu_c r}{2\pi a^2} I \qquad (3.52)$$

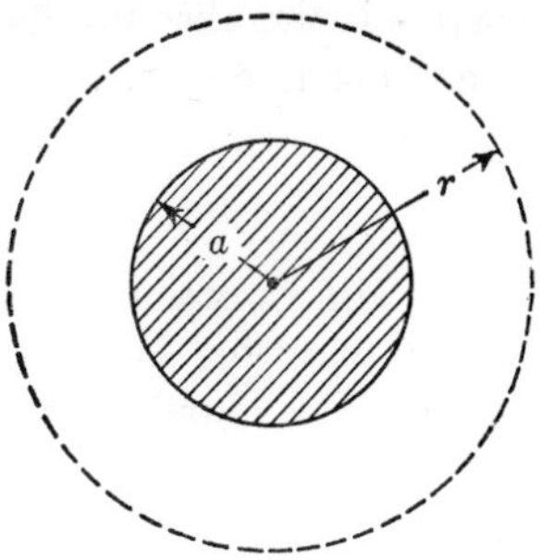

where μ_c is the absolute permeability of the conductor.   The foregoing relation will not hold, of course, when the skin effect is appreciable.   As the frequency is increased, the current will be forced to the surface of the conductor and the internal flux density will be reduced.   The effect of this on inductance was shown in Fig. 3.11.

FIG. 3.12.   Cross section of a cylindrical conductor.

The magnetic field intensity, H, is sketched in Fig. 3.13 for an isolated current-carrying conductor.   The straight-line variation shown within the conductor assumes a uniform current distribution.

In calculating capacitance, we shall need an expression for the electric field surrounding an isolated cylindrical conductor which carries an electrical charge of q coulombs per meter length.   From symmetry considerations, the electric flux will extend radially from the conductor.   The total

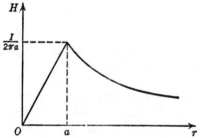

FIG. 3.13.   The magnetic field intensity about an isolated cylindrical conductor. The linear variation of H shown inside the conductor assumes a uniform current density.

electric flux is numerically equal to the charge in coulombs (the factor 4π is again absent in the mks rationalized system).   Therefore, at a radius r the electric flux density is

$$D = \frac{q}{2\pi r}$$

where q is the charge in coulombs per meter length. The electric field intensity is, therefore,

$$\mathcal{E} = \frac{D}{\epsilon} = \frac{q}{2\pi r \epsilon} \quad \text{volts/meter} \tag{3.53}$$

where ϵ is the absolute dielectric constant of the insulator surrounding the wire. For free space,

$$\epsilon_0 = \frac{1}{36\pi} \times 10^{-9} \text{ farad/meter}$$

3.8. The Constants of Parallel-wire Lines. A cross-sectional view of a parallel-wire line is shown in Fig. 3.14. A sketch of the magnetic and electric fields about the line was shown in Fig. 1.1a. We shall assume that

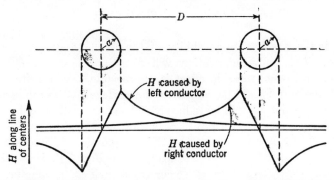

FIG. 3.14. Cross-sectional view of a parallel-wire line. The variation of H shown within the conductors assumes a uniform current distribution.

the spacing between wires, D, is considerably greater than the wire diameter, a, so that proximity effects can be neglected. Consider a 1-meter length of the complete circuit formed by both conductors. We shall use superposition and shall find the flux linkages created in the circuit by current flowing in only one conductor, and shall then multiply the result by 2 to take account of the current in both conductors. We shall consider separately the inductance caused by the flux which is outside the conductors and that caused by flux which is inside the conductors themselves (the "internal" inductance).

Consider first the external flux caused by current in the left-hand conductor. The portion of this flux which lies between the wires links the entire loop, while the portion which cuts through the right-hand conductor links only a portion of the complete circuit, and finally, the flux beyond the right-hand conductor does not create any linkages with the circuit at all. As an approximation, we shall consider the flux to the center of the right-hand conductor as linking the total loop and the remainder of the flux as not linking the circuit at all.   The external flux density at a radius $r > a$ is given by Eq. (3.51).   In an annular ring of thickness dr and unit axial length, the magnetic flux is

$$d\phi = B \, dr = \frac{\mu I}{2\pi r} \, dr \qquad \text{webers/meter length}$$

Between $r = a$ and $r = D$ this flux links one turn, and so the external flux linkage is

$$\int_a^D \frac{\mu I}{2\pi r} \, dr$$

Integrating this expression and multiplying by 2 to take account of both conductors, we obtain for the external flux linkage:

$$\psi_e = \frac{\mu I}{\pi} \log_e \frac{D}{a} \qquad \text{weber-turns/meter length} \qquad (3.54)$$

The inductance of the circuit is equal to the flux linkages per ampere; consequently, dividing ψ_e by I, we obtain the inductance per meter length caused by the external flux:

$$L_e = \frac{\mu}{\pi} \log_e \frac{D}{a} \qquad \text{henrys/meter} \qquad (3.55)$$

Assuming that μ has the free space value of $4\pi \times 10^{-7}$ henry/meter for the insulator, this becomes

$$L_e = 4 \times 10^{-7} \log_e \frac{D}{a} \qquad \text{henrys/meter} \qquad (3.56)$$

The inductance caused by the internal flux in a cylindrical wire has already been found in Sec. 3.6.   The internal inductance at any frequency, L_i, was given by Eq. (3.45), and a simplified formula for the internal inductance at very low frequencies, L_{i0}, was given by the second of Eqs. (3.46).   The ratio L_i/L_{i0} was plotted in Fig. 3.11 as a function of the ratio of wire radius to nominal depth of penetration.   In this section, we shall use a different method to derive the low-frequency internal inductance, L_{i0}.   If the frequency is low enough so that the current density is approximately uniform, the internal flux density can be found from Eq. (3.52).   Then, in an annular ring of unit length and thickness dr, the flux is

$$d\phi = B \, dr = \frac{\mu_c r I}{2\pi a^2} \, dr$$

where μ_c is the absolute permeability of the conductor. This flux links a fraction r^2/a^2 of the total current; therefore, the corresponding internal flux linkage is

$$d\psi_{i0} = \frac{r^2}{a^2} d\phi = \frac{\mu_c r^3 I}{2\pi a^4} dr$$

Integrating from $r = 0$ to $r = a$ and multiplying by 2 to take account of the two conductors, we have for the low-frequency internal flux linkage:

$$\psi_{i0} = \frac{\mu_c I}{4\pi} \qquad \text{weber-turns/meter length} \qquad (3.57)$$

Dividing ψ_{i0} by I, we obtain for the low-frequency internal inductance of the two wires:

$$L_{i0} = \frac{\mu_c}{4\pi} \qquad \text{henrys/meter length} \qquad (3.58)$$

This checks with the result obtained in Eq. (3.46), which gave the inductance as $\mu/8\pi$ henrys/meter for one wire. The foregoing result can be placed in more convenient form by multiplying and dividing by the permeability of free space, $\mu_0 = 4\pi \times 10^{-7}$ henry/meter:

$$L_{i0} = \left(\frac{\mu_c}{\mu_0}\right) \times \left(\frac{\mu_0}{4\pi}\right) = \left(\frac{\mu_c}{\mu_0}\right) \times 10^{-7} \qquad \text{henry/meter} \qquad (3.59)$$

in which the ratio (μ_c/μ_0) is the relative permeability of the conductor material (approximately unity for nonmagnetic materials).

As the frequency is increased, the current will be forced toward the surface of the conductor, and the internal flux and internal inductance will be reduced. The internal inductance at any frequency can be written as $L_i = (L_i/L_{i0})L_{i0}$, or, substituting from Eq. (3.59),

$$L_i = \left(\frac{L_i}{L_{i0}}\right)\left(\frac{\mu_c}{\mu_0}\right) \times 10^{-7} \qquad \text{henry/meter} \qquad (3.60)$$

The total inductance of the line per meter length will be the sum of the external and internal inductances. Using Eqs. (3.56) and (3.60), this can be expressed as

$$L = \left[4 \log_\epsilon \frac{D}{a} + \left(\frac{\mu_c}{\mu_0}\right)\left(\frac{L_i}{L_{i0}}\right) \right] \times 10^{-7} \qquad \text{henry/meter} \qquad (3.61)$$

At low frequencies the internal inductance ratio, L_i/L_{i0}, is approximately unity. The ratio at any frequency can be obtained from Fig. 3.11. When the skin effect is well developed, L_i/L_{i0} approaches zero and the internal inductance can be ignored.

To convert the foregoing relations to a mile basis, multiply by 1,609 meters/mile.

Next, we shall find the capacitance per unit length under the assumption that $D \gg a$. The potential difference between lines is obtained by integrating the electric field strength due to one charged conductor (Eq. 3.53) from $r = a$ to $r = D$, and then multiplying by 2 to take account of the fields of both conductors:

$$\text{Potential difference} = 2 \int_a^D \mathcal{E} \, dr = \frac{q}{\pi \epsilon} \log_\epsilon \frac{D}{a} \qquad \text{volts}$$

where q is the charge per meter length on each conductor (the charges on the two conductors are equal but of opposite sign). The capacitance is defined as the charge per unit potential difference; hence, dividing q by the potential difference and using

$$\epsilon = \left(\frac{\epsilon}{\epsilon_0} \right) \epsilon_0 = \left(\frac{\epsilon}{\epsilon_0} \right) \frac{10^{-9}}{36\pi} \qquad \text{farads/meter}$$

we can write

$$C = \frac{(\epsilon/\epsilon_0)}{36 \log_\epsilon (D/a)} \times 10^{-9} \qquad \text{farad/meter} \qquad (3.62)$$

In the above equation the quantity ϵ/ϵ_0 is the relative dielectric constant of the insulating medium. Most parallel-wire lines use air insulation, for which $\epsilon/\epsilon_0 \approx 1$.

For the high-frequency low-loss case, the characteristic impedance is $Z_0 \approx \sqrt{L/C}$ and the phase velocity is $v \approx 1/\sqrt{LC}$. Then, using Eqs. (3.61) and (3.62) and neglecting the internal inductance, we have for the two-wire line

$$Z_0 \approx \frac{120}{\sqrt{\epsilon/\epsilon_0}} \log_\epsilon \frac{D}{a} \qquad \text{ohms} \qquad (3.63)$$

and

$$v \approx \frac{3 \times 10^8}{\sqrt{\epsilon/\epsilon_0}} \qquad \text{meters/sec} \qquad (3.64)$$

The foregoing analysis assumed that the wire diameter was much smaller than the spacing between wires, and the resulting formulas are sufficiently accurate for most purposes. A more exact analysis, which does not make this assumption, shows that the capacitance is given more accurately by the relation[1]

$$C = \frac{(\epsilon/\epsilon_0) \times 10^{-9}}{36 \cosh^{-1}(D/2a)} \qquad \text{farads/meter} \qquad (3.65)$$

[1] See, for example, G. P. Harnwell, "Principles of Electricity and Electromagnetism," p. 40, McGraw-Hill Book Company, Inc., New York, 1938. The function abbreviated *cosh* is the hyperbolic cosine, which we will discuss in Sec. 4.3.

At high frequencies, with a well-developed skin effect, the corresponding expression for inductance is

$$L = 4 \times 10^{-7} \cosh^{-1} \frac{D}{2a} \qquad \text{henrys/meter} \qquad (3.66)$$

The phase velocity, $1/\sqrt{LC}$, remains the same as given by Eq. (3.64), but the high-frequency characteristic impedance is now given by

$$Z_0 = \frac{120}{\sqrt{\epsilon/\epsilon_0}} \cosh^{-1} \frac{D}{2a} \qquad \text{ohms} \qquad (3.67)$$

The resistance of the line at low frequencies can be found from wire tables or by the d-c formula $\rho \times$ length/area. When the skin effect becomes appreciable, the resistance can be found from Fig. 3.10. At frequencies so high that the depth of penetration is much smaller than the radius of the wire, the method developed in Sec. 3.4 can be used: the a-c resistance of the wires is equal to the d-c resistance of a pair of cylindrical tubes which have a radius a and a wall thickness equal to the nominal depth of penetration, $\sqrt{\rho/\pi f \mu}$. Then we obtain for the two-wire line:

$$R = \frac{2\rho}{2\pi a \delta} = \frac{1}{a} \sqrt{\frac{\rho f \mu_c}{\pi}} \qquad \text{ohms/meter length of line} \qquad (3.68)$$

where a is the wire radius in meters, ρ is the resistivity in ohm-meters, and μ_c is the permeability of the conductor material in henrys per meter. For copper, this formula becomes

$$R = 8.34 \times 10^{-6} \frac{\sqrt{f}}{a} \qquad \text{ohms/meter length} \qquad (3.69)$$

where a is now the wire radius in centimeters.

3.9. The Constants of Coaxial Lines. We shall develop the constants of coaxial lines only for the case of a well-developed skin effect and shall neglect entirely the internal inductance of the conductors. Because of the cylindrical symmetry, the field in the annular space has the same configuration as that about an isolated cylindrical conductor (see Fig. 1.1d). The magnetic flux density at a radius r is given by Eq. (3.51) as $\mu I/2\pi r$. Referring to Fig. 3.15, the flux in an annular ring of unit length and thickness dr is

$$d\phi = \frac{\mu I}{2\pi r} dr$$

Integrating this from $r = a$ to $r = b$, we obtain for the total magnetic flux in the annular space:

$$\phi = \frac{\mu I}{2\pi} \log_\epsilon \frac{b}{a} \qquad \text{webers/meter length}$$

This flux links one turn; hence, the inductance is obtained by dividing ϕ by I, resulting in

$$L = \frac{\mu}{2\pi} \log_\epsilon \frac{b}{a} \qquad \text{henrys/meter} \qquad (3.70)$$

Assuming the free-space value for μ, this becomes

$$L = 2 \times 10^{-7} \log_\epsilon \frac{b}{a} \qquad \text{henrys/meter} \qquad (3.71)$$

The electric field strength in the annular space is given by Eq. (3.53) as $q/2\pi r\epsilon$, where q is the charge per unit length. Integrating the field strength with respect to radius from $r = a$ to $r = b$, we obtain for the potential difference:

$$\text{Potential difference} = \frac{q}{2\pi\epsilon} \log_\epsilon \frac{b}{a}$$

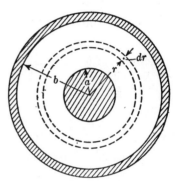

To find the capacitance per unit length we divide q by the potential difference and obtain

$$C = \frac{2\pi\epsilon}{\log_\epsilon (b/a)} \qquad \text{farads/meter} \qquad (3.72)$$

This can be written in terms of the free-space ϵ_0 as follows:

Fig. 3.15. Cross-sectional view of coaxial cable.

$$C = \frac{1}{18 \log_\epsilon (b/a)} \left(\frac{\epsilon}{\epsilon_0}\right) \times 10^{-9} \qquad \text{farads/meter} \qquad (3.73)$$

where ϵ/ϵ_0 is the relative dielectric constant of the insulator.

For the high-frequency case, the characteristic impedance is very nearly $\sqrt{L/C}$. Using the foregoing equations for L and C, we obtain

$$Z_0 \approx \frac{60}{\sqrt{\epsilon/\epsilon_0}} \log_\epsilon \frac{b}{a} \qquad \text{ohms} \qquad (3.74)$$

For air-insulated lines, the relative dielectric constant ϵ/ϵ_0 is, of course, practically unity.

In the high-frequency case, the phase velocity is approximately $1/\sqrt{LC}$. Then,

$$v \approx \frac{3 \times 10^8}{\sqrt{\epsilon/\epsilon_0}} \qquad \text{meters/sec} \qquad (3.75)$$

It should be observed that the comparatively high dielectric constant of a solid insulator will reduce the phase velocity appreciably below the free-space value and hence will shorten the wavelength at a given frequency.

The high-frequency resistance of a coaxial cable is equal to the d-c resistance of a circuit composed of two hollow conductors with radii a and b respectively, and with wall thicknesses equal to the nominal depth of penetration. The result is

$$R = \frac{1}{2} \sqrt{\frac{\rho f \mu}{\pi}} \left(\frac{1}{a} + \frac{1}{b} \right) \qquad \text{ohms} \qquad (3.76)$$

in which ρ is in ohm-meters, μ is in henries per meter, and a and b are measured in meters. Assuming copper conductors, this becomes

$$R = 4.2 \times 10^{-6} \sqrt{f} \left(\frac{1}{a} + \frac{1}{b} \right) \qquad \text{ohms/meter} \qquad (3.77)$$

where a and b are here measured in *centimeters*.

A flexible coaxial cable is made with a braided outer conductor and has solid insulation, often polyethylene, to give firmness to the cable and to support the inner conductor. At very high frequencies, the dielectric losses in the solid insulation may cause the conductance G to be appreciable. These losses are caused by a hysteresis effect of the molecules as they are polarized by the a-c field, but the result on the line is the same as that of a true conductivity. Consider for a moment an ordinary condenser of any shape, which is filled with a leaky dielectric. The admittance of the condenser will be of the form

$$Y = G + j\omega C = j\omega C \left(1 - j \frac{G}{\omega C} \right) \qquad (3.78)$$

The angle of the complex number in parentheses indicates the relative magnitude of the loss component of current. It is the amount by which the phase angle between voltage and current differs from 90° and is sometimes called the *loss angle*. The tangent of this angle, which is called the *loss tangent*, is often used to specify the loss properties of dielectrics. For small values of the loss angle, the loss tangent and the power factor are practically equal. From Eq. (3.78) the loss tangent is $G/\omega C$, and so for a transmission line with solid insulation we have

$$G = T_L \omega C \qquad (3.79)$$

where T_L signifies the loss tangent of the insulating material.

The high-frequency attenuation constant of a line is approximately

$\alpha \approx R/2Z_0 + GZ_0/2$. For a given outer radius b, there is a certain ratio b/a that will minimize the attenuation. The attenuation caused by dielectric loss is independent of the dimensions of the line, for we can write

$$\frac{G}{2}\sqrt{\frac{L}{C}} = \frac{T_L\omega C}{2}\sqrt{\frac{L}{C}} = \frac{T_L\omega\sqrt{LC}}{2}$$

and $\sqrt{LC}$ is the reciprocal of the phase velocity, which does not depend on the line dimensions. However, the part of the attenuation constant caused by conductor losses, $R/2Z_0$, does depend on the dimensions. If the inner conductor is very small, R will be high. On the other hand, if the inner conductor is nearly as large as the outer one, Z_0 will be very small. When the expression for α is minimized by the usual methods, it turns out that $b/a \approx 3.6$ for minimum attenuation. The corresponding value of Z_0 is approximately $77/\sqrt{\epsilon/\epsilon_0}$ ohms. However, the minimum is rather flat, and the value of b/a is not at all critical.

The maximum power that can be transmitted by a coaxial line may be limited by the dielectric strength of the insulator, or, in the case of solid insulation at the highest frequencies, by the heat generated by losses in the dielectric.[1] For a given permissible electric field intensity, and for a given ratio b/a, the power-handling capacity of a coaxial line varies as the square of its diameter. Therefore, high values of instantaneous power require cables of comparatively large diameter.

In the type of wave that we have been analyzing, the electric and magnetic fields are both transverse to the direction of propagation. If the diameter of a coaxial cable is made large enough compared with a wavelength, nontransverse waves can also be transmitted in the same fashion as in a hollow-pipe wave guide. The analysis of these waves is beyond the scope of this text; however, an approximate rule is that they may exist if the mean circumference of the annular space is greater than the wavelength; $i.e.$, if $(2\pi a + 2\pi b)/2 = \pi(a + b)$ is greater than the wavelength.[2] In coaxial cable transmission, the transverse electromagnetic wave, or "principal mode," is generally desired without the presence of higher modes. This tends to place an upper limit on the radius of the line, a limit which may be embarrassing at microwave frequencies, particularly if the attenuation is to be kept low or if large amounts of power are to be handled.

[1] See, for example, Willis Jackson, "High-frequency Transmission Lines," pp. 53–58, Methuen & Co., Ltd., London. Distributed in the United States by the Sherwood Press, Cleveland, Ohio.

[2] See S. Ramo and J. R. Whinnery, "Fields and Waves in Modern Radio," Chap. 9, John Wiley & Sons, Inc., New York, 1944.

3.10. The Constants of Parallel-strip Lines. A cross-sectional view of a parallel-strip line is shown in Fig. 3.16. If $b \gg a$ and the frequency is high enough so that the skin effect is well developed, most of the current will flow on the inner surfaces of the strips. Edge effects can be neglected, and the fields can be assumed to be uniform between the conductors. This assumption is similar to the one commonly made in computing the capacitance of a flat-plate condenser. Under these conditions, and assuming in addition that the insulator has a permeability equal to that of free space, the formulas for the line constants can be shown to be:

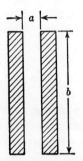

$$L = 4\pi \times 10^{-7} \frac{a}{b} \quad \text{henrys/meter} \quad (3.80)$$

$$C = \frac{1}{36\pi} \times 10^{-9} \frac{b}{a} \left(\frac{\epsilon}{\epsilon_0}\right) \quad \text{farads/meter} \quad (3.81)$$

$$Z_0 = \frac{120\pi}{\sqrt{\epsilon/\epsilon_0}} \frac{a}{b} \quad \text{ohms} \quad (3.82)$$

and

$$R = \frac{2}{b} \sqrt{\pi \rho f \mu} \quad \text{ohms/meter} \quad (3.83)$$

Fig. 3.16. Cross section of a parallel-strip transmission line.

Meter units are used throughout in the above formulas. The proof of the relations will be left to tne student.

It will be seen that, with $a \ll b$, the characteristic impedance is comparatively low. The attenuation constant is increased as the separation, a, is reduced.

PROBLEMS

1. Compute the nominal depth of penetration at 60 cps for

a. Silver, with $\mu = \mu_0 = 4\pi \times 10^{-7}$ henry/meter and $\rho = 1.63 \times 10^{-8}$ ohm-meter.

b. Brass, with $\mu = \mu_0$ and $\rho = 6.41 \times 10^{-8}$ ohm-meter.

c. Aluminum, with $\mu = \mu_0$ and $\rho = 2.83 \times 10^{-8}$ ohm-meter.

d. Iron, with $\mu = 250\mu_0$ and $\rho = 10.7 \times 10^{-8}$ ohm-meter. (The relative permeability of 250 is the "incremental permeability," which applies to small flux densities, and is considerably smaller than the maximum permeability of the iron.)

2. Compute the nominal depth of penetration for copper and express it in mils, for the following frequencies: 1,000 cps, 100 kc (k = kilo), 10 Mc (M = mega), and 1,000 Mc.

3. Show that, with direct current, the IR drop through a uniform wire of length l is equal to $J\rho l$, where J is the current density in the wire and ρ is the resistivity of the conductor material.

4. Compute ber 2 and bei 2 from the series expressions (3.37).

5. A copper wire with a diameter of 0.128 in. is to be used at a frequency of 15,000 cps. Determine the ratio of the current density at the center to the current density at the surface; also, find the phase difference between the two. Determine the ratio of the current density at $r = 0.032$ in. to the current density at the surface, and find the phase difference between them.

6. A two-wire line is to be made of copper wires which have a diameter of 0.128 in. Find the ratio of the a-c resistance to the d-c resistance for frequencies of 1,000 and 10,000 cps.

7. A copper wire with a diameter of 0.128 in. and a resistivity of 1.74×10^{-8} ohm-meter is to be used at a frequency of 15,000 cps.

a. Compute the internal impedance per meter length from Eq. (3.43) and Figs. 3.7 and 3.8. Compute the resistive and reactive components of this impedance.

b. Find R and L_i from Figs. 3.10 and 3.11, and compare with the results of part *a*.

8. An air-insulated two-wire line is composed of copper wires that have a diameter of 0.104 in. The wire spacing is 8 in. For a frequency of 10,000 cps, find R, L, C, and Z_0 of the line.

9. An air-insulated two-wire line is composed of two copper wires that have a diameter of 0.064 in. The spacing between conductors is 1 in. from center to center. For a frequency of 50 Mc, find R, L, C, and Z_0 of the line.

10. *a.* A high-frequency two-wire line is composed of two cylindrical copper conductors, each with a diameter of 0.125 in., and spaced 1 in. between centers. Compute the characteristic impedance from the logarithmic formula (3.63) and also from the hyperbolic formula (3.67). Compare the results.

b. Repeat part *a* for conductors with a diameter of 0.500 in. spaced 1 in. between centers.

11. An air-insulated coaxial line is composed of copper conductors and has the following dimensions: inside diameter of outer tube = 0.795 in., diameter of inner conductor = 0.250 in.

For a frequency of 100 Mc, compute Z_0, R, the attenuation constant assuming zero leakage conductance, and the attenuation in decibels per meter length.

12. A certain flexible coaxial line is filled with a dielectric material that has a relative dielectric constant of 2.25 and a loss tangent of 5×10^{-4} at 300 Mc. The copper conductors have the following dimensions: $a = 0.039$ in., $b = 0.143$ in. For a frequency of 300 Mc, find the capacitance of the cable per meter length. Also, determine Z_0, R, G, and the attenuation constant in both nepers per meter and in decibels per meter. Compute the phase velocity.

13. Prove Eqs. (3.80) through (3.83), which give the characteristics of parallel-strip lines.

14. A certain parallel-strip line is to be made of copper strips 2.50 in. wide. The spacing between strips is to be 0.200 in. and the frequency will be 200 Mc. The dielectric will be air. Find Z_0, R, and the attenuation constant assuming zero leakage conductance.

15. Some telegraph lines use a single wire, the other side of the circuit being the earth. If the earth is a good conductor, its surface can be considered as an equipotential plane in calculating the line constants. Furthermore, as indicated in Fig. P15, this plane can be replaced by an "image" conductor without affecting the field configuration. Show that, if the losses and the internal inductance of the wires are neglected, the characteristic impedance of such a line is given by

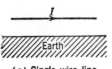

(a) Single-wire line

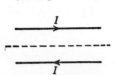

(b) Earth replaced by an image conductor

FIG. P15. A single-wire line with earth return.

$$Z_0 = 60 \log_\epsilon \frac{2h}{a} \qquad \text{ohms}$$

where a is the radius of the wire, and h is its height above the earth.

16. Prove Eq. (3.76), which gives the high-frequency resistance of a coaxial line.

17. Prove that, if the power-handling capacity of a coaxial cable is limited by the maximum electric field intensity, and if the ratio b/a is fixed, then the maximum permissible power is proportional to b^2.

18. Express the high-frequency attenuation constant of a coaxial line in terms of the dimensions a and b. Assuming that b is constant, minimize this with respect to a. Show that minimum attenuation is obtained when $b/a \approx 3.6$, and that the corresponding value of Z_0 is $77/\sqrt{\epsilon/\epsilon_0}$ ohms.

19. Use the series expressions for the ber and bei functions to show that, as the frequency approaches zero, the general expressions for resistance and internal inductance given by Eqs. (3.44) and (3.45) reduce to $R_0 = \rho/\pi a^2$ and $L_{i0} = \mu/8\pi$, respectively.

20. Use the series expressions for the ber and bei functions to show that, at low frequencies, the resistance of a cylindrical conductor is given approximately by

$$R = R_0 \left[1 + \frac{1}{48} \left(\frac{a}{\delta} \right)^4 \right]$$

where R_0 is the resistance at zero frequency, a is the radius of the conductor, and δ is the nominal depth of penetration.

21. Plot a graph of the characteristic impedance of a low-loss air-insulated two-wire line, as given by Eq. (3.63), vs. the ratio D/a over a range $2 < D/a < 200$. Use semilog paper, with D/a along the logarithmic scale. Superimpose on this a graph of the characteristic impedance as given by Eq. (3.67), and compare. Observe that the wires are in contact when $D/a = 2$.

CHAPTER 4

LINES WITH REFLECTIONS

4.1. Various Exponential Forms of the A-C Steady-state Solution. In Sec. 2.2 we solved the steady-state differential equations for the uniform transmission line and showed that the voltage and current could be written as

$$E = A_1 \epsilon^{-\gamma x} + A_2 \epsilon^{\gamma x} \tag{4.1}$$

and

$$I = \frac{1}{Z_0} (A_1 \epsilon^{-\gamma x} - A_2 \epsilon^{\gamma x}) \tag{4.2}$$

where the A's are constants with the dimensions of voltage, and x is the distance measured from the sending end. The quantities Z_0 and γ are, respectively, the characteristic impedance and the propagation constant of the line:

$$Z_0 = \sqrt{\frac{R + j\omega L}{G + j\omega C}}$$

$$\gamma = \sqrt{(R + j\omega L)(G + j\omega C)}$$

We saw that the real and imaginary parts of γ play entirely different roles in the propagation of the waves; the real part is the attenuation constant α, and the imaginary part is the phase constant β:

$$\gamma = \alpha + j\beta$$

To obtain the instantaneous values of voltage and current, we multiply Eqs. (4.1) and (4.2) by $\sqrt{2}\epsilon^{j\omega t}$ and take the real part of the result. Thus, the instantaneous voltage can be written as

$$e = \sqrt{2} \ \Re e[A_1 \epsilon^{-\alpha x} \epsilon^{j(\omega t - \beta x)} + A_2 \epsilon^{\alpha x} \epsilon^{j(\omega t + \beta x)}]$$

The first term within the brackets represents a wave that travels in the positive x direction with the phase velocity ω/β, as we saw in Sec. 2.4. Similarly, the second term in brackets represents a wave that travels in the negative x direction with the same speed. The wave traveling to the right is attenuated by the factor $\epsilon^{-\alpha x}$, and the wave traveling to the left is similarly attenuated in its direction of travel. In the latter case, the attenuation factor must appear as $\epsilon^{+\alpha x}$ because the wave travels in the direction of decreasing x.

In Sec. 1.6 we studied the propagation and reflection of waves on a lossless line and calculated the transients caused by a d-c emf. As will be shown in the next section, the a-c steady state can also be calculated in a similar way: one determines the initial outgoing wave from the generator and then considers all the reflections and rereflections of this wave. The results found in this way are identical with the steady-state solutions (4.1) and (4.2). The term $A_1 \epsilon^{-\gamma x}$ represents the net sum of all the individual waves that travel to the right, while the term $A_2 \epsilon^{\gamma x}$ represents the sum of all waves that travel to the left.

We shall call the first term in Eq. (4.1) the "incident" voltage and denote it by the symbol E^+. The second term we shall call the "reflected" voltage and denote it by the symbol E^-. Then we can write

$$E = E^+ + E^- \tag{4.3}$$

If we similarly write the total current as the sum of incident and reflected components, designated, respectively, by I^+ and I^-, we have

$$I = I^+ + I^- \tag{4.4}$$

Comparison with Eqs. (4.1) and (4.2) shows that

$$I^+ = \frac{E^+}{Z_0} \tag{4.5}$$

and

$$I^- = -\frac{E^-}{Z_0} \tag{4.6}$$

The characteristic impedance generally has only a small phase angle. Hence, I^+ is nearly in phase with E^+, while I^- is nearly 180° out of phase with E^-. We first saw this relation when we were analyzing transients on lossless lines. Physically, the explanation for the minus sign in Eq. (4.6) is the same as was given in Sec. 1.5 and Fig. 1.7: the current caused by a charge moving to the right is opposite to the current produced by a similar charge moving to the left, although the voltages have the same sign.

Next, we shall evaluate the constants A_1 and A_2 in the solutions (4.1) and (4.2). There are many equivalent forms of the solution, depending on the terminal quantities that are used. The more common forms use the following (see Fig. 4.1): I_R and Z_R; or I_s and the sending-end impedance Z_s; or Z_R, E_g, and Z_g. Often it is convenient to measure distances from the receiving end instead of from the sending end ($d = l - x$ as shown in Fig. 4.1).

To express the constants in terms of the sending-end quantities, set $E = I_s Z_s$ and $I = I_s$ at $x = 0$ in Eqs. (4.1) and (4.2). Then we have

$$I_s Z_s = A_1 + A_2$$
$$I_s Z_0 = A_1 - A_2$$

Solving for A_1 and A_2, we obtain

$$A_1 = \frac{I_s}{2}(Z_s + Z_0) \quad \text{and} \quad A_2 = \frac{I_s}{2}(Z_s - Z_0)$$

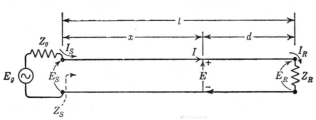

Fig. 4.1. Diagram showing notation.

Substituting back into (4.1) and (4.2) we obtain the solution in terms of sending-end quantities:

$$E = \frac{I_s}{2}[(Z_s + Z_0)\epsilon^{-\gamma x} + (Z_s - Z_0)\epsilon^{\gamma x}] \qquad (4.7)$$

and

$$I = \frac{I_s}{2Z_0}[(Z_s + Z_0)\epsilon^{-\gamma x} - (Z_s - Z_0)\epsilon^{\gamma x}] \qquad (4.8)$$

It is convenient for many purposes to express the constants in terms of the receiving-end quantities I_R and Z_R. For this, we go back to Eqs. (4.1) and (4.2) and substitute $E = I_R Z_R$ and $I = I_R$ at $x = l$. Then, solving for A_1 and A_2, we have

$$A_1 = \frac{I_R}{2}(Z_R + Z_0)\epsilon^{\gamma l} \quad \text{and} \quad A_2 = \frac{I_R}{2}(Z_R - Z_0)\epsilon^{-\gamma l}$$

Substituting these relations into Eqs. (4.1) and (4.2) and using $d = l - x$ for convenience, we obtain the solution in the form

$$E = \frac{I_R}{2}[(Z_R + Z_0)\epsilon^{\gamma d} + (Z_R - Z_0)\epsilon^{-\gamma d}] \qquad (4.9)$$

and

$$I = \frac{I_R}{2Z_0}[(Z_R + Z_0)\epsilon^{\gamma d} - (Z_R - Z_0)\epsilon^{-\gamma d}] \qquad (4.10)$$

The distance d is measured from the receiving end, as shown in Fig. 4.1.

The solution in terms of E_g, Z_g, and Z_R can be derived in a similar way by using $E = E_g - I_sZ_g$ at $x = 0$ and $E_R = I_RZ_R$ at $x = l$, but instead it will be found in a different way in the next section.

In each of the different forms of the solution, the first term represents the steady-state incident wave (E^+ and I^+), and the second term represents the steady-state reflected wave (E^- and I^-). Observe that the relations (4.5) and (4.6) apply to each.

The ratio of the reflected voltage to the incident voltage we shall call the *reflection coefficient*, k. From Eq. (4.9) this ratio at any distance from the receiving end is

$$k = \frac{E^-}{E^+} = \frac{Z_R - Z_0}{Z_R + Z_0}\epsilon^{-2\gamma d}$$

$$= \frac{Z_R - Z_0}{Z_R + Z_0}\epsilon^{-2\alpha d}\epsilon^{-j2\beta d} \tag{4.11}$$

The reflection coefficient is, in general, a complex number. At the receiving end it is

K_R

$$k_R = \frac{E_R^-}{E_R^+} = \frac{Z_R - Z_0}{Z_R + Z_0} \tag{4.12}$$

This is the same expression that we used in Sec. 1.6 for transients on a lossless line. The reflection coefficient at any point can be expressed in terms of k_R by dividing (4.11) by (4.12):

$$k = k_R\epsilon^{-2\gamma d} = k_R\epsilon^{-2\alpha d}\epsilon^{-j2\beta d} \tag{4.13}$$

Inspection of Eq. (4.10) shows that the reflection coefficient for current is the negative of that for voltage, i.e., at any point,

$$\frac{I^-}{I^+} = -k = \frac{Z_0 - Z_R}{Z_0 + Z_R}\epsilon^{-2\gamma d} \tag{4.14}$$

The impedance of a transmission line at any point is defined as the complex ratio of E to I at that point. Dividing Eq. (4.9) by (4.10), we obtain

$$Z = \frac{E}{I} = Z_0\frac{(Z_R + Z_0)\epsilon^{\gamma d} + (Z_R - Z_0)\epsilon^{-\gamma d}}{(Z_R + Z_0)\epsilon^{\gamma d} - (Z_R - Z_0)\epsilon^{-\gamma d}} \tag{4.15}$$

At $d = 0$ this impedance reduces, of course, to Z_R. The sending-end impedance of the line, Z_s, is obtained by placing $d = l$.

The transfer impedance of the line is defined as the ratio of the sending-end voltage to the receiving-end current. Using Eq. (4.9) and setting $E = E_s$ at $d = l$, we have

$$Z_{tr} = \frac{E_s}{I_R} = \frac{1}{2}[(Z_R + Z_0)\epsilon^{\gamma l} + (Z_R - Z_0)\epsilon^{-\gamma l}] \tag{4.16}$$

The ratio of the sending-end current to the receiving-end current can be obtained from Eq. (4.10) and is

$$\frac{I_s}{I_R} = \frac{(Z_R + Z_0)\epsilon^{\gamma l} - (Z_R - Z_0)\epsilon^{-\gamma l}}{2Z_0} \qquad (4.17)$$

In using the foregoing exponential relations, the student should recall that

$$\epsilon^{\pm \gamma d} = \epsilon^{\pm(\alpha+j\beta)d} = \epsilon^{\pm\alpha d}\epsilon^{\pm j\beta d} = \epsilon^{\pm\alpha d}(\cos \beta d \pm j \sin \beta d)$$

When a transmission line is terminated in its characteristic impedance $(Z_R = Z_0)$, the general relations derived above reduce to those of Sec. 2.3, as of course they should. First, the impedance (4.15) reduces to $Z = Z_0$ at all points on the line, and hence $Z_s = Z_0$. The solutions (4.7) and (4.8) become $E = E_s\epsilon^{-\gamma x}$, $I = I_s\epsilon^{-\gamma x}$, and the current ratio (4.17) reduces to $I_R/I_s = \epsilon^{-\gamma l}$.

As indicated in Fig. 4.1, the generator works into the impedance Z_s at its terminals. From the point of view of the generator, an equivalent sending-end circuit is that shown in Fig. 4.2. When Z_s is known, perhaps by use of Eq. (4.15) with $d = l$, the equivalent sending-end circuit can be solved for E_s, I_s, and the sending-end power. The solutions (4.7) and

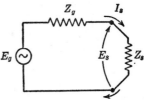

FIG. 4.2. Equivalent sending-end circuit.

(4.8) can then be used to find the current and voltage at any point on the line, or (4.16) or (4.17) can be used to find the receiving-end current.

Example. Given a telephone transmission line 100 miles long, with $Z_0 = 685 - j92$ ohms, $\alpha = 0.00497$ neper/mile, and $\beta = 0.0352$ rad/mile at 1,000 cps. The line is terminated in $Z_R = 2,000 + j0$ ohms. The generator has an emf of 10 volts rms and an internal impedance of 700 ohms resistance. Find the sending-end impedance, the sending-end current, voltage, and power, and the receiving-end voltage, current, and power.

For this line, $\beta l = 3.52$ rad $= 202°$. Then,

$$\epsilon^{\gamma l} = \epsilon^{0.497+j3.52} = \epsilon^{0.497}\underline{/202°}$$
$$= 1.64(\cos 202° + j \sin 202°)$$
$$= -1.52 - j0.615$$

Also,

$$\epsilon^{-\gamma l} = \epsilon^{-0.497}\underline{/-202°} = -0.566 + j0.229$$

Using Eq. (4.15), the sending-end impedance of the line is

$$Z_s = (685 - j92)\frac{(2,685 - j92)\epsilon^{\gamma l} + (1,315 + j92)\epsilon^{-\gamma l}}{(2,685 - j92)\epsilon^{\gamma l} - (1,315 + j92)\epsilon^{-\gamma l}}$$

$$= (691\underline{/-7.65°})\frac{5,060\underline{/14.4°}}{3,810\underline{/27.45°}} = 919\underline{/-20.7°}$$

$$= 861 - j325 \text{ ohms}$$

L
Z_0
α
β
f
Z_R
E_g
Z_g

A considerable part of the complex arithmetic necessary in the above computation can be avoided by use of the transmission-line chart described in Chap. 5, with, of course, some loss of potential accuracy.

The sending-end current can be found by solving the equivalent sending-end circuit of Fig. 4.2. If only magnitudes are desired, we have

$$|I_s| = \left|\frac{E_g}{Z_g + Z_s}\right| = \left|\frac{10}{700 + 861 - j325}\right|$$

$$= \frac{10}{1596} = 6.26 \times 10^{-3} \text{ rms amp}$$

The magnitude of the sending-end voltage is

$$|E_s| = |I_s| \cdot |Z_s| = 6.26 \times 10^{-3} \times 919 = 5.75 \text{ rms volts}$$

The sending-end power is $|I_s|^2 R_s$, where R_s is the resistive component of the sending-end impedance. Then,

$$P_s = (6.26 \times 10^{-3})^2 \times 861 = 33.8 \times 10^{-3} \text{ watt}$$

The variation of voltage and current along the line can now be written from Eqs. (4.7) and (4.8). The receiving-end current can be found from either (4.16) or (4.17). Using the latter and taking absolute values of numerator and denominator, we have

$$\frac{6.26 \times 10^{-3}}{|I_R|} = \frac{3,810}{2 \times 691}$$

in which we have made use of previous numerical results. Then,

$$|I_R| = 2.28 \text{ ma}$$

This flows through a 2000-ohm resistive load, and hence

$$|E_R| = 4.56 \text{ volts}, \qquad P_R = 10.4 \text{ mw}$$

4.2. Solution in Terms of E_g, Z_g, and Z_R. The equations for E and I along the line can be expressed in terms of E_g, Z_g, and Z_R by imposing the terminal relations $E_s = E_g - I_s Z_g$ and $E_R = I_R Z_R$ on Eqs. (4.1) and (4.2). However, we shall instead derive this form of the solution by a method analogous to that used in Sec. 1.6 for transients on a lossless line. In this analysis we shall ignore the possibility of local transients in the generator and load. Such transients may affect the manner in which the steady state builds up on the line, but they will not affect the final result or the general physical picture of the process.

When the generator is first connected to the line, it sees an impedance equal to Z_0 until the arrival of the first reflection from the load. Therefore, for the first few moments the sending-end voltage of the line will be $E_g Z_0/(Z_0 + Z_g)$. This wave, which we imagine to flow forevermore out of the generator, will be propagated down the line, and at a distance x will have the value

$$\frac{E_g Z_0}{Z_0 + Z_g} \epsilon^{-\gamma x} \tag{4.18}$$

As this wave reaches the load, it will be reflected with the reflection coefficient [see Eq. (1.25) or (4.12)]:

$$k_R = \frac{Z_R - Z_0}{Z_R + Z_0} \tag{4.19}$$

and thus will give rise to a second wave which will travel back toward the generator. To find an expression for the second wave, we take (4.18) and set $x = l$ to find the value of the wave at the load end, then we multiply by k_R to find the reflected voltage at that point, and finally we multiply by the propagation factor $\epsilon^{-\gamma d} = \epsilon^{-\gamma(l-x)}$ to allow for its travel backward. The result is

$$\frac{E_g Z_0}{Z_0 + Z_g} \epsilon^{-\gamma l} k_R \epsilon^{-\gamma(l-x)} \tag{4.20}$$

This wave reaches the generator with a magnitude obtained by setting $x = 0$ in the above expression and is reflected there with the coefficient

$$k_g = \frac{Z_g - Z_0}{Z_g + Z_0} \tag{4.21}$$

The new reflection is propagated to the right as indicated by the propagation factor $\epsilon^{-\gamma x}$; hence, combining these factors, we get an expression for the new reflection:

$$\frac{E_g Z_0}{Z_0 + Z_g} \epsilon^{-2\gamma l} k_R k_g \epsilon^{-\gamma x} \tag{4.22}$$

Continuing in this way, we find an infinite series of reflections, each smaller than the last. The sum is convergent and represents the steady state which is finally established. The infinite series can be written as

$$E = \frac{E_g Z_0}{Z_0 + Z_g} [\epsilon^{-\gamma x} + k_R \epsilon^{-2\gamma l} \epsilon^{\gamma x} + k_R k_g \epsilon^{-2\gamma l} \epsilon^{-\gamma x} + \cdots]$$

or

$$E = \frac{E_g Z_0}{Z_0 + Z_g} \left\{ (\epsilon^{-\gamma x} + k_R \epsilon^{-2\gamma l} \epsilon^{\gamma x})[1 + (k_R k_g \epsilon^{-2\gamma l}) + (k_R k_g \epsilon^{-2\gamma l})^2 + \cdots] \right\}$$

The part in brackets is an infinite geometric series of the form $1 + a + a^2 + \cdots$, which is known to converge to the value $1/(1 - a)$ for $|a| < 1$.

Therefore, we can write

$$E = \frac{E_g Z_0}{(Z_0 + Z_g)(1 - k_R k_g \, \epsilon^{-2\gamma l})} \left(\epsilon^{-\gamma x} + k_R \epsilon^{-2\gamma l} \, \epsilon^{\gamma x} \right) \tag{4.23}$$

This expression should be compared with the previous forms of the solution, for all are identical in form and meaning. The equation we have just derived gives explicitly the values of the two integration constants A_1 and A_2 in terms of E_g, Z_g, and Z_R. The result is again of the form $E^+ + E^-$, where E^+ is the net incident wave composed of the sum of all the individual waves that travel toward the load, and E^- is the net reflected wave composed of all the individual waves that travel toward the generator. The steady-state ratio E^-/E^+ was defined in Eq. (4.11) as the reflection coefficient k, and is equal to $k_R \epsilon^{-2\gamma l} \epsilon^{2\gamma x} = k_R \epsilon^{-2\gamma d}$. This ratio is determined only by the load and the line, not by the generator. It is completely unaffected by the quantity $k_g = (Z_g - Z_0)/(Z_g + Z_0)$, which is the reflection coefficient seen by an individual backward-traveling wave as it reaches the generator terminals. As indicated by Eq. (4.23), the latter affects the steady-state solution only through its influence on the sending-end voltage, i.e., through its influence on the magnitude of the entire solution (also see Prob. 14 of Chap. 2).

It can be seen from Eq. (4.23) that the reflected wave in the vicinity of the sending end will be very small whenever k_R is small, or, regardless of k_R, whenever the total attenuation is so great that $k_R \epsilon^{-2al} \ll 1$. Then, so far as the generator is concerned, the line is equivalent to one of infinite length.

In deriving the foregoing equation, we have assumed that the generator could be represented accurately by an emf in series with an effective internal impedance. If the line is driven from a complicated network, Thévenin's theorem can be used to find an equivalent series circuit. It should be realized, however, that not all sources can be represented accurately by an emf of constant amplitude in series with a fixed impedance. This is particularly true where the line is driven directly from an oscillator, for the loading effect of the line may affect both the intensity of the oscillation and the frequency. The simple series representation shown in Figs. 4.1 and 4.2 should be regarded as the nearest linear equivalent of whatever generator is actually used.

It is often convenient to express Eq. (4.23) in terms of the distance d from the load. Substituting $x = l - d$, we obtain

$$E = \frac{E_g Z_0 \epsilon^{-\gamma l}}{(Z_0 + Z_g)(1 - k_R k_g \epsilon^{-2\gamma l})} \left(\epsilon^{\gamma d} + k_R \epsilon^{-\gamma d} \right) \tag{4.24}$$

The expression for current corresponding with Eq. (4.24) can be obtained by use of Eqs. (4.5) and (4.6): the voltage is divided by Z_0 and the sign of the reflected component is reversed. Then we have

$$I = \frac{E_g \epsilon^{-\gamma l}}{(Z_0 + Z_g)(1 - k_R k_g \epsilon^{-2\gamma l})} (\epsilon^{\gamma d} - k_R \epsilon^{-\gamma d}) \tag{4.25}$$

The reflection coefficient for current is again seen to be the negative of that for voltage.

4.3. Hyperbolic Functions. In the next section we shall express the transmission-line solutions in terms of hyperbolic functions, and so in this section we shall consider these functions briefly.

Euler's formulas for the exponential with an imaginary exponent are

and
$$\left. \begin{array}{l} \epsilon^{jz} = \cos z + j \sin z \\[2mm] \epsilon^{-jz} = \cos z - j \sin z \end{array} \right\} \tag{4.26}$$

Adding and subtracting these equations, we obtain the familiar exponential expressions for the sine and cosine:

$$\left. \begin{array}{l} \cos z = \dfrac{\epsilon^{jz} + \epsilon^{-jz}}{2} \\[4mm] \sin z = \dfrac{\epsilon^{jz} - \epsilon^{-jz}}{2j} \end{array} \right\} \tag{4.27}$$

As is well known, the sine and cosine can be defined for real values of z by means of a circle, as indicated in Fig. 4.3. For this reason these quantities are sometimes called *circular functions*. The angle z in radians is equal to the arc s divided by the radius a, or alternatively, is equal to $2A/a^2$, where A is the area of the shaded circular sector shown in Fig. 4.3.

It is frequently convenient to define a somewhat different combination of exponentials named the *hyperbolic cosine* and the *hyperbolic sine*, abbreviated *cosh* and *sinh*:

$$\left. \begin{array}{l} \cosh z = \dfrac{\epsilon^{z} + \epsilon^{-z}}{2} \\[4mm] \sinh z = \dfrac{\epsilon^{z} - \epsilon^{-z}}{2} \end{array} \right\} \tag{4.28}$$

For real values of z, the exponents are real and the exponential expressions for the hyperbolic functions are simpler than those for the circular functions. If Eqs. (4.28) are added and subtracted, we obtain hyperbolic relations analogous to (4.26):

$$\epsilon^{\pm z} = \cosh z \pm \sinh z \tag{4.29}$$

These functions have the name "hyperbolic" because, for real values of z, they can be represented geometrically on a rectangular hyperbola as shown in Fig. 4.4.   The argument, z, is equal to $2A/a^2$, where A is the area of the shaded hyperbolic sector.

Other hyperbolic functions are defined in a manner similar to the circular functions; for example,

$$\tanh z = \frac{\sinh z}{\cosh z} \tag{4.30}$$

The hyperbolic sine, cosine, and tangent, together with the exponentials ϵ^z and ϵ^{-z}, are plotted in Fig. 4.5 for real values of the argument.[1]   The

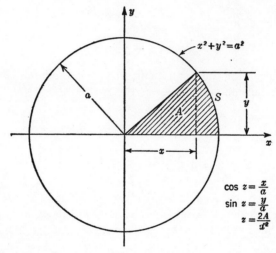

FIG. 4.3.   Geometrical representation of circular functions.

relationship of the hyperbolic functions to the exponentials should be noted particularly.   It can be shown from the exponential definitions that the following differentiation formulas apply:

$$\left.\begin{aligned}\frac{d}{dz} \sinh z &= \cosh z\\[1em]\frac{d}{dz} \cosh z &= \sinh z\\[1em]\frac{d}{dz} \tanh z &= \operatorname{sech}^2 z\end{aligned}\right\} \tag{4.31}$$

[1] Tables of hyperbolic functions are available in various handbooks.   For example, see H. B. Dwight, "Mathematical Tables," pp. 148–178, McGraw-Hill Book Company, Inc., New York. 1941.

Furthermore, identities similar to those of the circular functions can be proved. For example,

$$\cosh^2 z - \sinh^2 z = 1 \tag{4.32}$$

$$\sinh (x \pm y) = \sinh x \cosh y \pm \cosh x \sinh y \tag{4.33}$$

$$\cosh (x \pm y) = \cosh x \cosh y \pm \sinh x \sinh y \tag{4.34}$$

Comparison of Eqs. (4.27) and (4.28) show that the hyperbolic functions

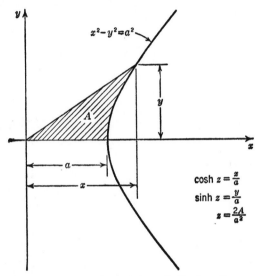

Fig. 4.4. Geometrical representation of hyperbolic functions on the rectangular hyperbola $x^2 - y^2 = a^2$.

of an imaginary argument bear a simple relation to the circular functions of a real argument, namely,

$$\cosh (jx) = \cos x$$
$$\sinh (jx) = j \sin x \tag{4.35}$$

and, similarly,

$$\cos (jx) = \cosh x$$
$$\sin (jx) = j \sinh x \tag{4.36}$$

For a complex argument $z = a \pm jb$, we can either use the definitions (4.28) directly, or we can express the result in terms of functions of a real argument by placing $x = a$, $y = jb$ in (4.33) and (4.34), and then use (4.35). The resulting formulas are

$$\left. \begin{array}{l} \sinh (a \pm jb) = \sinh a \cos b \pm j \cosh a \sin b \\ \cosh (a \pm jb) = \cosh a \cos b \pm j \sinh a \sin b \end{array} \right\} \tag{4.37}$$

4.4. Hyperbolic Form of the Solution. The form of the transmission-line solutions is somewhat simplified by the use of hyperbolic functions. The result is compact and lends itself to algebraic manipulation. However, because the arguments are in general complex, the hyperbolic form is not necessarily easier to use for numerical computations, except possibly

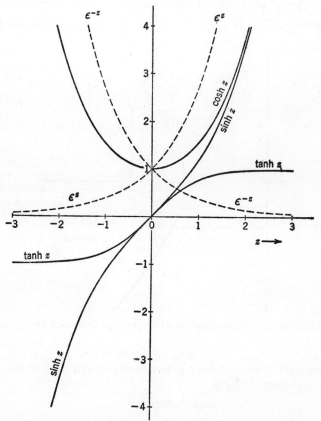

FIG. 4.5.   The hyperbolic functions cosh z, sinh z, and tanh z.

when charts or tables of hyperbolic functions of complex arguments are available.[1]

Equations (4.7) and (4.8) can be written in the form

$$E = E_s \left[\left(\frac{\epsilon^{\gamma x} + \epsilon^{-\gamma x}}{2} \right) - \frac{Z_0}{Z_s} \left(\frac{\epsilon^{\gamma x} - \epsilon^{-\gamma x}}{2} \right) \right]$$

[1] See A. E. Kennelly, "Tables of Complex Hyperbolic and Circular Functions," Harvard University Press, Cambridge, Mass., 1927; also, "Chart Atlas of Complex Hyperbolic and Circular Functions," Harvard University Press, Cambridge, Mass.. 1924.

and

$$I = I_s \left[\left(\frac{\epsilon^{\gamma x} + \epsilon^{-\gamma x}}{2} \right) - \frac{Z_s}{Z_0} \left(\frac{\epsilon^{\gamma x} - \epsilon^{-\gamma x}}{2} \right) \right]$$

The quantities in parentheses are seen to be the hyperbolic cosine and hyperbolic sine, respectively, with the complex argument $\gamma x = \alpha x + j\beta x$. Hence, we can write

$$E = E_s \left(\cosh \gamma x - \frac{Z_0}{Z_s} \sinh \gamma x \right) \qquad (4.38)$$

and

$$I = I_s \left(\cosh \gamma x - \frac{Z_s}{Z_0} \sinh \gamma x \right) \qquad (4.39)$$

where x is measured from the sending end.

Similarly, Eqs. (4.9) and (4.10) can be expressed in the form

$$E = E_R \left(\cosh \gamma d + \frac{Z_0}{Z_R} \sinh \gamma d \right) \qquad (4.40)$$

and

$$I = I_R \left(\cosh \gamma d + \frac{Z_R}{Z_0} \sinh \gamma d \right) \qquad (4.41)$$

where d is measured from the receiving end.

The line impedance can be obtained by dividing E by I. The result can be expressed as

$$Z = Z_0 \frac{Z_R + Z_0 \tanh \gamma d}{Z_0 + Z_R \tanh \gamma d} \qquad (4.42)$$

When d is placed equal to the line length, this expression yields the sending-end impedance, Z_s.

The transfer impedance can be written from (4.16) as

$$Z_{tr} = \frac{E_s}{I_R} = Z_R \cosh \gamma l + Z_0 \sinh \gamma l \qquad (4.43)$$

The ratio of sending-end current to receiving-end current can be written from (4.17):

$$\frac{I_s}{I_R} = \cosh \gamma l + \frac{Z_R}{Z_0} \sinh \gamma l \qquad (4.44)$$

The foregoing expressions are complex, in general. The relations (4.37) are useful in making numerical computations.

Example. Consider the line specified in the example of Sec. 4.1, for which the following data apply:

$$l = 100 \text{ miles}$$
$$f = 1,000 \text{ cps}$$
$$Z_0 = 685 - j92 \text{ ohms}$$
$$\gamma = 0.00497 + j0.0352 \text{ per mile}$$
$$Z_R = 2,000 + j0 \text{ ohms}$$
$$E_g = 10 \text{ volts rms}$$
$$Z_g = 700 + j0 \text{ ohms}$$

Find the sending-end impedance of the line, the sending-end current, and the receiving-end voltage, using the hyperbolic form of the line equations.

First, we compute $\sinh \gamma l$ and $\cosh \gamma l$. Making use of Eq. (4.37), we have

$$\sinh \gamma l = \sinh (0.497 + j3.52) = \sinh 0.497 \cos 3.52 + j \cosh 0.497 \sin 3.52$$

and, since 3.52 radians are equivalent to 202°, we obtain

$$\sinh \gamma l = -0.518 \times 0.927 - j1.126 \times 0.375$$
$$= -0.480 - j0.422$$

Also, from (4.37) we have

$$\cosh \gamma l = \cosh 0.497 \cos 3.52 + j \sinh 0.497 \sin 3.52$$
$$= -1.126 \times 0.927 - j0.518 \times 0.375$$
$$= -1.042 - j0.194$$

We compute $\tanh \gamma l$ to be

$$\tanh \gamma l = \frac{\sinh \gamma l}{\cosh \gamma l} = \frac{-0.480 - j0.422}{-1.042 - j0.194}$$
$$= 0.523 + j0.311$$

The sending-end impedance can now be computed from Eq. (4.42), using $d = l$. Then,

$$Z_s = (685 - j92) \left[\frac{2,000 + (685 - j92)(0.523 + j0.311)}{685 - j92 + (2,000)(0.523 + j0.311)} \right]$$
$$= 859 - j325 \text{ ohms}$$

This checks quite well with the value computed by means of the exponential form of the equations $(861 - j325 \text{ ohms})$.

The sending-end current will be

$$|I_s| = \left| \frac{E_g}{Z_g + Z_s} \right| = \frac{10}{|700 + 859 - j325|} = 6.26 \times 10^{-3} \text{ amp}$$

The receiving-end current can be found by use of Eq. (4.44):

$$\frac{6.26 \times 10^{-3}}{|I_R|} = \left| -1.042 - j0.194 + \left(\frac{2,000}{685 - j92}\right)(-0.480 - j0.422) \right|$$

from which

$$\frac{6.26 \times 10^{-3}}{|I_R|} = 2.76$$

or

$$|I_R| = 2.27 \times 10^{-3} \text{ amp}$$

The receiving-end voltage is

$$|E_R| = |I_R| \cdot |Z_R| = 2.27 \times 10^{-3} \times 2,000$$

$$= 4.54 \text{ volts}$$

4.5. Interference and Standing-wave Patterns. The preceding sections have been concerned with the equations for the steady-state a-c current and voltage along a transmission line in the general case where a portion of the incident energy is reflected from the load. It was shown in Chap. 2 that, when there is no reflection from the load, the rms voltage and current decrease exponentially from the generator toward the load. However, when the load reflects part of the incident energy, it is found that the rms voltage and current vary almost periodically along the line. This effect is caused by the interference between the incident and reflected waves, and the resultant variation is called a *standing wave*.

The effect is shown in a qualitative way in Figs. 4.6 to 4.8 for the extreme case of a low-loss transmission line which is loaded in such a way that there is complete reflection; *i.e.*, the reflected wave is equal in magnitude to the incident wave. This will occur whenever the load can dissipate no energy; *i.e.*, when the termination is a short circuit, an open circuit, or a pure reactance. Figure 4.6 shows the traveling waves of current and voltage at two successive instants of time a quarter cycle apart. Both of the component waves have a sinusoidal form; however, they travel in opposite directions, so that at one instant they reinforce each other, while a quarter of a cycle later they interfere and cancel. The net instantaneous voltage or current is the sum of the incident and reflected components, as shown by the dashed curves. Observe that e^+ and i^+ are in phase, and that e^- and i^- are opposite in phase.

Figure 4.7 shows the net voltage and current plotted against distance for successive instants of time. Observe that at any particular point the voltage and current vary sinusoidally with time, but that the amplitude of oscillation is different at different points on the line. Furthermore, there are points spaced a half wavelength apart at which, in this extreme case, the voltage is always zero (neglecting losses). These are called the *nodes* of the standing wave. The points of maximum oscillation are called *antinodes*. The pattern between two nodes is called a *loop*. Adjacent loops in this extreme standing wave are 180° out of phase. At any particular point on the line, comparison of the numbers on the diagrams show that the current and voltage are in time quadrature. This corresponds with our expectation of a reactive impedance associated with zero average power flow.

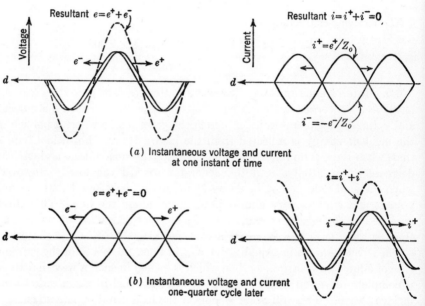

(*a*) Instantaneous voltage and current
at one instant of time

(*b*) Instantaneous voltage and current
one-quarter cycle later

Fig. 4.6.   Instantaneous current and voltage on the interior of a line when the load gives complete reflection ($| k_R | = 1$). Attenuation is neglected.

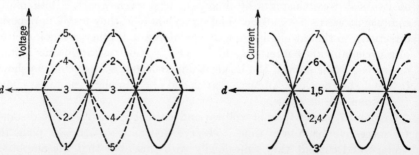

Fig. 4.7.   Net instantaneous values of voltage and current plotted for successive instants of time in the sequence 1, 2, 3, ... Graph 1 corresponds to Fig. 4.6*a*; graph 3 to Fig. 4.6*b*.

Figure 4.8 shows the rms voltage and current plotted against distance for the case of complete reflection. The voltage at any point oscillates sinusoidally with time, and hence the rms value is equal to the amplitude at that point, as shown in Fig. 4.7, divided by $\sqrt{2}$. With complete reflection, the rms standing-wave pattern consists of a succession of sine loops, with the voltage nodes located at the current antinodes and the current nodes located at the voltage antinodes.

The method of determining standing wave patterns by means of instantaneous values as shown in Figs. 4.6 to 4.8 is inaccurate and clumsy. A more practical method is to use vectors to represent the incident and reflected current and voltage at any given point on the line. This method is illustrated in Fig. 4.9. An arbitrary length is laid off to represent the

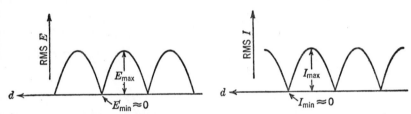

FIG. 4.8. Standing-wave patterns of rms voltage and current plotted against distance for a line with complete reflection.

incident voltage at the receiving end, E_R^+. The reflection coefficient for the receiving end is then computed (see Eq. 4.12):

$$k_R = \frac{Z_R - Z_0}{Z_R + Z_0}$$

Then the length and direction of E_R^- are calculated from the relation

$$E_R^- = k_R E_R^+$$

and E_R^- is laid off to scale. The receiving-end voltage is represented by the vector sum $E_R^+ + E_R^-$, usually to an unknown scale, of course. The construction is shown in Fig. 4.9a for the particular case $k_R = 0.6\underline{/-45°}$, corresponding to a load with resistance and capacitance: $Z_R = (1.25 - j1.66)Z_0$.

The incident and reflected components of the receiving-end current can be drawn similarly. From Eq. (4.5),

$$I^+ = \frac{E^+}{Z_0}$$

Therefore, the angle between E_R^+ and I_R^+ must be that of the characteristic impedance Z_0. In Fig. 4.9 the angle is drawn at 10°, with the current leading. It is often convenient to choose the length of $I_R{}^+$ to be equal to that of $E_R{}^+$.

The reflected component of current can be determined from the relation (4.6), which was

$$I^- = -\frac{E^-}{Z_0}$$

The vector sum $I_R{}^+ + I_R{}^-$ represents the receiving-end current, as shown in the illustration.

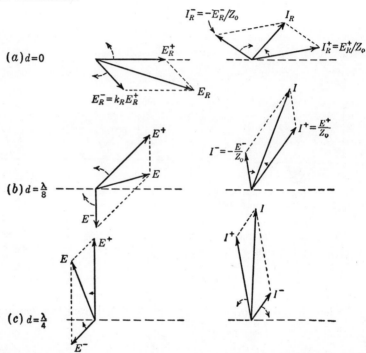

FIG. 4.9. Vector diagrams of voltage and current at three different positions on a transmission line. Drawn for $k_R = 0.6\underline{/-45°}$. The angle of Z_0 is taken to be $-10°$. The curved arrows show the directions in which the vectors shift in phase as d is increased.

The relative voltage and current at any point on the line can now be found. As the distance from the receiving end increases, the reflected vectors E^- and I^- shift in the lagging direction through an angle equal to 360° for every wavelength. In addition, they are attenuated in magnitude by the factor $\epsilon^{-\alpha d}$. Conversely, the incident vectors E^+ and I^+ shift in the leading direction as the distance from the receiving end increases, and their magnitudes grow by the factor $\epsilon^{\alpha d}$. At any point, the vectors representing the actual line voltage and current are given by

$$E = E^+ + E^-$$

and

$$I = I^+ + I^-$$

The construction is shown in Fig. 4.9 for $d = \lambda/8$ and $d = \lambda/4$, assuming an attenuation of 0.4 neper per wavelength, or 3.47 db per wavelength. This might correspond to an open-wire telephone line operating at a frequency of several thousand cycles per second.

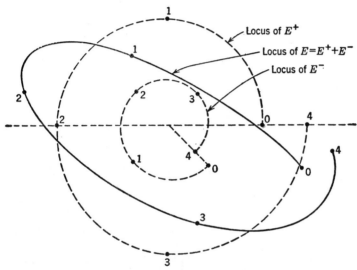

Locus of E^+

Locus of $E = E^+ + E^-$

Locus of E^-

FIG. 4.10.   Polar plot showing the loci of E^+, E^-, and the line voltage E over a span of 1 wavelength. The numerals indicate the number of quarter wavelengths from the load. Drawn for $k_R = 0.6/\underline{-45°}$ and an attenuation of 0.4 neper/wavelength.

The loci of E^+, E^-, and E are shown in Fig. 4.10 for one wavelength of the line illustrated previously.   Figure 4.11 shows the standing-wave patterns of voltage and current for several wavelengths.   It can be seen that the attenuation of the traveling waves causes the interference effect to diminish as the distance from the load increases.   At a sufficient distance from the load, the reflected voltage is negligible, and the current and voltage variation along the line is substantially exponential as in the case of a line with no reflection.

The magnitude and angle of the line impedance at any point can be found from the vector diagrams of Fig. 4.9.   Since E and I are shown in their correct phase relationships, the angle between them is that of the line impedance.   If the lengths representing $E_R{}^+$ and $I_R{}^+$ were made equal at the start, then the magnitude of the line impedance can be found from the relation

$$|Z| = \frac{\text{length representing } E}{\text{length representing } I} \times |Z_0| \qquad (4.45)$$

which follows directly from the fact that $E^+ = I^+ Z_0$ and $E^- = -I^- Z_0$ (see Prob. 10).

Referring to Fig. 4.9, the vector I_R is seen in this case to be smaller than E_R and to lead in phase; hence, Z_R is capacitive and is greater in magnitude than Z_0. This, of course, corresponds with the assumed load impedance: $Z_R = (1.25 - j1.66)Z_0$. The diagrams for $d = \lambda/8$ and $\lambda/4$ show the

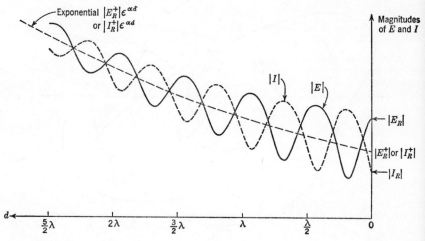

FIG. 4.11. Magnitudes of E and I as functions of distance from the load, corresponding to Figs. 4.9 and 4.10.

impedance decreasing below $|Z_0|$ and turning inductive. Further diagrams would show the impedance increasing again, becoming capacitive, and repeating cyclically. The variation in line impedance is shown in Fig. 4.12 for the data of the preceding example. This variation is typical of the interference effect on all transmission lines. Because of attenuation, the line impedance approaches Z_0 at a sufficient distance from the load.

4.6. The Crank Diagram. In the case of a line with low losses, the standing-wave patterns of voltage and current along the line can be determined most easily by means of the "crank diagram" shown in Fig. 4.13. This diagram corresponds to the vector diagrams shown in Fig. 4.9 except for two differences: (1) the crank diagram is most useful when $\alpha d \ll 1$, in which case the vectors do not change appreciably in size; and (2) in the crank diagram the incident E^+ vector is arbitrarily held stationary, so that to maintain the correct relative phase positions between E^- and E^+ $2\beta d$ the reflected vector must be rotated through twice the angle βd.

In constructing the diagram, a horizontal line of arbitrary length is first laid off to represent the incident voltage vector E^+.   Next, the voltage

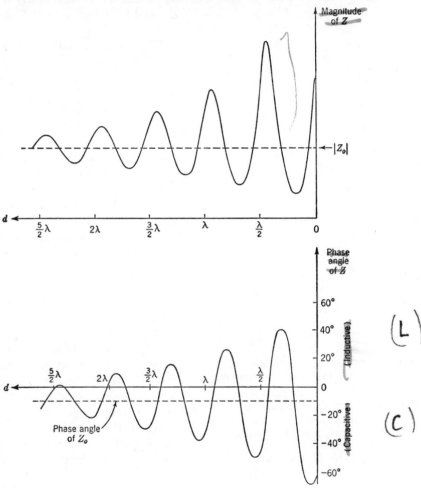

Fig. 4.12. Magnitude and phase angle of the line impedance, corresponding to Figs. 4.9 to 4.11.

reflection coefficient for the receiving end is calculated and the vector $E_R^- = k_R E_R^+$ is laid off from the tip of E^+.   The vector sum $E^+ + E_R^-$ represents the receiving-end voltage, E_R.   To find the voltage at a distance d from the receiving end, the vector E^- is rotated clockwise through the angle $2\beta d$. The new vector sum $E^{+} + E^{-}$ correctly represents the relative magnitude of the voltage at that point. The vector representing the sending-end voltage is obtained by making $d = l$. If the magnitude of E_s is a known quantity, this fixes the voltage scale of the entire diagram.

For a low-loss line, the characteristic impedance is very nearly a pure resistance; hence, I^{+} is in phase with E^{+}, and I^{-} is 180° out of phase with E^{-}. This leads to the diagram of Fig. 4.13b which represents current as well as voltage. The lengths of I^{+} and I^{-} are made equal to those of E^{+} and E^{-}, respectively, so that Eq. (4.45) for the line impedance will apply. As the distance from the load increases, the "crank" revolves once for each half wavelength, and the vectors drawn from the origin to the opposite ends of the revolving diameter represent, respectively, the rms current and voltage on the line. The magnitude of the line impedance can be found by means of Eq. (4.45), and its angle can, of course, be measured directly from the diagram by a protractor.

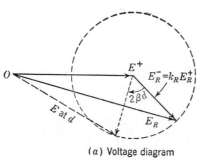

(a) Voltage diagram

Rotate through angle $2\beta d$

(b) Voltage and current diagram

Fig. 4.13. Crank diagram for determining the voltage and current variation along a low-loss line. The "crank" makes one revolution per half wavelength.

The crank diagram shows clearly that at a point where E is maximum on a low-loss line, I is minimum; where E is minimum, I is at a maximum. The line impedance shows the same type of cyclic variation observed in the preceding section and is a pure resistance at the maximum and minimum points.

When the line has appreciable losses, the inconvenience of changing the lengths of the vectors removes one of the principal charms of the crank diagram, namely, its simplicity.

The student may wonder whether the standing-wave pattern of E (or I) can be represented as a simple function of d. In the limiting case of no reflection, we have $k_R = 0$ and the pattern reduces to a straight horizontal line (the "flat" line). In the other limiting case, that of complete reflection, we have $|k_R| = 1$, and the standing-wave pattern becomes a succession of loops of a sine wave. Between these two extremes, the pattern has a rather peculiar shape with the valleys sharper than the crests. It is not hard to prove by the trigonometry of the crank diagram that the *square* of the voltage has a sinusoidal variation down the line:

$$\left|\frac{E}{E^+}\right|^2 = 1 + K_R^2 + 2K_R \cos(2\beta d + \theta_R) \tag{4.46}$$

where K_R is the magnitude of the reflection coefficient at the load and θ_R is its angle. The proof of this will be left to the student.

4.7. Measurement of the Characteristics of Lines. A common problem is the determination of the characteristic impedance and propagation constant of a given transmission line which is physically available for measurement and test. These quantities can be found by measurement of the input impedance of the line under two conditions: with the far end short-circuited, and with the far end open-circuited.

The sending-end impedance can be obtained from (4.42) by placing $d = l$. Then, with the far end shorted, we have

$$Z_{s(s)} = Z_0 \tanh \gamma l \tag{4.47}$$

and with the far end open,

$$Z_{s(o)} = \frac{Z_0}{\tanh \gamma l} \tag{4.48}$$

Inspection of the foregoing equations shows that Z_0 can be computed from the short- and open-circuit measurements by the relation

$$Z_0 = \sqrt{Z_{s(s)}Z_{s(o)}} \tag{4.49}$$

The length of the line makes no theoretical difference; however, it does make a practical difference in the individual measurements of $Z_{s(s)}$ and $Z_{s(o)}$, particularly when the over-all attenuation is small. To examine this, take the case of negligible attenuation, where $\alpha \approx 0$. Then, using (4.37), we have $\sinh \gamma l \approx j \sin \beta l$, $\cosh \gamma l \approx \cos \beta l$, and therefore $\tanh \gamma l \approx j \tan \beta l$. The short- and open-circuit impedances are now reactances:

$$Z_{s(s)} \approx jZ_0 \tan \beta l \tag{4.50}$$

and

$$Z_{s(o)} \approx \frac{-jZ_0}{\tan \beta l} \tag{4.51}$$

If the line is almost an odd number of quarter wavelengths long, the angle βl will be nearly an odd integer times $\pi/2$ radians; then $Z_{s(s)}$ will approach an open circuit and $Z_{s(o)}$ will approach a short circuit. This makes accurate measurements difficult. On the other hand, if the line is almost an even number of quarter wavelengths long, $Z_{s(o)}$ will be very large and $Z_{s(s)}$

will be small. It is convenient to have both impedances of the same order of magnitude, and when the length of line can be selected at will, we can arrange to have tan βl approximately equal to $1/\tan \beta l$. This occurs when the line is an odd number of eighth wavelengths long. If the over-all attenuation is high, the variation of impedance with length of line is not so violent, and it is not so necessary to select the length with care.

If we divide Eq. (4.47) by Eq. (4.48), we obtain an expression for the propagation constant:

$$\tanh \gamma l = \sqrt{\frac{Z_{s(s)}}{Z_{s(o)}}} \tag{4.52}$$

or

$$\alpha l + j\beta l = \tanh^{-1} \sqrt{\frac{Z_{s(s)}}{Z_{s(o)}}} \tag{4.53}$$

There are several equivalent methods for computing γl from (4.53). A simple method is to use the identity

$$\tanh^{-1} u = \frac{1}{2} \log_\epsilon \frac{1 + u}{1 - u} \tag{4.54}$$

and then use the following expression for the logarithm of a complex number:

$$\log_\epsilon A \underline{/\theta} = \log_\epsilon A + j(\theta + n2\pi) \tag{4.55}$$

where A is the magnitude of the complex number, θ is its angle, and n is any integer. Since the imaginary part of the last expression is infinitely many valued, we will be left in doubt as to which value to choose for βl. This matter can be settled only by having additional information; for example, if we can guess an approximate value for the velocity of propagation, we can estimate β and thus select the proper value.

It will be found that the foregoing method of determining γl will not yield accurate results when the over-all losses are so great that both $Z_{s(s)}$ and $Z_{s(o)}$ are nearly equal to Z_0.

The constants R, L, G, and C can be determined by making open- and short-circuit impedance measurements on a section of line which is much shorter than a quarter wavelength. For $\gamma l \ll 1$ we have $\tanh \gamma l \approx \gamma l$. Using (4.47) for the sending-end impedance with the far end short-circuited, we obtain for a small length of line,

$$Z_{s(s)} \approx Z_0 \gamma l = \sqrt{\frac{Z}{Y}} \sqrt{ZY}\, l$$

or

$$Z_{s(s)} \approx Zl = (R + j\omega L)l \tag{4.56}$$

This result is what we expect for a loop of wire which is so short that the small-circuit theory applies.

Similarly, using (4.48) for the open-circuit impedance, we obtain for a short section,

$$\frac{1}{Z_{s(o)}} \approx Yl = (G + j\omega C)l \qquad (4.57)$$

Again, this is the result that would be anticipated from small-circuit theory, for here we have two conductors much shorter than a quarter wavelength, separated by a dielectric medium. The effect is that of a condenser which has, perhaps, some leakage.

4.8. Equivalent Four-terminal Networks. It is sometimes convenient to have a lumped-constant equivalent circuit which will simulate the

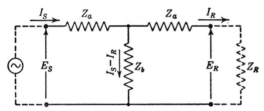

FIG. 4.14. T-connected four-terminal network which is to be equivalent to a transmission line in steady state at a single frequency.

terminal characteristics of a transmission line at a given frequency. When the line is electrically long, the values of the elements in the equivalent circuit will change rapidly with frequency; therefore, such a circuit is of use mainly when one is considering the steady state at a single frequency. A transmission line is essentially a four-terminal network. In most applications, it can be more accurately called a "two-terminal pair," for two of the terminals are designated as the input and the other two as the output without any outside interconnection between the two ends. We shall show that, for steady-state conditions at a single frequency, the transmission line can be represented by either a T or a π two-terminal pair.

Figure 4.14 shows a T-connected two-terminal pair whose terminal characteristics are, if possible, to be made equivalent to those of a given transmission line. Since a uniform line can be turned end for end without affecting the results, we expect that the network will be symmetrical and that the two series arms will have equal impedances. We shall now set up the equations for the two-terminal pair and shall show that these can be made equivalent to the equations for the smooth transmission line. The phase shift through the network will be the same as the phase shift caused by the traveling waves on the actual line, and the attenuation in the line will be simulated by loss in the network.

If we write a voltage equation around each of the two meshes in the equivalent circuit, we obtain

$$(Z_a + Z_b)I_s - Z_bI_R = E_s \tag{4.58}$$

and

$$-Z_bI_s + (Z_a + Z_b)I_R = -E_R \tag{4.59}$$

To obtain corresponding equations for the transmission line, use Eq. (4.39) with $x = l$ and Eq. (4.41) with $d = l$. Then, we have

$$I_R = I_s \cosh \gamma l - \frac{E_s}{Z_0} \sinh \gamma l \tag{4.60}$$

and

$$I_s = I_R \cosh \gamma l + \frac{E_R}{Z_0} \sinh \gamma l \tag{4.61}$$

These equations can be rearranged into the form of (4.58) and (4.59) immediately. The result is

$$(Z_0 \coth \gamma l)I_s - \left(\frac{Z_0}{\sinh \gamma l}\right) I_R = E_s \tag{4.62}$$

and

$$-\left(\frac{Z_0}{\sinh \gamma l}\right) I_s + (Z_0 \coth \gamma l)I_R = -E_R \tag{4.63}$$

By comparison with Eqs. (4.58) and (4.59), we obtain

$$Z_b = \frac{Z_0}{\sinh \gamma l} \tag{4.64}$$

and

$$Z_a + Z_b = Z_0 \coth \gamma l \tag{4.65}$$

Eliminating Z_b between (4.64) and (4.65) and utilizing the identity $(\cosh z - 1)/\sinh z = \tanh z/2$, we find for Z_a:

$$Z_a = Z_0 \tanh \frac{\gamma l}{2} \tag{4.66}$$

Referring to Fig. 4.14 and using Eqs. (4.64) and (4.66), we obtain the equivalent circuit of Fig. 4.15. It should be emphasized again that a given network is equivalent to the actual transmission line only at one frequency, for a change in frequency will alter the value of γ and may also change Z_0 somewhat.

The corresponding π-connected two-terminal pair can be derived from the T-connected network by means of the well known T-π, or wye-delta, transformation. The result is shown in Fig. 4.16.

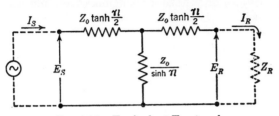

FIG. 4.15. Equivalent T network.

It is interesting to draw the approximate equivalent circuit for a transmission line which is much shorter than a quarter wavelength; *i.e.*, when $\gamma l \ll 1$. Then, in Eq. (4.64) we can write $\sinh \gamma l \approx \gamma l$. In Eq.

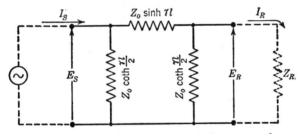

FIG. 4.16. The π form of the equivalent network.

(4.66) we have, similarly, $\tanh \gamma l/2 \approx \gamma l/2$. Recalling that $Z_0 = \sqrt{(R + j\omega L)/(G + j\omega C)}$ and $\gamma = \sqrt{(R+j\omega L)(G+j\omega C)}$, we can write

$$Z_a \approx \frac{(R + j\omega L)}{2} l \tag{4.67}$$

and

$$Y_b = \frac{1}{Z_b} \approx (G + j\omega C)l \tag{4.68}$$

The corresponding equivalent circuit is shown in Fig. 4.17. As we should expect for a short line, this circuit corresponds with one section of the lumpy line with which we started in Chap. 1 (compare Fig. 1.3). A 60-cps power line which is very much shorter than a quarter wavelength can be analyzed in this manner. A short section of high-frequency line can be considered in the same way, and, in addition, the losses can often be neglected, leaving only the inductance and capacitance. In high-impedance circuits the shunting effect of the capacitance is most important, and the inductance can often be neglected. An example of this is the connection of a cathode-ray oscilloscope to an amplifier through a section of coaxial cable, in which case the shunt capacitance of the cable may load the circuit at the higher frequencies. In very-low-impedance circuits the shunt capacitance may have a negligible effect, and the voltage drop in the inductance may become important.

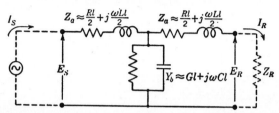

FIG. 4.17. Approximate equivalent circuit for a line which is much shorter than a quarter wavelength.

4.9. Insertion Ratio and Insertion Loss. Quantities called the *insertion ratio* and the *insertion loss* are used to describe the effect of inserting a four-terminal network (perhaps a transmission line) between a generator and a load, as compared with a direct connection between the two. Consider Fig. 4.18a in which, with the four-terminal network inserted, the receiving-end voltage is E_R. On the other hand, if a direct connection were used between generator and load as in Fig. 4.18b, the receiving-end voltage would be E_R'. We then define

$$\text{Voltage insertion ratio} = \frac{E_R'}{E_R} \qquad (4.69)$$

Since we are considering two voltages across the same impedance, the insertion ratio for current must be the same as that for voltage. The insertion ratio for power, sometimes called the power-loss ratio, will be equal to the square of the magnitude of (4.69).

The term *insertion loss* is used to denote the magnitude of the insertion effect as expressed in nepers or decibels, *i.e.*,

$$\left.\begin{array}{c} \text{Insertion loss} = \log_\epsilon \left|\dfrac{E_R'}{E_R}\right| \qquad \text{nepers} \\[2em] \text{Insertion loss} = 20 \log_{10} \left|\dfrac{E_R'}{E_R}\right| \qquad \text{db} \end{array}\right\} \qquad (4.70)$$

or

Referring to Fig. 4.18b, the load voltage with a direct connection is given simply by

$$E_R' = \frac{Z_R}{Z_R + Z_g} E_g \qquad (4.71)$$

The voltage at the receiving end of a uniform transmission line can be obtained from Eq. (4.24) with $d = 0$. The result is

$$E_R = \frac{E_g Z_0 \epsilon^{-\gamma l}(1 + k_R)}{(Z_0 + Z_g)(1 - k_R k_g \epsilon^{-2\gamma l})} \qquad (4.72)$$

where

$$k_R = \frac{Z_R - Z_0}{Z_R + Z_0} \qquad (4.73)$$

$$k_g = \frac{Z_g - Z_0}{Z_g + Z_0}. \qquad (4.74)$$

Dividing (4.71) by (4.72), we obtain for the voltage insertion ratio of a uniform transmission line,

$$\frac{E_R{}'}{E_R} = \epsilon^{\gamma l}\left[\frac{Z_R(Z_0 + Z_g)}{Z_0(Z_R + Z_g)(1 + k_R)}\right](1 - k_R k_g \epsilon^{-2\gamma l}) \qquad (4.75)$$

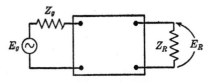

(a) Four-terminal network inserted between generator and load

The middle factor in this expression can be simplified by expressing Z_R and Z_g in terms of k_R and k_g, respectively. Solving Eq. (4.73) for Z_R and Eq. (4.74) for Z_g, we obtain

$$Z_R = \frac{1 + k_R}{1 - k_R} Z_0$$

and

$$Z_g = \frac{1 + k_g}{1 - k_g} Z_0$$

(b) With direct connection

Fig. 4.18. Illustrating the meaning of the voltage insertion ratio $E_R{}'/E_R$.

Upon substitution of these relations, the middle factor of Eq. (4.75) simplifies into $1/(1 - k_R k_g)$, whereupon the insertion ratio can be written as

$$\frac{E_R{}'}{E_R} = \epsilon^{\gamma l}\left(\frac{1}{1 - k_R k_g}\right)(1 - k_R k_g \epsilon^{-2\gamma l}) \qquad (4.76)$$

The last factor will be nearly unity for a long lossy line. Furthermore, if either $Z_R = Z_0$ or $Z_g = Z_0$, the last two factors both reduce to unity and the insertion ratio becomes simply $\epsilon^{\gamma l} = \epsilon^{\alpha l}\epsilon^{j\beta l}$, i.e., it has a magnitude $\epsilon^{\alpha l}$ and a phase angle βl radians.

The magnitude of (4.76) is

$$\left|\frac{E_R{}'}{E_R}\right| = \epsilon^{\alpha l}\left|\frac{1 - k_R k_g \epsilon^{-2\gamma l}}{1 - k_R k_g}\right| \qquad (4.77)$$

The insertion loss is defined as the logarithm of (4.77), and so, in nepers, we have

Insertion loss $= \alpha l - \log_\epsilon | 1 - k_R k_g |$

$$+ \log_\epsilon | 1 - k_R k_g \epsilon^{-2\gamma l} | \qquad \text{nepers} \qquad (4.78)$$

The corresponding expression in decibels is

Insertion loss $= 8.686\alpha l - 20 \log_{10} | 1 - k_R k_g |$

$$+ 20 \log_{10} | 1 - k_R k_g \epsilon^{-2\gamma l} | \qquad \text{db} \qquad (4.79)$$

The last term is nearly zero for a long lossy line. If either $Z_R = Z_0$ or $Z_g = Z_0$, only the first term remains and the insertion loss becomes simply αl nepers or $8.686 \ \alpha l$ db.

PROBLEMS

1. Given an open-wire telephone line with $Z_0 = 717 - j139$ ohms and $\gamma = 0.00718 + j0.0358$ per mile at 1,000 cps. The line is 100 miles long and is terminated in $Z_R = 2,000 + j0$ ohms.

a. Compute the complex values of $\epsilon^{\gamma l}$ and $\epsilon^{-\gamma l}$.

b. Compute the transfer impedance $Z_{\text{tr}} = E_s/I_R$.

c. If $E_s = 10.0/\underline{0°}$ volts rms, find the magnitude and phase of I_R.

2. Compute the sending-end impedance of the line of Prob. 1. Determine the magnitudes of the sending-end current and the sending-end power.

3. An open-wire telephone line has $Z_0 = 649 - j82.9$ ohms and $\gamma = 0.00539 + j0.0353$ per mile at 1,000 cps. The line is driven by a generator with $E_g = 10.0$ volts rms and $Z_g = Z_0$. It is terminated in an open circuit 125 miles from the sending end. Compute the magnitude of the voltage across the receiving end.

4. From the exponential definitions of the hyperbolic functions, compute the values sinh 0.6, cosh 0.6, and tanh 0.6.

5. Prove that $\cosh^2 z - \sinh^2 z = 1$.

6. From the exponential definitions of the hyperbolic functions, prove that $\sinh (x + y) = \sinh x \cosh y + \cosh x \sinh y$ and that $\cosh (x + y) = \cosh x \cosh y + \sinh x \sinh y$.

7. Show from the exponential definitions that $\tanh jb = j \tan b$.

8. Prove Eqs. (4.40) and (4.41) from Eqs. (4.9) and (4.10).

9. Given a line 100 miles long, with $Z_0 = 717 - j139$ ohms and $\gamma = 0.00718 + j0.0358$ per mile at 1,000 cps.

a. Compute $\sinh \gamma l$, $\cosh \gamma l$, and $\tanh \gamma l$.

b. Compute the sending-end impedance with the receiving end open.

c. Compute the sending-end impedance with the receiving end short-circuited.

10. Suppose that, in a vector diagram similar to that of Fig. 4.9, the length of any voltage vector is denoted by l_E and the length of the corresponding current vector is denoted by l_I. Denote the scale factors of the diagram by S_E = voltage scale in volts per unit length, S_I = current scale in amperes per unit length.

Show that, if $l_E{}^+ = l_I{}^+$, then $S_E/S_I = | Z_0 |$ and $| Z | = (l_E/l_I) | Z_0 |$.

11. Given a telephone line with $Z_0 = 690/\underline{-11.8°}$ ohms and $\gamma = 0.0075 + j0.0351$ per mile at 1,000 cps. The line is terminated in a short circuit at the receiving end.

a. Draw vector diagrams similar to those of Fig. 4.9 for the following distances from the receiving end: d = zero, and d = 65 miles.

b. If the magnitude of the receiving-end current is 5.0×10^{-3} amp, find the voltage and current scales of the diagram in volts per inch and amperes per inch, respectively.

c. From the diagram, find the line impedance in both magnitude and phase angle at d = 65 miles. Determine the line voltage and current at this position.

12. Given a high-frequency transmission line which has $Z_0 = 400 + j0$ ohms and $Z_R = 150 - j200$ ohms. The line is $1\frac{1}{8}$ wavelengths long. Neglect losses. The sending-end voltage is 100 volts rms.

a. Compute the complex reflection coefficient at the receiving end.

b. Draw a crank diagram for the line, and indicate the points which represent the voltage and current at the two ends of the line. Find the voltage and current scales in volts per inch and amperes per inch, respectively.

c. Using the crank diagram, plot the standing-wave patterns of voltage and current along the line. Determine the magnitudes of the receiving-end current and voltage. Compute the power transmitted by the line.

13. A high-frequency low-loss line with $Z_0 = 400 + j0$ ohms is terminated in $Z_R = -j400$ ohms. Construct a crank diagram for this line and find the line impedance at the following distances from the receiving end: $d = \lambda/16, \lambda/8, 3\lambda/16, \lambda/4, 3\lambda/8$, and $\lambda/2$.

14. A low-loss line with $Z_0 = 400 + j0$ ohms is operating at a wavelength of 3 meters. Using a crank diagram, sketch the standing-wave patterns of voltage and current over a distance of 5 meters from the receiving end, for the following two conditions:

a. Receiving end short-circuited, $|I_R|$ = 1.0 amp rms.

b. Receiving end open, $|E_R|$ = 400 volts rms.

15. Prove Eq. (4.46), which gives the variation of $|E|^2$ along a low-loss line.

16. Compute the magnitude and angle of the reflection coefficient for each of the following receiving-end impedances, and mark the points on a sheet of polar-coordinate paper: Z_R = zero, $Z_0/3$, Z_0, $3Z_0$, and infinity. Also, $Z_R = -jZ_0$, jZ_0, $-j2Z_0$, and $2Z_0$.

17. The following impedance measurements were made at the sending end of a 70-mile open-wire telephone line: (*a*) With the receiving end short-circuited, the sending-end impedance was $469\underline{/-66.9°}$ ohms at 1,000 cps; (*b*) With the receiving end open-circuited, the sending-end impedance was $681\underline{/55.1°}$ ohms at 1,000 cps. Determine the characteristic impedance of the line, also the attenuation and phase constants per mile.

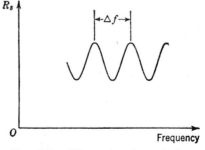

FIG. P18.   Effect of an irregularity.

18. If there is an irregularity on a transmission line, the input impedance of the line will fluctuate up and down as the frequency is increased. A typical graph of the resistive component of input impedance is shown in Fig. P18. Show that the distance to the irregularity is given by

$$x = \frac{v}{2\Delta f}$$

where v is the phase velocity and Δf is the frequency in cycles per second between adjacent peaks of impedance.

19. Starting with the T equivalent network of Fig. 4.15, and using the wye-delta transformation, show that the π network of Fig. 4.16 is a correct equivalent. (Formulas for the transformation will be found in Sec. 10.9.) Note the following hyperbolic identities:

$$\tanh \frac{z}{2} = \frac{\sinh z}{\cosh z + 1} = \frac{\cosh z - 1}{\sinh z}$$

20. Find the T equivalent network at 1,000 cps for the transmission line whose constants are specified in the example of Sec. 4.4. Using this equivalent circuit, compute the input impedance of the line for the condition of an open circuit at the receiving end. Also, find the magnitude of the voltage across the open end if the sending-end voltage is 5.0 volts rms.

21. Find the π equivalent network at 1,000 cps for the transmission line whose constants are specified in the example of Sec. 4.4. Using this equivalent circuit, compute the magnitude of the receiving-end current if the receiving end is short-circuited and if a voltage of 5 volts is impressed across the sending end.

22. Compute the insertion ratio, and the insertion loss in decibels, for the example of Sec. 4.1.

23. Given a transmission line which is terminated so as to have complete reflection at the receiving end ($|E_R^+| = |E_R^-|$).

a. Show that the envelope curves for the standing-wave pattern of voltage are

$$\text{Upper envelope} = 2|E_R^+| \cosh \alpha d$$

$$\text{Lower envelope} = 2|E_R^+| \sinh \alpha d$$

where d is the distance measured from the receiving end.

b. Show that the envelope curves for the standing-wave pattern of current are like those for the voltage except divided by $|Z_0|$.

c. Derive the envelope curves for the magnitude of the line impedance.

24. Use the results of Prob. 23 to sketch the standing-wave patterns of E and I vs. d for a distance of 300 miles from the receiving end, for a line operating under the following conditions: $\alpha = 0.00576$ neper/mile, $\lambda = 90.0$ miles, $|E_R^+| = 7.0$ volts rms, $|Z_0| = 700$ ohms, and with the

a. Receiving end short-circuited,

b. Receiving end open-circuited.

25. A certain tuned r-f amplifier has a parallel LC circuit with $L = 125 \ \mu h$ and a capacitance which resonates the circuit at 2.0 Mc. For purposes of measurement, one end of a coaxial cable is connected across the LC circuit. The other end of the cable is connected to a measuring instrument whose input impedance is effectively infinite. According to the manufacturer's specifications, the cable has a characteristic impedance of 52 ohms and a capacitance of 28.5 $\mu\mu$f/ft. The cable is $2\frac{1}{2}$ ft long. How much is the resonant frequency of the circuit changed by the connection of the cable?

26. A 75-mile transmission line is operating on direct current. Its constants are $R = 86.0$ ohms/mile and $G = 2.5 \times 10^{-6}$ mho/mile. The sending-end voltage is 24.0 volts. The receiving end is short-circuited. Using the hyperbolic form of the line solution, find the sending-end impedance, the sending-end current, and the receiving-end current.

27. If the line of Prob. 26 is open-circuited at the receiving end, find the sending-end impedance and the receiving-end voltage, using the hyperbolic form of the line solution.

CHAPTER 5

TRANSMISSION-LINE CHARTS

5.1. Introduction. The student who has calculated the characteristics of some transmission lines using the relations developed in the preceding chapter will appreciate the statement that a considerable amount of labor is often involved. Furthermore, after one set of calculations has been made, it is still difficult to see just how the results will be affected by a change in the system. In an effort to make calculation easier, various charts have been developed. The accuracy obtainable with these is, in general, not so high as can be obtained with a slide rule; however, it is sufficiently good for many purposes. The charts can be used for both lossy and lossless lines, and they facilitate many calculations which otherwise would be quite difficult.

Among the earliest sets of tables and charts for transmission-line problems were those prepared by A. E. Kennelly (see the footnote on page 104). These charts gave the values of hyperbolic and circular trigonometric functions over a range of complex arguments and were used in connection with the hyperbolic solutions given by Eqs. (4.38) to (4.44). However, the charts were rather bulky, and both the charts and the tables are now out of print.

In the sections that follow, we shall develop two simple charts which are in wide use and which are generally available.

5.2. The Reflection Coefficient and the Line Impedance. In Eq. (4.11) we defined the reflection coefficient for any point on a transmission line as the steady-state ratio of the reflected voltage to the incident voltage at that point:

$$k = \frac{E^-}{E^+} \tag{5.1}$$

This ratio is determined entirely by the load and the line. At the receiving end it is given by the complex number

$$k_R = \frac{Z_R - Z_0}{Z_R + Z_0} \tag{5.2}$$

As was shown in Eq. (4.13), the variation of the ratio along the line is

$$k = k_R \epsilon^{-2\gamma d} = k_R \epsilon^{-2\alpha d} \epsilon^{-j2\beta d} \tag{5.3}$$

126

where d is the distance from the receiving end. If we denote magnitudes by K and phase angles by θ, the quantity k_R can be written as

$$k_R = K_R \epsilon^{j\theta_R} = K_R\underline{/\theta_R}$$

Equation (5.3) can then be expressed as

$$k = K\underline{/\theta} \tag{5.4}$$

where

$$K = K_R \epsilon^{-2\alpha d} \tag{5.5}$$

and

$$\theta = \theta_R - 2\beta d \quad \text{radians} \tag{5.6}$$

As we go away from the load, the magnitude K shrinks by the factor $\epsilon^{-2\alpha d}$, the doubled exponent being caused by the shrinkage in E^- and the corresponding increase in E^+. The angle θ decreases, or shifts in the lagging direction, by the amount $2\beta d$ rad, which amounts to 360° in every half wavelength. This is to be expected from the fact that E^+ and E^- come back in phase with each other in this distance. On a lossless line, where $\alpha = 0$, the magnitude of k remains constant and only the phase angle changes.

Next, we shall show the connection between the reflection coefficient and the line impedance. We have already defined the line impedance at any point as the complex ratio of voltage to current at that point:

$$Z = \frac{E}{I} \tag{5.7}$$

By using the relations $E = E^+ + E^-$ and $I = I^+ + I^- = E^+/Z_0 - E^-/Z_0$, we can write

$$Z = \frac{E^+ + E^-}{E^+/Z_0 - E^-/Z_0}$$

or

$$\frac{Z}{Z_0} = \frac{1 + E^-/E^+}{1 - E^-/E^+} \tag{5.8}$$

We recognize the ratio E^-/E^+ as the complex reflection coefficient, k. The ratio Z/Z_0 will appear frequently in the pages that follow, and it will be convenient to consider it as an entity in itself. We shall call it the *normalized* or per-unit impedance and shall denote it by a lower-case z, with the real and imaginary parts r and x:

$$\frac{Z}{Z_0} = z = r + jx \tag{5.9}$$

Normalizing an impedance by dividing it by Z_0 amounts to a change of scale such that the unit of impedance is Z_0 ohms rather than 1 ohm.

The relation given by Eq. (5.8) can now be expressed in terms of the normalized line impedance and the reflection coefficient as

$$z = \frac{1+k}{1-k} \tag{5.10}$$

This can be expressed the other way around by solving for k:

$$k = \frac{z-1}{z+1} \tag{5.11}$$

The connection between the complex reflection coefficient on the one hand and the complex normalized impedance on the other, as indicated by the foregoing two equations, is presented in a graphical way by the two charts that we shall describe in the next section. This relation, together with the known variation of k along the line, simplifies calculations to a surprising extent.

5.3. The Rectangular and Circular Transmission-line Charts. One of the most common transmission-line charts is a plot of normalized impedance z in rectangular coordinates, with lines corresponding to k superimposed on it (see Fig. 5.1). Another much-used diagram, often called the Smith chart,[1] plots k in polar coordinates and has lines corresponding to z superimposed upon it (see Fig. 5.2). The Smith chart is, in effect, a generalized form of the crank diagram shown in Fig. 4.13. More accurately drawn charts are given in Figs. 5.3 and 5.4. Accurate printed charts are commercially available.[2]

The circular chart is drawn with an outside radius which represents unit magnitude of the reflection coefficient. The rectangular chart is drawn for positive values of the real part, r, of the normalized impedance and so, in principle, should occupy an infinite half plane. The foregoing ranges take care of most practical problems; however we may note in passing that, with lossy lines, it is possible for points to fall outside these regions. This occurs whenever the difference between the angles of Z and Z_0 is greater than 90°. For example, for $Z = j1$ and $Z_0 = 0.866 - j0.500$, we have $z = -0.500 + j0.866$ and $k = j1.73 = 1.73\ \underline{/90°}$. This difficulty arises only rarely, however.

[1] P. H. Smith, Transmission Line Calculator, *Electronics*, January, 1939, p. 29, and An Improved Transmission Line Calculator, *Electronics*, January, 1944, p. 130.

[2] Copies of the Smith chart, printed on 8½- by 11-in. paper, can be purchased from the Emeloid Company, Arlington, N.J. One form of the rectangular transmission-line chart can be obtained from the Addison-Wesley Press, Inc., Cambridge, Mass.

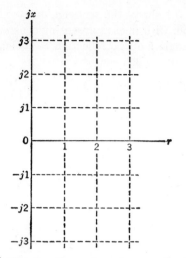

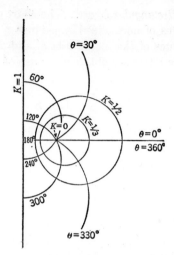

(a) Showing the rectangular coordinates of normalized impedance, $z = Z/Z_0$.

(b) The lines of constant K and θ ($k = Ke^{j\theta}$) which are to be superimposed on the coordinates shown at the left.

FIG. 5.1. The two sets of lines of the rectangular impedance chart. A completed chart is given in Fig. 5.3.

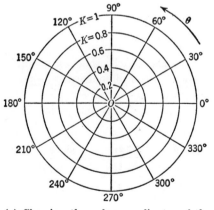

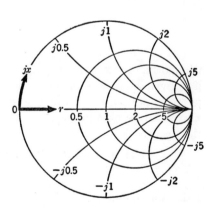

(a) Showing the polar coordinates of the reflection coefficient $k = Ke^{j\theta}$.

(b) The lines of constant r and x which are to be superimposed on the coordinates shown at the left.

FIG. 5.2. The two sets of lines which compose the circular (Smith) transmission-line chart. Part b above should be compared with Fig. 5.1a, and part a should be compared with Fig. 5.1b. A completed chart is given in Fig. 5.4.

Rectangular Chart.   The chart of Fig. 5.1 shows the rectangular coordinates of normalized impedance z, together with lines representing constant values of the magnitude K and phase angle θ of the reflection coefficient. The latter are obtained by use of Eq. (5.10), which can be written as

$$z = \frac{1 + K\underline{/\theta}}{1 - K\underline{/\theta}} = \frac{1 + K\cos\theta + jK\sin\theta}{1 - K\cos\theta - jK\sin\theta} \qquad (5.12)$$

For example, for $K = 0.5$ and $\theta = 30°$, we have

$$z = \frac{1 + 0.5 \times 0.866 + j0.5 \times 0.5}{1 - 0.5 \times 0.866 - j0.5 \times 0.5} = 1.95 + j1.31$$

If K is held constant while θ is varied, the corresponding locus of z turns out to be a circle which encloses the point $1 + j0$.   On the other hand, if θ is held constant while K is varied, the corresponding locus is another circle which crosses the foregoing one at right angles and goes through the point $1 + j0$.   This leads to two families of circles on the z plane: one for constant values of K, and another, everywhere at right angles to the first, for constant values of θ.   The intercepts on the r axis can be obtained by placing θ equal, respectively, to 0 and 180° which yields

$$r_i = \frac{1 + K}{1 - K}$$

and

$$r_i' = \frac{1 - K}{1 + K}$$

$$(5.13)$$

Obviously, the two intercepts are the reciprocals of each other; for example, if a given K circle intercepts the r axis at $\frac{1}{3}$, the diametrically opposite intercept will be located at $r = 3$.   The intercepts of the θ circles on the x axis can be obtained by setting $K = 1$ in Eq. (5.12) and rationalizing the result:

$$z = \frac{(1 + \cos\theta + j\sin\theta)(1 - \cos\theta + j\sin\theta)}{(1 - \cos\theta - j\sin\theta)(1 - \cos\theta + j\sin\theta)} = \frac{j\sin\theta}{1 - \cos\theta}$$

Using the trigonometric identity $(1 - \cos\theta)/\sin\theta = \tan\theta/2$, we obtain for the intercept on the x axis:

$$x_i = \frac{1}{\tan\theta/2} \qquad (5.14)$$

The continuation of any given θ circle after it passes through the point $1 + j0$ can be shown from the geometry of the figure to be associated with

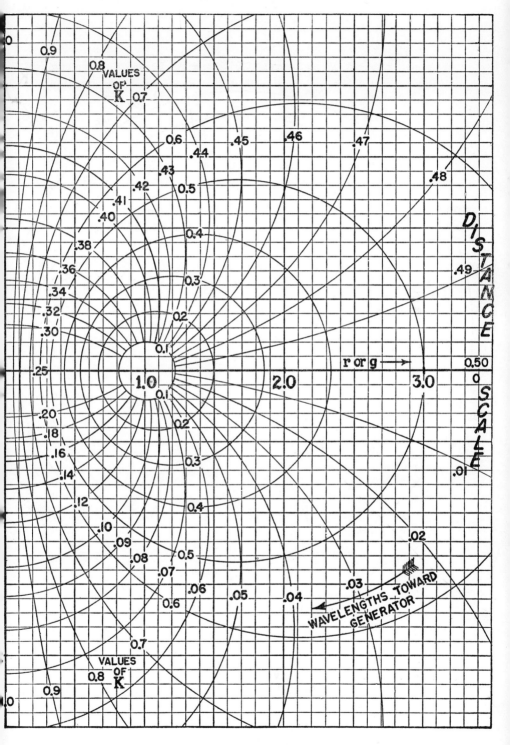

FIG. 5.3 The rectangular transmission-line chart. $|E^-/E^+| = K = K_R \epsilon^{-2\alpha d}$.

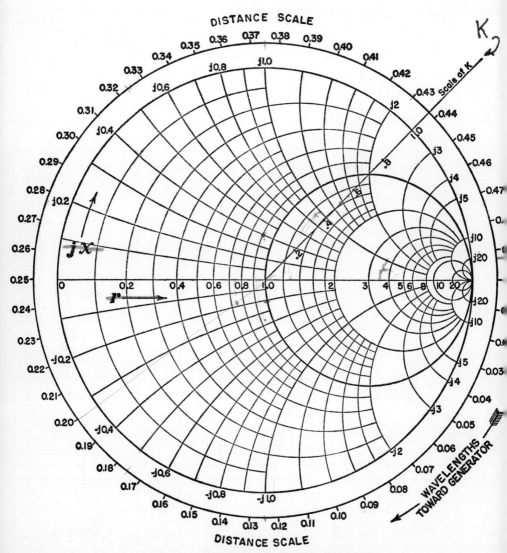

Fig. 5.4. The circular transmission-line chart.

the angle $\theta + 180°$; hence, replacing θ by $\theta + 180°$ in Eq. (5.14) we obtain for the other intercept,

$$x_i' = \frac{1}{\tan (\theta/2 + 90°)} = \frac{1}{\cot \theta/2} = -\tan \frac{\theta}{2} \quad (5.15)$$

In the completed chart of Fig. 5.3, the θ scale has been replaced by a scale of distances along the line, expressed in terms of wavelength. As shown by Eq. (5.6), the angle θ starts at the value θ_R and decreases along the line by the amount $2\beta d$, or 360° in every half wavelength. Thus, a change of $-7.2°$ on the scale of θ always corresponds to a change of 0.01λ on the distance scale. However, we cannot arrange to have the zero of the distance scale start at the position on the chart that corresponds to the load, for θ_R may have any value. The zero is placed arbitrarily, and the distance scale must therefore be used in a relative way: distances along the line are given by the differences between two readings on the scale. In the chart of Fig. 5.3 the zero of the scale has arbitrarily been located at $\theta = 0$. The scale undergoes a jump from $d = 0.5\lambda$ to $d = 0$, and this must be taken into account when passing over this portion of the scale.

The value of θ corresponding to any given reading on the distance scale can be found by use of Eq. (5.6) with θ_R set equal to our arbitrarily chosen value of zero. If we express this equation in degrees and, for convenience, take θ_R to be 360° instead of zero, we have the relation

$$\theta = \left(1 - 2\frac{d}{\lambda}\right) \times 360° \quad (5.16)$$

where d/λ represents the reading on the distance scale.

Circular Chart. The circular reflection-coefficient, or Smith, chart is plotted within a circle of unit radius (see Fig. 5.2). The magnitude K of the reflection coefficient is represented by the radius from the center. Superimposed on this chart are lines representing constant values of r and x, respectively, as obtained from Eq. (5.11).[1] The point $z = 1 + j0$ is at the center of the chart ($k = 0$). The lines of constant r and x form two orthogonal families of circles. The advantages of the Smith chart are that the lines are less crowded in the low-impedance region, and also that all the usual values of k and z are contained within a finite area. On the other hand, the values of z are much easier to read when plotted on rectangular coordinates, and certain graphical constructions that we shall describe later are easier to follow on the rectangular chart.

In the completed chart of Fig. 5.4, the θ scale has been replaced by a

[1] The reader who has studied complex function theory will recognize one chart as the conformal map of the other, with Eqs. (5.10) and (5.11) as the transforming functions.

distance scale expressed in terms of wavelengths. This is obtained from the θ scale in the same way that we obtained the distance scale for the rectangular chart, and it is used in the same manner.

The zero of the distance scale has arbitrarily been located at $\theta = 0$ in Fig. 5.4. Some forms of this chart have the zero located at $\theta = 180°$. Since the scale is used in a relative manner, with actual distances represented by the differences between two readings, the location of the zero is of no essential importance.

The value of θ corresponding to any given reading on the distance scale can be determined by use of a protractor, or Eq. (5.16) can be used.

5.4. Computation of Line Impedance. The charts that we have described show the relation between normalized line impedance and the complex ratio of reflected to incident voltage, k. The θ scale has been replaced by a scale of distances, but the magnitude of the reflection coefficient, K, has been retained. As shown by Eq. (5.5), this magnitude varies as

$$K = K_R \epsilon^{-2\alpha d} \tag{5.17}$$

If we desired, we could replace the scale of K by a scale of αd in either nepers or decibels, similar to the replacement of the θ scale by a scale of wavelengths. The computation (5.17) is not laborious, however, and we shall continue to use K.

If the load impedance and the line constants are known, the impedance at any other point can be found from either chart as follows:

1. Compute the normalized receiving-end impedance, $z_R = Z_R/Z_0$, and enter either chart at this point. Note the corresponding readings on the distance and K scales.

2. To find the impedance at a distance d measured toward the generator, decrease the value of K by the factor $\epsilon^{-2\alpha d}$ and increase the reading of the distance scale by the proper number of wavelengths. This locates a new point on the chart. (If losses can be neglected, the quantity K is maintained at its original value and only the reading of the distance scale is changed.) The distance scale repeats itself every half wavelength, and any multiple of 0.5λ can be subtracted without changing the result.

3. At the point on the chart located as described above, read the value of normalized impedance, z. Convert to complex ohms by the relation $Z = zZ_0$. The sending-end impedance is obtained by using d equal to the length of the line.

Some of the available rectangular charts do not have the values of K marked on them, but the value of K associated with any circle can be found from its right-hand intercept r_i by use of Eq. (5.13), which becomes

$$K = \frac{r_i - 1}{r_i + 1}$$

)n the circular chart, the magnitude K can be read as a radial distance, taking the radius of the chart to be unity.

Example. Find the complex sending-end impedance of a 100-mile telephone ine which has $Z_0 = 685 - j92$ ohms, $\alpha = 0.00497$ neper/mile, and $\beta = 0.0352$ rad/mile at 1,000 cps. The line is terminated in $Z_R = 2,000 + j0$ ohms.
First, compute the normalized receiving-end impedance:

$$z_R = \frac{2,000 + j0}{685 - j92} = 2.87 + j0.385$$

Enter either chart with this value of z and read $K_R = 0.491$; also read 0.492λ on the distance scale (see Fig. 5.5).
Next, compute K at the sending end and the length of the line in wavelengths:

$$K_s = K_R\epsilon^{-2\alpha l} = 0.491\epsilon^{-0.994} = 0.182$$

and

$$\frac{l}{\lambda} = \frac{\beta}{2\pi}l = \frac{0.0352 \times 100}{2\pi} = 0.561$$

Adding the length of the line to the initial reading on the wavelength scale, we find that the point representing the sending-end impedance is located at $(0.492 + 0.561)\lambda = 1.053\lambda$. The distance scale repeats every half wavelength and so, subtracting two half wavelengths, we read the sending-end impedance on the chart at 0.053λ and $K_s = 0.182$. As indicated in Fig. 5.5, the result is $z_s = 1.30 - j0.31$. Then, in ohms,

$$Z_s = (1.30 - j0.31)(685 - j92)$$
$$= 861 - j332 \text{ ohms}$$

This compares with an answer of $861 - j325$ ohms obtained by the use of the exponential solution in the example of Sec. 4.1 and has been obtained with considerably less labor.

5.5. The Calculation of Currents and Voltages. As was shown in the preceding section, the charts simplify the calculation of line impedance. When Z_s is known, the sending-end voltage, current, and power can be computed by using the equivalent sending-end circuit of Fig. 4.2. The calculation of currents and voltages along the line can then proceed in any one of several ways. One can use the exponential or hyperbolic relations given in Chap. 4. Or, if the line has negligible losses, the conservation of power can be used: The sending-end power is computed and is set equal to the power at any other desired point. From this and the impedance at the second point, the voltage and current can be calculated. This procedure will be illustrated in Example 2 of this section. Still another method is to determine the forward-traveling component of voltage at the sending end, and then compute the attenuation and phase shift of this wave as it travels down the line, is reflected at the load, and travels back again. The total voltage'at any point is obtained as the vector sum of the incident and reflected components. We shall illustrate this method in Example 1 of this section, but first we shall derive a relation by which the forward-traveling component of voltage can be separated from the whole

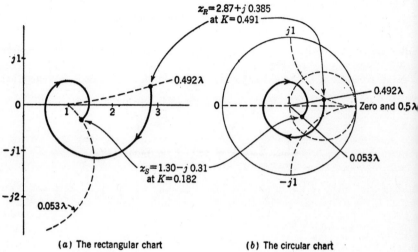

(a) The rectangular chart　　　　　　(b) The circular chart

FIG. 5.5.　Loci of line impedance for the example.

We wish to find a convenient expression for the ratio E^+/E. To do this, we write

$$\frac{E^+}{E} = \frac{E^+}{E^+ + E^-}$$

$$= \frac{1}{1 + (E^-/E^+)}$$

$$= \frac{1}{1 + k}$$

Substituting the value for k from Eq. (5.11) and simplifying, we obtain

$$\frac{E^+}{E} = \frac{1}{2}\left(1 + \frac{1}{z}\right) = \frac{1}{2}\left(1 + \frac{Z_0}{Z}\right) \tag{5.18}$$

This formula is especially convenient when the transmission-line chart is being used, as the normalized impedance $z = (Z/Z_0)$ can be read directly from the chart.

In a similar manner, it can be shown that the ratio of the forward-traveling voltage to the net current at any point is given by

$$\frac{E^+}{I} = \frac{1}{2}(Z + Z_0) \tag{5.19}$$

When the total voltage or current is known at any point, *e.g.*, at the sending end, Eq. (5.18) or (5.19) can be used to find E^+ at that point. The forward-traveling voltage can then be "followed" down the line by shrinking the vector by the factor $\epsilon^{-\alpha x}$ and rotating it in the lagging direction through the angle βx. At the receiving end, the vector E_R^+ is reflected by the factor k_R, giving rise to the backward-traveling component E_R^-. The sum $E_R^+ + E_R^-$ gives the net voltage at the receiving end, and from E_R and Z_R one can calculate the receiving-end current. If desired, the reflected component, E^-, can be "followed" back up the line for a calculation of the voltage and current at any point.

Example 1: A Lossy Line. Consider the 100-mile telephone line given as an example in Sec. 5.4, which had $Z_0 = 685 - j92$ ohms, $\gamma = 0.00497 + j0.0351$ rad/mile, and $Z_R = 2{,}000 + j0$ ohms. The sending-end impedance was calculated to be $z_s = 1.30 - j0.31$ per unit, or $Z_s = 861 - j332$ ohms.

Now assume that the line is driven at the sending end with a generator which has an open-circuit emf of 10 volts rms and an internal impedance of 700 ohms pure resistance. We shall compute the receiving-end voltage, current, and power by finding the magnitudes of the traveling waves of voltage, starting our computation at the sending end.

The generator works into the impedance Z_s; hence, the magnitude of the sending-end current is

$$|I_s| = \left|\frac{E_g}{Z_g + Z_s}\right|$$
$$= \left|\frac{10}{1{,}561 - j332}\right| = \frac{10}{1{,}597}$$
$$= 6.26 \times 10^{-3} \text{ amp rms}$$

The magnitude of the sending-end voltage is

$$|E_s| = |I_s| \cdot |Z_s|$$
$$= 6.26 \times 10^{-3} \sqrt{(861)^2 + (332)^2}$$
$$= 5.79 \text{ volts rms}$$

We now use Eq. (5.18) to find the magnitude of the forward-traveling component of voltage at the sending end:

$$E_s^+ = \frac{E_s}{2}\left(1 + \frac{1}{z_s}\right)$$
$$|E_s^+| = \frac{5.79}{2} \times \left|1 + \frac{1}{1.30 - j0.31}\right| = 5.03 \text{ volts rms}$$

Next, we compute the magnitude of the incident voltage at the receiving end, taking into account the attenuation factor $\epsilon^{-\alpha l}$:

$$| E_R{}^+ | = | E_s{}^+ | \epsilon^{-\alpha l}$$
$$= 5.03 \times 0.608$$
$$= 3.06 \text{ volts rms}$$

We can compute the reflection coefficient at the receiving end from the relation $k = (z - 1)/(z + 1)$, or we can obtain it from the chart. Reference to either chart shows that $z_R = 2.87 + j0.385$ corresponds to $K = 0.491$. The angle of k_R can be measured on the circular chart with a protractor as 6°. Therefore,

$$k_R = 0.491\underline{/6°}$$

Arbitrarily taking $E_R{}^+$ to be horizontal, we compute

$$E_R{}^- = k_R E_R{}^+$$
$$= (0.491\underline{/6°})(3.06\underline{/0°})$$
$$= 1.50\underline{/6°} \text{ volts}$$

Therefore, the resultant voltage at the receiving end is

$$E_R = E_R{}^+ + E_R{}^- = 3.06\underline{/0°} + 1.50\underline{/6°}$$
$$= 4.55 + j0.157 \text{ volts}$$

or

$$| E_R | = 4.55 \text{ volts rms}$$

The magnitude of the receiving-end current is

$$| I_R | = \left| \frac{E_R}{Z_R} \right| = \frac{4.55}{2,000} = 2.28 \times 10^{-3} \text{ amp rms}$$

This current is in phase with the receiving-end voltage, since the load is a pure resistance. The power absorbed by the load is

$$P_R = | I_R |^2 R_R = (2.28 \times 10^{-3})^2 \times 2,000$$
$$= 10.4 \times 10^{-3} \text{ watt}$$

For comparison, we shall compute the power which enters the sending end of the line. The magnitude of the sending-end current is

$$| I_s | = \left| \frac{E_s}{Z_s} \right| = \frac{5.79}{| 861 - j332 |} = 6.26 \times 10^{-3} \text{ amp rms}$$

The sending-end power is, therefore,

$$P_s = | I_s |^2 R_s = (6.26 \times 10^{-3})^2 \times 861 = 33.8 \times 10^{-3} \text{ watt}$$

The foregoing answers are identical with those computed in the example of Sec. 4.1.

The difference between P_s and P_R is obviously the power lost in the line. If the line were terminated in Z_0 at the receiving end, the power loss would be equivalent to αl nepers, or 8.686 αl db. However, the above loss is somewhat greater than this because of the power lost by the reflected wave as it travels back toward the generator.

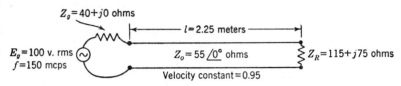

FIG. 5.6. Circuit for Example 2. Losses are assumed to be negligible ($\alpha l \ll 1$).

Example 2: Lossless Line. In this example we shall find the receiving-end voltage, current, and power for the line shown in Fig. 5.6, using the transmission-line chart. The over-all losses are small ($\alpha l \ll 1$), and so we can use the principle of the conservation of power.

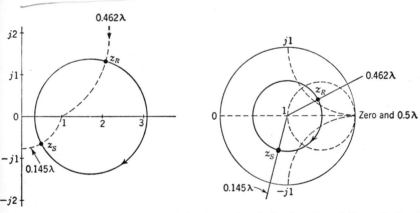

FIG. 5.7. Illustrating the use of the charts for the lossless line of Example 2.

First, we express the length of the line in terms of wavelengths. The velocity constant is given in the illustration as 0.95, and so the phase velocity is

$$v = 0.95 \times 3 \times 10^8 = 2.85 \times 10^8 \text{ meters/sec}$$

The wavelength is

$$\lambda = \frac{v}{f} = \frac{2.85 \times 10^8}{150 \times 10^6} = 1.90 \text{ meters}$$

The length of the line in wavelengths is

$$\frac{l}{\lambda} = \frac{2.25}{1.90} = 1.183$$

Next, we normalize the receiving-end impedance and obtain

$$z_R = \frac{Z_R}{Z_0} = \frac{115 + j75}{55} = 2.09 + j1.36$$

Now we enter either of the transmission-line charts with this value of normalized impedance, which, we find, corresponds to a reading of 0.462λ on the distance scale (see Fig. 5.7).

To find z_s, we rotate clockwise an amount equivalent to 1.183λ, keeping to the same K circle because losses are negligible. Now, a rotation of λ/2 merely brings us back to the starting point on the chart; hence, the net rotation on the chart will merely be 0.183λ. Therefore, the rotation stops at (0.462 + 0.183)λ = 0.645λ. Taking into account the break in the scale of distances at 0.5λ, this is equivalent to (0.645 − 0.5)λ = 0.145λ on the chart. At this point, either chart provides the following result for sending-end impedance:

$$z_s = 0.49 - j0.65$$

or

$$Z_s = (0.49 - j0.65) \times 55 = 27 - j35.8 \text{ ohms}$$

This is the impedance into which the generator works; hence, the sending-end current is

$$I_s = \frac{E_g}{Z_g + Z_s} = \frac{100}{67 - j\,35.8}$$

or

$$|I_s| = \frac{100}{75.8} = 1.32 \text{ amp rms}$$

The sending-end power is

$$P_s = |I_s|^2 \times R_s = (1.32)^2 \times 27$$
$$= 47 \text{ watts}$$

Since the losses in the line are negligible, the receiving-end power is equal to the sending-end power. Therefore, we can compute the receiving-end current as follows:

$$P_R = |I_R|^2 R_R$$

or

$$47 = |I_R|^2 \times 115$$

from which

$$|I_R| = 0.639 \text{ amp rms}$$

The receiving-end voltage is given by

$$|E_R| = |I_R| \times |Z_R|$$
$$= 0.639 \times |115 + j75|$$
$$= 0.639 \times 137$$
$$= 87.5 \text{ volts rms}$$

The circular chart fulfills all the requirements of the crank diagram described in Sec. 4.6 and Fig. 4.13 and can be used as a generalized form of this diagram to find the standing-wave patterns of voltage and current along a low-loss line.   On the circular chart, the complex reflection coefficient k is measured from the center, taking the outside radius to be unity. The corresponding quantity in the crank diagram is the vector $E^- = kE^+$, which is plotted from the tip of the E^+ vector.   To use the circular chart as a crank diagram, we regard the radius extending from $z = 0$ to the center of the chart as the vector representing E^+ and I^+, as shown in Fig. 5.8.   The vector originally representing k on the chart now represents

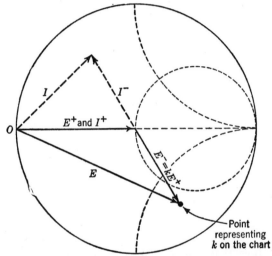

FIG. 5.8. Use of the circular chart as a crank diagram for low-loss lines.

kE^+, which is the reflected voltage E^-.   The reflected current, I^-, is the negative of E^-.   The total voltage and current, E and I, are measured from the extreme left edge of the chart to the tips of E^- and I^-, respectively.   As the distance d is increased, the points representing E^- and I^- revolve clockwise through the angle $2\beta d$ rad, or 360° for every half wavelength.

Example 2 Continued.   To determine the variation of E and I along the lossless line of Example 2, we can use the circular chart of Fig. 5.7 as a crank diagram in the manner illustrated by Fig. 5.8.   The receiving-end voltage is represented in Fig. 5.7 by a line joining the point $z = 0$ with the point indicated as z_R.   The receiving-end current is represented by a line extending from $z = 0$ to a point on the K circle diametrically opposite z_R.   As the distance from the load is increased, the tips of the E and I vectors rotate uniformly about the center of the chart, one revolution for each half wavelength.

We observe that, for this particular line, the tip of the E vector will cross the horizontal axis (at $z = 3.1 + j0$) at a distance of $(0.500 - 0.462)\lambda = 0.038\lambda$ from the receiving end. This is a position of maximum voltage and minimum current. If d is increased by another quarter wavelength, the tip of the E vector crosses the horizontal axis at $z = 1/3.1$. This is a position of minimum voltage and maximum current. Increasing d by another quarter wavelength, we find another voltage maximum. The sending end is located 0.645λ beyond this second maximum. The standing-wave patterns of E and I for the line are shown in Fig. 5.9.

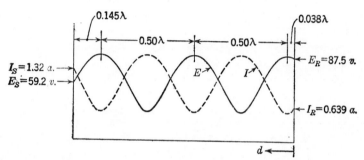

Fig. 5.9. Standing-wave patterns of E and I for the lossless line of Example 2.

5.6. Line Admittance. The Inversion of Complex Numbers. If we return to Eq. (5.12), which expresses the normalized impedance z in terms of the reflection coefficient at any point, we see that

$$z = \frac{1 + K\underline{/\theta}}{1 - K\underline{/\theta}} \tag{5.20}$$

We shall now show that the complex number associated with the same value of K but with a new angle given by $\theta' = \theta \pm 180°$ is the complex reciprocal of z. To prove this, replace θ by $\theta \pm 180°$ in Eq. (5.20) and examine the result.

$$\frac{1 + K\underline{/\theta} \pm 180°}{1 - K\underline{/\theta} \pm 180°}$$

The addition of 180° to the phase angle of a complex number is equivalent to taking the negative of the number; hence, we can write the foregoing relation as

$$\frac{1 - K\underline{/\theta}}{1 + K\underline{/\theta}}$$

This is the reciprocal of (5.20) and must represent the complex number $1/z$, which was to be proved. If we define

$$\frac{1}{Z_0} = Y_0 \quad \text{and} \quad \frac{1}{Z} = Y \tag{5.21}$$

then the reciprocal of the normalized impedance is

$$\frac{1}{z} = \frac{Z_0}{Z} = \frac{Y}{Y_0}$$

which we define as the normalized admittance

$$y = \frac{1}{z} = \frac{Y}{Y_0} \tag{5.22}$$

The foregoing development shows that, whenever we have located an impedance on the transmission-line chart, we can find the corresponding admittance by rotating on a fixed K circle to a new value of θ which is

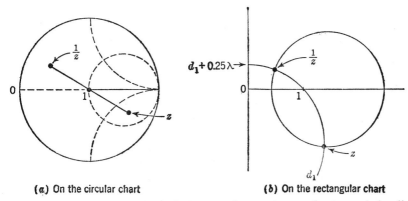

(a) On the circular chart   (b) On the rectangular chart

FIG. 5.10. Obtaining the reciprocal of a complex number on the transmission-line charts.

either 180° larger or smaller than the one at which we started. This corresponds to a change of 0.25λ on the wavelength scale. On the circular transmission-line chart the second point is diametrically opposite the first, as indicated in Fig. 5.10. On the rectangular chart, the complex reciprocal is located on the extension of the θ circle through the point $1 + j0$, as shown in Fig. 5.10b. As the circles are hard to follow through this point, it is best to note the original reading on the wavelength scale and add or subtract 0.25λ.

At this point, it should be clear that, as the impedance of a lossy line winds around a spiral, as shown in Fig. 5.5, or revolves around a circle of fixed radius for a lossless line, the normalized line admittance follows the impedance around with the same value of K, but at a value of θ greater by 180°. On the circular chart, the two points y and z are on the opposite ends of a diameter. Thus, all the methods for using the charts for impedances will apply equally well to admittances also. We can find the characteristic admittance of the line, Y_0, and compute the normalized receiving-end admittance by use of Eq. (5.22). We can then enter the transmission-line chart with this value of normalized admittance and re-volve around the chart in exactly the same fashion as when we were using normalized impedance, shrinking the value of K by the factor $\epsilon^{-2\alpha d}$ to take account of the losses in the line. Certain problems which involve parallel loads are easiest to solve on an admittance basis.

It should also be clear that we can use either of the charts to invert any complex number. In order to make the number fall on a convenient portion of the chart, it is often necessary to remove some simple factor such as a power of 10. One then enters the chart with the complex number and revolves on a circle of constant K to a new point on the distance scale which is either 0.25λ larger or smaller than the initial reading. The result is the reciprocal of the complex number with which the chart was entered, and any factor previously removed must now be inserted. On the circular chart one merely locates the original number and goes to a point diametrically opposite to find its complex reciprocal.

PROBLEMS

1. A certain transmission line has a characteristic impedance of $70 + j0$ ohms and is terminated in $Z_R = 90 + j65$ ohms. The line is 1.20 wavelengths long and has $\alpha l = 0.35$ neper. Determine the sending-end impedance of the line.

2. A 200-mile telephone line has $Z_0 = 645 - j95$ ohms and $\gamma = 0.00525 + j0.0351$ per mile at 1,000 cps. The receiving end is open-circuited, and the sending-end voltage is 10 volts rms. Find the sending-end impedance and the magnitude of the receiving-end voltage.

3. The line of Prob. 2 is terminated in a short circuit. The sending-end voltage is 10 volts rms. Determine the sending-end impedance and the magnitude of the receiving-end current.

4. A type of coaxial cable is made with a purposely high attenuation by using a Nichrome wire for the center conductor. The characteristic impedance is 53 ohms, and the phase velocity is 1.98×10^8 meters/sec. At a frequency of 120 Mc, the attenuation constant is 0.061 neper/meter. A section of this line $2\frac{1}{8}$ wave-lengths long is terminated in an open circuit. The sending-end voltage is 100 volts rms. Determine the sending-end impedance and the magnitude of the re-ceiving-end voltage.

5. A high-frequency line with a characteristic impedance of 70 ohms is termi-nated in $Z_R = 300 + j0$ ohms. The over-all attenuation is negligible. The line is driven with a generator that has $E_g = 10$ volts rms and $Z_g = 70$ ohms. Find the sending-end impedance, the magnitude of the receiving-end voltage, and the re-ceiving-end power for the following lengths of line:

    a. $l = \lambda$.
    b. $l = 1\frac{1}{8}\lambda$.
    c. $l = 1\frac{1}{4}\lambda$.

6. A certain high-frequency line has a characteristic impedance of 52 ohms and a phase velocity of $(2.00 \pm 0.02) \times 10^8$ meters/sec. Its length is 10.0 ± 0.1

meters. The frequency is 300 Mc. Neglect losses. Find the sending-end impedance of this line for the following receiving-end impedances:

   a. $Z_R = Z_0$.

   b. $Z_R = 4Z_0$. Explain any difficulties encountered.

   7. Prove Eq. (5.19).

   8. Compute the following quantities for Example 1 of Sec. 5.5: The efficiency of power transmission for the line itself under the conditions given in the example; the power loss of the line in decibels, based on the power input and power output of the line; the quantity αl expressed in decibels; the insertion loss of the line in decibels. Note the differences in meaning among the latter three quantities.

   9. Given a high-frequency line with negligible losses and with $Z_0 = 70$ ohms. The line is terminated in an open circuit. Determine the sending-end impedance for the following lengths of line: $\lambda/8$, $\lambda/4$, $3\lambda/8$, $\lambda/2$, and $5\lambda/8$.

   10. Given a high-frequency line with negligible losses and with $Z_0 = 70$ ohms. The line is terminated in $Z_R = -j140$ ohms. What is the shortest length of line for which Z_s is zero? For which $Z_s = j70$ ohms? For which Z_s is infinite?

   11. For the line given in Example 2 of Sec. 5.5, compute the rms voltage and current at a voltage maximum on the standing-wave pattern. Compute the rms voltage and current at a voltage minimum.

   12. Given a high-frequency line with $Z_0 = 150$ ohms, terminated in $Z_R = 70 + j0$ ohms. The line is 0.680 wavelength long. Neglect losses. Using a transmission-line chart on an admittance basis, determine the reactance which, if connected across the sending end of the line, will make the input impedance of the combination a pure resistance. Under this condition, what is the input impedance of the combination?

   13. Two transmission lines are connected in parallel at their sending ends. One line is $5\lambda/8$ long, has $Z_0 = 400$ ohms, and is terminated in a resistance of 800 ohms. The other is $3\lambda/8$ long, has $Z_0 = 400$ ohms, and is terminated in an impedance of $400 + j400$ ohms. Using a transmission-line chart on an admittance basis, determine the input impedance of the parallel combination of the two lines.

   14. Use one of the transmission-line charts to invert the following complex numbers:

   a. $2 + j1$.

   b. $100 + j150$.

   c. $1.47 - j2.05$.

   d. $0.062 + j0.120$.

CHAPTER 6

SPECIAL CONSIDERATIONS FOR RADIO-FREQUENCY LINES

6.1. Introduction. This chapter is devoted to a consideration of the special problems of low-loss lines; *i.e.*, lines for which the attenuation per wavelength is much less than 1 neper. Most radio-frequency lines fall in this category, and so the theory based on this approximation has wide usefulness. In addition, many phenomena which are of importance in connection with lossy lines are brought into prominence by a consideration of lines with small loss.

In Sec. 2.7 we derived the following relations for high-frequency lines where $\omega L \gg R$ and $\omega C \gg G$:

$$Z_0 \approx \sqrt{\frac{L}{C}} \qquad \text{ohms} \tag{6.1}$$

$$\alpha_0 \approx \frac{R}{2Z_0} + \frac{G}{2} Z_0 \qquad \text{nepers per unit length} \tag{6.2}$$

and

$$\beta \approx \omega\sqrt{LC} \qquad \text{radians per unit length} \tag{6.3}$$

The phase velocity ω/β is, therefore,

$$v \approx \frac{1}{\sqrt{LC}} \tag{6.4}$$

If the line is uniformly insulated with a material having a relative dielectric constant ϵ/ϵ_0 and a permeability equal to that of free space, Eq. (6.4) reduces to

$$v \approx \frac{3 \times 10^8}{\sqrt{\epsilon/\epsilon_0}} \qquad \text{meters/sec}$$

The wavelength on the line is

$$\lambda = \frac{v}{f} \tag{6.5}$$

We shall now show that a line for which $\omega L \gg R$ and $\omega C \gg G$, has a low attenuation per wavelength. The attenuation in nepers per wavelength is equal to $\alpha\lambda$. Using Eqs. (6.2), (6.4) and (6.5), this can be expressed as

$$\alpha\lambda = \left(\frac{R}{2Z_0} + \frac{G}{2}\, Z_0\right)\left(\frac{1}{f\sqrt{LC}}\right)$$

$$= \frac{R}{2fL} + \frac{G}{2fC}$$

or

$$\alpha\lambda = \pi\left(\frac{R}{\omega L} + \frac{G}{\omega C}\right) \text{ nepers per wavelength}$$

Therefore, the conditions $\omega L \gg R$ and $\omega C \gg G$ are equivalent to a small attenuation per wavelength.

Both of these conditions are generally satisfied by practical transmission lines at radio frequencies. When the skin effect is well developed, the effective resistance of the conductors increases as $\sqrt{\omega}$, and the inequality $\omega L \gg R$ therefore increases with frequency. As was shown in Chap. 3, the conductance of a uniformly insulated line can be expressed as $G = \omega C T_L$, where T_L is the loss tangent of the dielectric material. The inequality $\omega C \gg G$ thus requires only that $T_L \ll 1$, a condition that is satisfied by reasonably good insulating materials.

The use of transmission-line charts for the calculation of line impedance was discussed in Chap. 5, and Example 2 of Sec. 5.5 demonstrated the use of the charts for a line with negligible losses. When $\alpha l \ll 1$, the magnitude of the reflection coefficient, K, remains constant over the length of the line, and the locus of the line impedance is merely a circle of constant K. This simplifies impedance (or admittance) calculations. If the current and impedance at one point are known, the magnitude of the current at another point can be found from the conservation of power:

$$|I_1|^2 R_1 = |I_2|^2 R_2$$

where I_1 and I_2 are the currents at the two points, and R_1 and R_2 are the resistive components of the impedances at those points. The voltage at any point can be found from the current and the impedance. Thus, the charts can be used to simplify voltage and current calculations. Other uses will be developed in the sections that follow.

Even when losses are small, the variation of K with distance may be important if K is nearly unity; i.e., if the reflected wave is nearly equal in magnitude to the incident wave. The K circle will cross the resistance axis near zero on one side and at a very large resistance on the other, and if good accuracy is desired at these maximum and minimum points, the charts are not generally satisfactory. Appropriate formulas for this condition will be developed in Sec. 6.8.

6.2. Standing-wave Ratio. The relative magnitude of the reflected wave on a low-loss line is generally expressed by means of the standing-wave ratio ρ, which is defined as[1]

$$\rho = \frac{|E_{\max}|}{|E_{\min}|} \qquad \text{VSWR} \qquad (6.6)$$

where $|E_{\max}|$ is the rms value of the voltage at the highest point on the standing-wave pattern, and $|E_{\min}|$ is the rms value of the voltage at the lowest point on the pattern (see Fig. 6.1). The standing-wave ratio is widely used because it is one of the more easily measured quantities on a

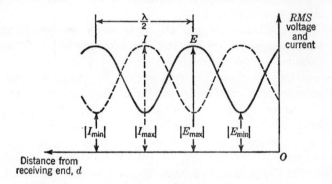

FIG. 6.1. Standing-wave patterns on a low-loss line.

line. A flat line, which has no reflected wave, has $|E_{\max}| = |E_{\min}|$ and a standing-wave ratio of unity. The ratio becomes larger without limit as complete reflection is approached.

The instruments used to measure standing-wave patterns at the higher frequencies are frequently of a type which give an indication proportional to the square of the voltage, and, for this reason, a ratio that is the square of Eq. (6.6) is sometimes used for convenience. To distinguish the two, the definition (6.6) is often given the name "voltage standing-wave ratio," abbreviated $VSWR$, while its square is called the "power standing-wave ratio," and is abbreviated $PSWR$. The latter name is somewhat misleading, however, for the transmitted power is the same at all points on the line. The name refers, not to the transmitted power, but to the power absorbed by the detecting device, for this is proportional to the square of the voltage across which the detector is connected.

[1] This symbol is also the conventional one for resistivity and was so used in Chap. 3. The context will make the usage clear, and no confusion should arise.

The standing-wave ratio cannot be defined when the standing-wave pattern changes its form markedly from one loop to another, as it will when losses are large. On a low-loss line, the ratio will remain constant for at least a few wavelengths and may be defined for any one region. When losses are negligible, the standing-wave ratio remains the same throughout the length of the line.

The standing-wave ratio for current $|I_{max}|/|I_{min}|$ is the same as that for voltage. This can be seen geometrically from the crank diagram of Fig. 6.2 (see also Sec. 4.6). The quantities E_{max} and I_{max} are represented by the same length on the diagram, although they occur a quarter wavelength apart. Furthermore, E_{min} and I_{min} are represented by the same length. Therefore, we have $|E_{max}|/|E_{min}| = |I_{max}|/|I_{min}|$.

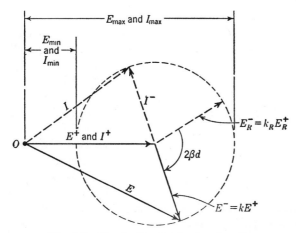

Fig. 6.2. Crank diagram for low-loss line.

The voltage standing-wave ratio bears a simple relation to the magnitude of the reflection coefficient, and its value can be read directly from the transmission-line chart. We observe from the crank diagram that a maximum in voltage occurs when E^+ and E^- add directly, and a minimum occurs when E^- subtracts directly from E^+. Therefore, we can write

$$|E_{max}| = |E^+| + |E^-| \qquad (6.7)$$

and

$$|E_{min}| = |E^+| - |E^-| \qquad (6.8)$$

The standing-wave ratio can then be expressed as

$$\rho = \frac{|E_{max}|}{|E_{min}|} = \frac{|E^+| + |E^-|}{|E^+| - |E^-|} = \frac{1 + |E^-/E^+|}{1 - |E^-/E^+|}$$

But $|E^-/E^+|$ is the magnitude of the reflection coefficient, K, and so

$$\rho = \frac{1 + K}{1 - K} \tag{6.9}$$

This relation is useful in itself, and also it indicates the method of determining standing-wave ratio from the transmission-line charts. Note that Eq. (6.9) is precisely the same as Eq. (5.13), which gave the value of normalized resistance at which a K circle crosses the r axis to the right of the point $1 + j0$. Therefore, to find the standing-wave ratio from the charts, we follow the K circle around to its right-hand intersection with the horizontal axis. The normalized resistance at this point is numerically equal to the voltage standing-wave ratio ρ.

If the receiving end is terminated in a pure resistance ($Z_R = R_R + j0$), the corresponding normalized impedance is the real number $R_R/Z_0 + j0$. If this is greater than unity, it represents the above-mentioned right-hand intersection and is equal to the standing-wave ratio. But if R_R/Z_0 is smaller than unity, it corresponds to the left-hand intersection of the K circle with the horizontal axis. The right-hand intercept, which is ρ, is the reciprocal of this, or Z_0/R_R. Therefore, with a purely resistive load, the standing-wave ratio is either R_R/Z_0 or Z_0/R_R, whichever is greater than unity.

The crank diagram for a low-loss line shows that E_{max} and I_{min} occur at the same position on the line and, furthermore, are in phase at this point. This gives a maximum impedance which is purely resistive and which corresponds with the right-hand intersection of the impedance locus with the r axis. The normalized resistance here is equal to the standing-wave ratio ρ, and so, in ohms,

$$Z_{max} = \rho Z_0 + j0 \tag{6.10}$$

Similarly, E_{min} and I_{max} occur together, a quarter wavelength away from E_{max} and I_{min}. This is a position of minimum impedance—again a resistance—and it corresponds with the left-hand intersection of the impedance locus with the r axis. The normalized resistances at the two intersections are the reciprocals of each other; the maximum is ρ and so the minimum is $1/\rho$. This gives

$$Z_{min} = \frac{Z_0}{\rho} + j0 \tag{6.11}$$

The ratio Z_{max}/Z_{min} is obviously equal to ρ^2, which gives another physical interpretation to the so-called "power standing-wave ratio."

Example. A low-loss line with $Z_0 = 70$ ohms is terminated in an impedance $Z_R = 115 - j80$ ohms. The wavelength on the line is 2.5 meters. Find the standing-wave ratio and the maximum and minimum line impedances. Also, find the distance between the load and the first voltage minimum.

The normalized receiving-end impedance is

$$z_R = \frac{115 - j80}{70} = 1.64 - j1.14$$

As shown in Fig. 6.3, we enter either transmission-line chart at this impedance and find that it corresponds with a reading of 0.052λ on the distance scale. Rotating on a circle of constant $K(= 0.46)$, the locus intersects the horizontal axis at $z = 0.37 + j0$; this is the first position of minimum impedance, minimum voltage, and maximum current. The distance from the load is $(0.250 - 0.052)\lambda = 0.198 \times 2.5 = 0.495$ meter, and the impedance is $Z_{\min} = 0.37 \times 70 = 25.9 + j0$ ohms.

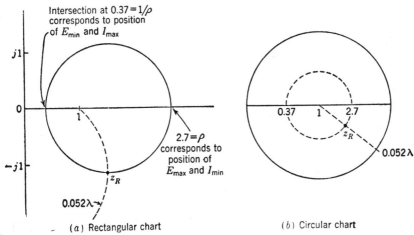

(a) Rectangular chart   (b) Circular chart

Fig. 6.3. The use of the charts for the example.

A quarter wavelength farther along, the locus intersects the horizontal axis at $z = 2.7 + j0$; this is the first position of maximum impedance, maximum voltage, and minimum current. The standing-wave ratio is given by the normalized resistance at this intersection: $\rho = 2.7$. The distance from the load to the first voltage maximum is $0.495 + \lambda/4 = 1.12$ meters, and the impedance at this point is $Z_{\max} = \rho Z_0 = 2.7 \times 70 = 189 + j0$ ohms.

6.3. Extreme Values on the Standing-wave Pattern. Power. For a low-loss line, the relation between the maxima of current and voltage on the standing-wave pattern is an especially simple one, even though the two quantities occur a quarter wavelength apart. As in Eq. (6.7), we write

$$|E_{\max}| = |E^+| + |E^-|$$

But $|E^+| = |I^+|Z_0$ and $|E^-| = |\varGamma|Z_0$, and so we have

$$|E_{\max}| = \big[|I^+| + |\varGamma|\big]Z_0$$

The bracketed portion is the magnitude of I_{max}, and thus we have the simple relation

$$|E_{max}| = |I_{max}|Z_0 \qquad (6.12)$$

In a similar fashion, we can show that the minima of current and voltage, although they occur a quarter wavelength apart on the line, are related simply by

$$|E_{min}| = |I_{min}|Z_0 \qquad (6.13)$$

The power transmitted by a line can be expressed in a number of equivalent ways. The computation is easiest at a point of maximum or minimum voltage, where the impedance is purely resistive. At a point of maximum voltage and minimum current, we have $E_{max} = I_{min}Z_{max} = I_{min}\rho Z_0$, and so the power can be expressed as

$$P = |E_{max}| \cdot |I_{min}| = \frac{|E_{max}|^2}{\rho Z_0} \qquad (6.14)$$

An equivalent expression for the power can be written at a point of minimum voltage and maximum current, where $E_{min} = I_{max}Z_{min} = I_{max}Z_0/\rho$. Here we have

$$P = |E_{min}| \cdot |I_{max}| = \frac{\rho|E_{min}|^2}{Z_0} \qquad (6.15)$$

A more symmetrical expression is obtained by solving (6.13) for $|I_{min}|$ and substituting into (6.14):

$$P = \frac{|E_{max}| \cdot |E_{min}|}{Z_0} \qquad (6.16)$$

If we use Eq. (6.16) and express E_{max} and E_{min} in terms of E^+ and E^-, the power can be written as

$$P = \frac{(|E^+| + |E^-|)(|E^+| - |E^-|)}{Z_0}$$

or

$$P = \frac{|E^+|^2}{Z_0} - \frac{|E^-|^2}{Z_0} \qquad (6.17)$$

We can regard the first term in this expression as the power associated with the forward-traveling wave, and the second term as the reflected power. This simple separation of power into two components, each associated with one of the traveling waves, can be done only when the characteristic impedance is a pure resistance. Otherwise, the interaction of the two waves gives rise to a third component of power. Thus, the concept applies to low-loss lines and to distortionless lines, but not to lossy lines in general.

Example. The line in the example of the preceding section had $Z_0 = 70$ ohms and $Z_R = 115 - j80$ ohms, and was found to have a standing-wave ratio of 2.7. If the line is to transmit a power of 50 watts, find the magnitudes of the maximum and minimum voltage and current, also the magnitude of the receiving-end voltage. The maximum and minimum impedances are purely resistive, and so we can write for the maximum point:

$$P = \frac{|E_{\max}|^2}{Z_{\max}} = \frac{|E_{\max}|^2}{\rho Z_0}$$

or

$$50 = \frac{|E_{\max}|^2}{2.70 \times 70}$$

from which

$$|E_{\max}| = 97.3 \text{ volts}$$

To find the minimum voltage along the standing-wave pattern, we use the relation

$$|E_{\max}| = \rho |E_{\min}|$$

or

$$97.3 = 2.7 |E_{\min}|$$

from which

$$|E_{\min}| = 36.0 \text{ volts}$$

For the maxima and minima of the current, we use Eqs. (6.12) and (6.13):

$$|I_{\max}| = \frac{|E_{\max}|}{Z_0} = \frac{97.3}{70} = 1.39 \text{ amp}$$

and

$$|I_{\min}| = \frac{|E_{\min}|}{Z_0} = \frac{36.0}{70} = 0.514 \text{ amp}$$

The receiving-end current can be found from the power relation

$$P = |I_R|^2 R_R$$

where R_R is the resistive component of the receiving-end impedance. Then

$$50 = |I_R|^2 \times 115$$

from which we obtain

$$|I_R| = 0.660 \text{ amp}$$

The receiving-end voltage is then

$$|E_R| = |I_R Z_R| = 0.660\sqrt{(115)^2 + (80)^2}$$

$$= 92.5 \text{ volts}$$

The standing-wave pattern is shown in Fig. 6.4. The wavelength is 2.5 meters, and the distance from the load to the first voltage minimum was found in the preceding section to be 0.495 meter.

6.4. The Impedance of Lossless Lines. The variation of impedance along a lossless line can, of course, be found from the transmission-line charts, and it has been shown previously that the locus for negligible losses simply follows a circle of constant K. For some purposes it is, however, convenient to have a formula for the impedance, perhaps to obtain better

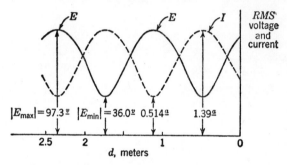

FIG. 6.4. Standing-wave pattern for the example.

accuracy than can be realized graphically, but more often because the formulas for certain special cases turn out to be especially simple and revealing. We return to Eq. (4.42), which gave the impedance of a general lossy line in terms of hyperbolic functions, and set the attenuation constant equal to zero; *i.e.*, we use $\gamma d = j\beta d$. Now we note from the identities (4.35) that $\tanh j\beta d = j \tan \beta d$, and so we can write Eq. (4.42) for the lossless case as

$$Z = Z_0 \frac{Z_R + jZ_0 \tan \beta d}{Z_0 + jZ_R \tan \beta d}$$

$$= Z_0 \frac{Z_R + jZ_0 \tan 2\pi d/\lambda}{Z_0 + jZ_R \tan 2\pi d/\lambda} \tag{6.18}$$

The angle $\beta d = 2\pi d/\lambda$ is expressed in radians. The sending-end impedance of the line, Z_s, is obtained from the foregoing expression by setting $d = l$.

Example. Find the sending-end impedance of a line with negligible losses when $Z_0 = 55$ ohms, $Z_R = 115 + j75$ ohms, and $l = 1.183\lambda$. We use Eq. (6.18), and first express the angle $2\pi d/\lambda$ in degrees. The angle is $2\pi \times 1.183 = 7.43$ rad, or $426°$, which is equivalent to $66°$. Then,

$$Z_s = 55\frac{(115 + j75) + j55\tan 66°}{55 + j(115 + j75)\tan 66°}$$

$$= 26.5 - j36.0 \text{ ohms}$$

This compares with $27 - j35.8$ ohms which was obtained with the transmission-line chart, using the same data, in Example 2 of Sec. 5.5.

6.5. Half-wavelength and Quarter-wavelength Lines. A distance of a half wavelength along a low-loss line is represented on a transmission-line chart by a complete rotation on a circle of constant K. Thus, the same impedance is repeated at half-wavelength intervals. This can also be seen in Eq. (6.18), for the tangent function repeats at intervals of π rad. Although impedances are repeated at intervals of $\lambda/2$, both the voltage and the current undergo a phase shift of $180°$ in this distance.

Because of the impedance-repeating effect of half-wavelength sections, bead-type insulators in a coaxial line should not be placed a half wavelength apart, as this is equivalent to connecting them all in parallel. If the beads are placed a quarter wavelength apart, the reflections from them will cancel except for their loss components.

A quarter-wavelength section of low-loss line acts as an "impedance inverter"; for example, if the terminating impedance is greater than Z_0, the impedance at the other end will be smaller than Z_0. Moreover, an inductive load will cause the sending-end impedance to be capacitive, and vice versa. To show this property, use Eq. (6.18): divide both numerator and denominator by $\tan 2\pi d/\lambda$, and let d approach $\lambda/4$. The sending-end impedance of the quarter-wave section becomes

$$Z_s = \frac{Z_0^2}{Z_R} \tag{6.19}$$

This can be written as $Z_s/Z_0 = Z_0/Z_R$, or, in terms of normalized impedances,

$$z_s = \frac{1}{z_R} \tag{6.20}$$

Thus, the complex normalized impedance is inverted; *i.e.*, its magnitude is inverted and its phase angle is reversed. An important application of this property is in the so-called "quarter-wave transformer," described in Sec. 7.8. Reference to Eq. (6.18) shows that any section an odd number of quarter wavelengths long will have the same impedance-inverting property, provided, of course, that the over-all losses are small. This can be regarded as the result of adding one or more half-wavelength impedance-repeating sections to a quarter-wave section.

6.6. Short Sections as Circuit Elements. As the frequency is increased and the wavelength becomes smaller, the conventional inductance coils and condensers become less and less useful as circuit elements. The distributed capacitance that exists between the turns of a coil offers a lower impedance as the frequency is raised, until finally the coil no longer behaves as a lumped element with a constant inductance but acts more as a very complex and almost unpredictable transmission line. In addition, the losses are needlessly high because of skin and proximity effects and often because of the radiation of energy. Condensers have similar difficulties because of a distributed inductance which is caused by the magnetic flux associated with the flow of current, although the range of usefulness of the condenser can be extended considerably by using small plates of simple geometry. As the distributed effects of supposedly lumped elements become troublesome, it becomes desirable to utilize the distributed constants in a simple geometry which permits easier prediction of the effects and which keeps losses to a minimum. This leads to the use of short sections of transmission line with completely reflecting terminations. They are useful as inductive and capacitive reactances and as resonant circuits at wavelengths below a few meters, where the required line sections are short enough to be convenient. The short lines are often called "stubs." Parallel-tube or parallel-strip lines are frequently used, but at the shorter wavelengths the coaxial line is preferred because it is self-shielded. However, precautions must be taken at the terminations to keep the energy within the annular space and prevent it from being transmitted along the outside of the outer conductor. At wavelengths of 10 cm and below, hollow metal boxes are used as resonators. Some extreme types of cavity resonators can be viewed as ordinary transmission lines but, in general, they are outside the realm of the distributed circuit and must be analyzed by the more general methods of electromagnetic theory. This places the hollow-box resonator beyond the scope of this book. We shall treat only those reactive and resonant devices which can be analyzed by conventional transmission-line theory.

The completely reflecting termination required to produce a low-loss element can be obtained with a short circuit, an open circuit, or a pure reactance. The short-circuit termination is most frequently used because it is easier to produce complete reflection in this way when the wavelength is small. Open-circuited lines may radiate energy off the open ends unless the spacing between conductors is quite small compared with a quarter wavelength. On an open-wire line, a conducting plate erected at right angles to the line and electrically connected to the wires is an excellent reflector. A coaxial cable can be completely short-circuited by closing off the annular space with a conducting plunger.

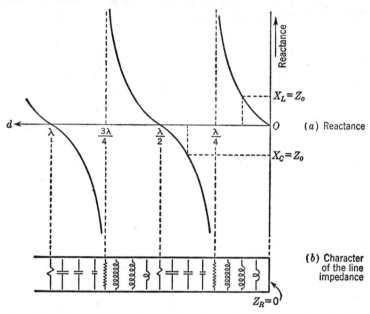

FIG. 6.5. The variation of reactance along a lossless short-circuited line.

Short-circuit Termination. To find the impedance of a lossless short-circuited line, use Eq. (6.18) with $Z_R = 0$. The result is

$$Z = jZ_0 \tan \frac{2\pi d}{\lambda} \qquad (6.21)$$

where the angle $2\pi d/\lambda$ is in radians. With the angle expressed in degrees, we have

$$Z = jZ_0 \tan \left(\frac{d}{\lambda} 360° \right) \qquad (6.22)$$

This impedance is a pure reactance, and its variation with d is sketched in Fig. 6.5. The distance from the load, d, is shown increasing to the left to make it correspond with our usual sketch of the line. When the lossless line is an integral number of half wavelengths long, the sending-end impedance is zero. Line losses cause the impedance actually to be a very small resistance here. A line of this length is said to be *resonant* and can be compared with a lumped LC series circuit at resonance. When the lossless line is an odd number of quarter wavelengths long, the sending-end impedance is theoretically infinite (inversion of the receiving-end impedance). The actual impedance, considering losses, is a very large resistance, and the line is said to be *antiresonant*. It can be compared with a lumped LC parallel circuit when $\omega L = 1/\omega C$. The effect of losses will be considered in Sec. 6.8.

Reference to Fig. 6.5 shows that the reactance of a short-circuited stub is inductive for lengths smaller than $\lambda/4$, and is capacitive for lengths between $\lambda/4$ and $\lambda/2$. The magnitude of the reactance can be adjusted by changing the length of the stub. When $l = \lambda/8$, so that $\beta l = \pi/4$ rad

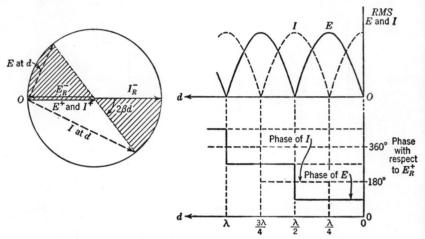

FIG. 6.6. Crank diagram, standing-wave pattern, and phase graph for a lossless short-circuited line.

or 45°, we obtain an inductive reactance numerically equal to the characteristic impedance: $Z = jZ_0$. Similarly, when $l = 3\lambda/8$, so that $\beta l = 3\pi/4$ rad or 135°, we obtain $Z = -jZ_0$.

The standing-wave pattern along the short-circuited line can be obtained from a crank diagram, as shown in Fig. 6.6. The reflection coefficient for voltage at the receiving end is -1. The pattern consists of a succession of sine loops, with $E_{\min}$ ($=0$) and $I_{\max}$ occurring at the receiving end. This diagram can be compared with the one of Fig. 4.8. The pattern can easily be drawn from memory, starting with $E = 0$ and $I = I_{\max}$ at the short circuit.

We recall from Sec. 4.6 that the crank diagram is drawn with E^+ arbitrarily held stationary, whereas E^+ should really be rotated counterclockwise through the angle βd, and E^- should be rotated oppositely through the same angle. Therefore, to obtain the actual phase angles of E and I with respect to a fixed reference, say $E_R{}^+$, it is necessary to give the whole crank diagram a counterclockwise rotation through the angle βd. The resulting phase angles for the present problem are shown in Fig. 6.6. A phase shift of 180° occurs abruptly at a node; i.e., adjacent loops are 180° out of phase with each other. In the first quarter wavelength I lags E by 90°, in the next quarter wavelength I leads E by 90°, and so on in a cyclic variation.

Figure 6.5 shows the variation of impedance with distance from the load, but the variation with frequency for a given length of line is of equal

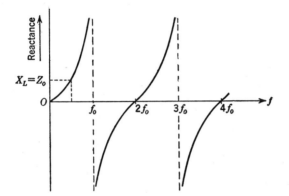

FIG. 6.7. Variation of input impedance with frequency; short-circuited lossless line.

importance. If we return to Eq. (6.21) and use $\lambda = v/f$, we obtain for the sending-end impedance of a short-circuited stub of length l:

$$Z_s = jZ_0 \tan \frac{2\pi l}{v} f$$

$$= jZ_0 \tan \frac{\pi f}{2f_0} \tag{6.23}$$

where f_0 is the frequency that makes the line a quarter wavelength long. This is plotted in Fig. 6.7, and is similar in form to Fig. 6.5. A uniform line with a short-circuited receiving end has an infinite number of resonant and antiresonant frequencies spaced at uniform intervals.

Various applications of low-loss short-circuited lines are based on the properties just described. A stub with a movable short circuit is commonly used as an adjustable reactance and receives considerable application as an impedance-matching element.

In the vicinity of an antiresonant frequency, the low-loss line behaves in the same manner as a high-Q parallel-resonant circuit and can be used for similar purposes. The variation of reactance with frequency for the two systems is compared in Fig. 6.8a. Similarly, in the vicinity of a resonant frequency, a low-loss line behaves like a high-Q series-resonant

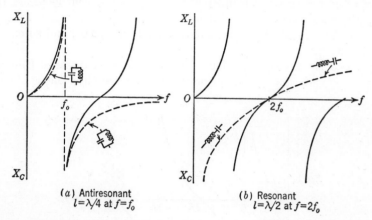

(a) Antiresonant
$l=\lambda/4$ at $f=f_0$

(b) Resonant
$l=\lambda/2$ at $f=2f_0$

FIG. 6.8.   The short-circuited lossless line as a resonant circuit.   The dashed curves show the reactance of lumped LC circuits for comparison.

circuit, as illustrated in Fig. 6.8b.   The Q of a resonant or antiresonant short-circuited line is considered in Sec. 6.9.

A quarter-wave short-circuited stub is often used as a low-loss insulator at short wavelengths.   Two examples of this application are shown in

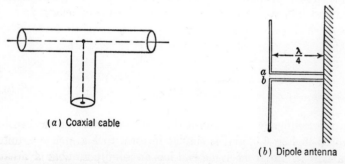

(a) Coaxial cable

(b) Dipole antenna

FIG. 6.9.   Short-circuited quarter-wave lines used as insulating supports.

Fig. 6.9.   The quarter-wave stub is sometimes used as the support for the center conductor of rigid coaxial cable, for at the shorter wavelengths it introduces less reflection than a dielectric bead.   The stub support, being an antiresonant structure, is quite sensitive to changes in frequency, and

n its simplest form is suitable only for narrow-band operation. It can,
however, be broad-banded by a simple modification of its construction.[1]

A somewhat similar application is the line balance converter, sometimes
called a *balun*, which is used as a transition between an unbalanced coaxial
system and a balanced two-wire line or dipole antenna. At lower frequencies, the balance-to-unbalance transition could be made with a conventional transformer. In so far as high-frequency alternating currents
are concerned, the outer conductor of a coaxial cable is generally at or near
the potential of any large nearby conducting surfaces, either because of an
intentional connection between the two, or simply because of the capacitance existing between them. A two-wire line, on the other hand, should
be balanced with respect to ground so that the conductors carry equal and
opposite currents. Unbalance will result in the induction of currents on
nearby conducting surfaces. In addition, if a coaxial cable is connected

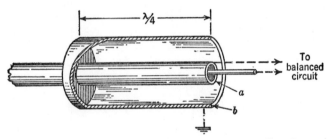

FIG. 6.10. Cutaway view of quarter-wave decoupling sleeve.

to an unbalanced two-wire line, a wave will be transmitted along the outside of the coaxial cable. The currents so produced generally serve no
useful purpose and lead to unnecessarily high losses. The simple balun
shown in Fig. 6.10 is sometimes called a decoupling sleeve or a "bazooka."
It consists of a quarter-wave sleeve at the end of the coaxial cable. The
end at the point of transition is open; the other end is closed. The sleeve
and the outside surface of the cable form a quarter-wave coaxial line
which is short-circuited at one end; this produces a high impedance between points *a* and *b*. Point *a* is now isolated from ground and can be
connected to the balanced circuit. The balun shown in Fig. 6.10 is useful
over only a narrow range of frequencies. Other types, with a greater
band width, have been devised.[2]

[1] See G. L. Ragan, "Microwave Transmission Circuits," Sec. 4.4, McGraw-Hill
Book Company, Inc., New York, 1948, and T. Moreno, "Microwave Transmission
Design Data," pp. 88–93, McGraw-Hill Book Company, Inc., New York, 1948.
[2] See "Very High-frequency Techniques," by the Radio Research Laboratory
Staff of Harvard University, Vol. I, Secs. 3–13 to 3–16, McGraw-Hill Book Com-
pany, Inc., New York, 1947.

If a lumped load is used at the end of a high-frequency line, the line can be extended a quarter wavelength beyond the load and there terminated in a short circuit. The impedance of this extension will be extremely large as viewed from the load and will not alter the performance of the system as calculated by circuit theory. However, viewed from the field standpoint, the short circuit will reflect any energy that gets past the load and will send the reflection back to the load so that it arrives with the proper phase relationship to reinforce the next half cycle of the wave arriving from the generator.

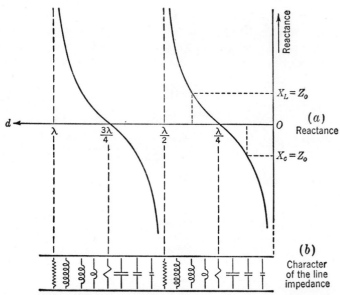

FIG. 6.11. The variation of reactance along a lossless open-circuited line.

Open-circuit Termination. The impedance of a lossless open-circuited line can be found from Eq. (6.18) by letting Z_R approach infinity. Then we obtain

$$Z = \frac{-jZ_0}{\tan 2\pi d/\lambda} \tag{6.24}$$

or, with the argument in degrees,

$$Z = \frac{-jZ_0}{\tan [(d/\lambda)360°]} \tag{6.25}$$

This has the variation sketched in Fig. 6.11. The character of the line impedance at various points is shown symbolically. The impedance is very small a quarter wavelength from the termination (resonance), and is very large a half wavelength from the open-circuited end (antiresonance). A capacitive reactance is obtained with an open-ended stub shorter than a quarter wavelength, an inductive reactance is obtained when the length is between $\lambda/4$ and $\lambda/2$, and so on cyclically.

Figure 6.12 shows the standing-wave pattern and the variation in phase of E and I along the line. These can be obtained from a crank diagram, as was done in Fig. 6.6. Again the pattern can easily be drawn from memory, for $I = 0$ and $E = E_{\max}$ at the open end.

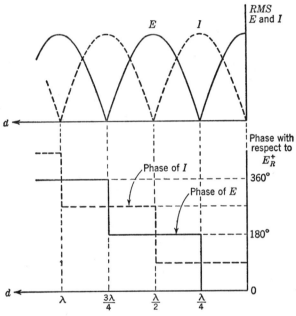

FIG. 6.12. Standing-wave pattern and phase graph for a lossless open-circuited line.

Reactive Terminations. Complete reflection can be produced not only by an open circuit and by a short circuit, but also by any reactive termination which absorbs no energy. The variation of impedance along the line can be found, of course, from Eq. (6.18), and a crank diagram can be used to find the standing-wave pattern. However, it is convenient for many purposes to visualize the effect of the reactive load by replacing it with a fictitious open- or short-circuited extension of the line. The length of the fictitious extension is chosen so that the reactance seen at the actual receiving end is the same as before. The standing-wave pattern and the impedances at various points can then be found for the new system composed of the actual line plus the extension, after which the fictitious portion can be discarded.

Figure 6.13 shows two examples of this procedure, together with the resulting standing-wave patterns. In Fig. 6.13a the inductive load has been replaced by a short-circuited extension, the length of which is selected so as to provide an impedance at its terminals of $j600$ ohms. To find the

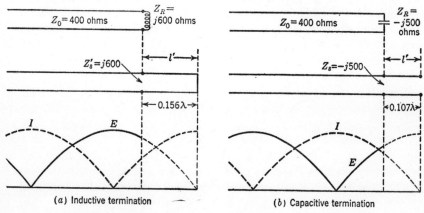

(a) Inductive termination   (b) Capacitive termination

FIG. 6.13. Two examples of reactive loads on a lossless line.

length of the extension, we use Eq. (6.22) with $Z_s' = j600$ ohms. Then we have

$$j600 = j400 \tan \left(\frac{l'}{\lambda} 360° \right)$$

Hence,

$$\tan \left(\frac{l'}{\lambda} 360° \right) = 1.50$$

from which

$$\frac{l'}{\lambda} 360° = 56.3°$$

or

$$l' = 0.156\lambda$$

The length of the equivalent open-circuited extension in Fig. 6.13b is found by use of Eq. (6.25):

$$-j500 = \frac{-j400}{\tan \left(\frac{l'}{\lambda} 360° \right)}$$

from which we obtain

$$l' = 0.107\lambda$$

An inductive load is equivalent to a short-circuited extension with a length smaller than $\lambda/4$. Since E is zero at the end of the fictitious shorted extension, this places a voltage maximum less than $\lambda/4$ from an inductive load. Larger inductances are equivalent to longer extensions, and the voltage maximum moves toward the load until, at infinite inductance, the effect is that of an open circuit.

Similarly, a capacitive load is equivalent to an open-ended extension with a length smaller than $\lambda/4$. This leads to a voltage minimum less than $\lambda/4$ from the load. Larger capacitances are equivalent to longer extensions, thus causing the voltage minimum to move closer to the load. A very large capacitance gives the effect of a short circuit, with $E = 0$ at the load.

6.7. Other Resonant Systems. The resonant and antiresonant properties of short sections of low-loss line were discussed in the preceding section from the point of view of the impedance presented to the generator. Other viewpoints are often useful, and these will be discussed in connection with systems of slightly greater complexity. The word "resonant" is often used as a generic term to denote the state of a system when it is driven at one of its natural frequencies of oscillation; when the word is used in this way, it includes both phenomena that we have previously called resonance and antiresonance. The context will make the meaning clear.

Consider Fig. 6.14, in which the two boxes A and B represent lossless networks whose input impedances at a given frequency are reactances of equal magnitude but of opposite sign. When driven by a generator G which maintains a voltage E at the terminals, the currents taken by the networks are respectively $I_1 = E/jX$ and $I_2 = -E/jX$. The generator current, being the sum of these, is zero. The impedance seen by the generator is infinite, and the two networks represent an antiresonant system at the given frequency.

If, on the other hand, the generator were connected in series with one of the wires joining the two networks, the generator would see the sum of the two impedances, or zero impedance, at the given frequency. The system would then be resonant, as distinguished from antiresonant. In this section we shall be interested chiefly in the properties of antiresonant systems, and from here on we shall use the word "resonant" in its broader sense.

In a practical situation, the networks of Fig. 6.14 will have some loss, and the generator will be able to maintain the circulating current between the networks by supplying an in-phase current to balance the losses. In general, a resonant system can be maintained in oscillation by supplying it at some point with energy at the proper frequency. The amplitude of oscillation will, in steady state, adjust itself so that all the supplied energy is consumed in losses.

When losses are small, the generator current in Fig. 6.14 will be much smaller than the circulating current I_1 or I_2, and the impedance seen by the generator will be much greater in magnitude than X ohms. Therefore, unless the generator is connected across a very-low-impedance part of the system, the input impedance will be quite large.

Figure 6.15 shows a resonant system which consists of a line that is short-circuited at both ends. If this system is driven by a generator that has a sinusoidally varying voltage, a large standing wave will be set up on the system at a frequency which makes the length equal to $\lambda/2$. This

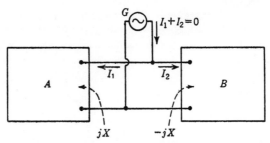

FIG. 6.14. Equal and opposite reactive impedances joined to form an antiresonant system.

is a natural frequency of oscillation of the system, for it allows a standing wave to satisfy the boundary conditions by having a voltage node at each short-circuited end, as shown in Fig. 6.15a. The result is similar to the vibration of a stretched elastic string which is fastened at the ends. One can imagine two oppositely traveling waves, each continuously reflected from the end upon which it is incident and, upon its reflection, being the "source" of the other wave. This oscillation, caused by a continuous back-and-forth reflection, can be sustained at a constant amplitude by supplying it with enough energy from the outside to balance the losses. If a generator is connected across the line as shown in Fig. 6.15, the line and generator voltages must be equal at the point of connection. If the generator is not placed too near a node, this voltage will be of the same order of magnitude as the greatest voltage on the line. Then the generator will be able to supply the required power with only a small current, and it will see the system as an antiresonant one. On the other hand, a generator connected in series with the line would see a low impedance and would view the system as resonant rather than antiresonant.

Considering the standing wave as a whole, the current is maximum at the moment when the voltage is everywhere zero, and at this instant all the energy in the system is stored in the magnetic field. A quarter cycle later the current is zero everywhere and the voltage is maximum, and the stored energy has been transferred to the electric field.

If the frequency of the driving generator is increased beyond the first resonance, a second resonant frequency is found at double the first one. This frequency makes the line a wavelength long and again allows a standing wave to satisfy the boundary conditions by having a node at both short-circuited ends. The resulting standing-wave pattern for the second mode of oscillation is shown in Fig. 6.15b. This mode has a node of voltage in the center of the line. A generator connected across the line at the nodal point will not be able to supply energy to the mode and will not be able to excite or sustain this particular oscillation.

Still higher modes of oscillation are found at integral multiples of the first resonant frequency. If the voltage of the generator contains harmonics, two or more of the modes may be excited at once. Several frequencies will then be present on the system at the same time,

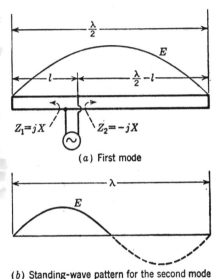

(a) First mode

(b) Standing-wave pattern for the second mode

FIG. 6.15. A low-loss antiresonant system and the standing-wave patterns for the first two modes.

and the standing-wave pattern, as measured by an rms voltage indicator, will have a complex shape.

Reasoning on an energy basis shows that a parallel-connected generator will see a high impedance, but we shall demonstrate this with the transmission-line equations for the first mode of Fig. 6.15a. Neglecting losses and using Eq. (6.21), the impedance of the left-hand section as viewed from the generator terminals is

$$Z_1 = jZ_0 \tan \frac{2\pi l}{\lambda}$$

The impedance of the right-hand section, viewed from the same terminals, is

$$Z_2 = jZ_0 \tan \frac{2\pi}{\lambda} \left(\frac{\lambda}{2} - l \right)$$

$$= jZ_0 \tan \left(\pi - \frac{2\pi l}{\lambda} \right)$$

$$= -jZ_0 \tan \frac{2\pi l}{\lambda}$$

which is the negative of Z_1. The generator sees the two impedances in parallel, the result being an infinite impedance, independently of the length l. In practice, of course, the impedance is finite at resonance

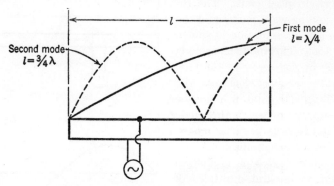

FIG. 6.16. Another antiresonant system.

although much larger than Z_1. If the point of attachment of the generator is moved toward the node of voltage, the impedance Z_1 is reduced and finally becomes zero at the node ($l = 0$). The impedance presented to the generator is similarly reduced toward zero as the node is approached, until, at the node, no energy can be supplied by the generator.

Another example of an antiresonant transmission-line system is shown in Fig. 6.16. The natural modes of oscillation of this system must have maximum voltage at the open end and a node at the short circuit. The first mode makes the line a quarter wavelength long, and its standing-wave pattern is shown by the solid curve. The second mode has a frequency three times the first; it makes the line $3\lambda/4$ in length and has the standing-wave pattern shown by the dashed curve. This system, operated in its lowest mode, is frequently used in practice. When the generator is connected at the open end of the line, the system reduces to the simpler one discussed in the preceding section (Figs. 6.5 to 6.8). When the line is driven at an intermediate point, it can be used as a tuned autotransformer, as shown in Fig. 6.17. The standing-wave pattern shown beside the line assumes that the output operates into a very high impedance.

The arrangement shown in Fig. 6.18 is often encountered in practical applications. The condenser C may represent the interelectrode capacitance of a vacuum tube. The system is resonant when the input line impedance Z is a reactance equal and opposite to that of the condenser. Using Eq. (6.21) with the angle $2\pi l/\lambda$ expressed as $\omega l/v$, this condition is

$$Z_0 \tan \frac{l\omega}{v} = \frac{1}{\omega C} \tag{6.26}$$

The antiresonant frequencies are given by the intersections of the hyperbola $1/\omega C$ with the function $Z_0 \tan l\omega/v$, as shown in Fig. 6.18b, and these

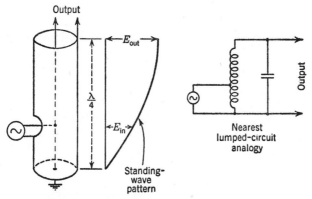

FIG. 6.17. A quarter-wave line used as an autotransformer.

are not harmonically related to each other. The standing-wave patterns of voltage for the first two modes are sketched roughly in Fig. 6.18a.

6.8. The Impedance near Resonance and Antiresonance. Because of the neglect of losses, the equations derived in Sec. 6.6 indicate that the input impedance is infinite at antiresonance and zero at resonance. In this section, we shall take the losses into account and investigate the behavior of the impedance near the critical frequencies. We shall consider only the simpler resonant systems which are driven at one end and are connected at the other to a perfectly reflecting termination.

Input Admittance near Antiresonance. If the termination is perfectly reflecting, the incident and reflected components of voltage will be equal in magnitude at the receiving end. Denote this magnitude by the letter A, i.e.,

$$|E_R^+| = |E_R^-| = A \tag{6.27}$$

Then the magnitude of the current components at the receiving end will be

$$| I_R{}^+ | \, = \, | I_R{}^- | \, = \frac{A}{Z_0} \qquad (6.28)$$

At the sending end, the reflected voltage and current will have been shifted $-\beta l$ rad from their phase positions at the receiving end and will

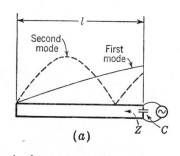

(a)

have been shrunk by the factor $\epsilon^{-\alpha l}$ (see Fig. 6.19). For αl much less than 1 neper, this factor will be very nearly equal to $1 - \alpha l$. Similarly, the forward-traveling components will be βl rad ahead of their phase positions at the load and will be larger by the factor $\epsilon^{\alpha l}$ $\approx 1 + \alpha l$.

Near an antiresonant frequency, the two voltage vectors will be nearly lined up in direct addition, and the two current components will be nearly in phase opposition, as indicated in Fig. 6.19. The amount by which βl differs from the antiresonant value will be denoted by $\Delta\theta$.

Now we add the voltage components to obtain the sending-end voltage. The vertical components of the vectors are very small and nearly cancel, and so they can be neglected in comparison with the large horizontal components. Then, since $\cos \Delta\theta \approx 1$, we have

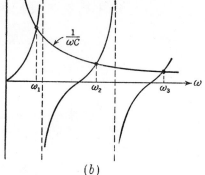

(b)

FIG. 6.18. An antiresonant line loaded with capacitance at one end.

$$E_s = E_s{}^+ + E_s{}^- \approx A(1 + \alpha l) + A(1 - \alpha l) = 2A \qquad (6.29)$$

Resolving the current vectors into real and imaginary parts and using $\cos \Delta\theta \approx 1$, $\sin \Delta\theta \approx \Delta\theta$, we have

$$I_s = I_s{}^+ + I_s{}^- \approx \frac{A(1 + \alpha l)}{Z_0}(1 + j\,\Delta\theta) + \frac{A(1 - \alpha l)}{Z_0}(-1 + j\,\Delta\theta)$$

or

$$I_s = \frac{2A}{Z_0}(\alpha l + j\,\Delta\theta) \qquad (6.30)$$

Dividing I_s by E_s, we obtain for the admittance near antiresonance,

$$Y_s = \frac{I_s}{E_s} = \frac{1}{Z_0} (\alpha l + j \,\Delta\theta) \qquad (6.31)$$

The quantity $\Delta\theta$ is the amount by which the angle βl differs from the antiresonant value. Since $\beta = \omega/v$, we can write the angle as

$$\beta l = \frac{l\omega}{v} \qquad (6.32)$$

If the length of the line is fixed and the frequency is variable, then

$$\Delta\theta = \frac{l}{v} \,\Delta\omega \qquad (6.33)$$

where $\Delta\omega$ is the amount by which the angular frequency ω is greater than

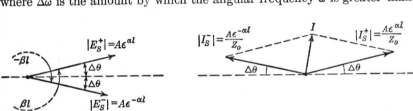

FIG. 6.19. The incident and reflected components of voltage and current near antiresonance.

the antiresonant value. On the other hand, if the frequency is fixed and the length is variable, we have

$$\Delta\theta = \frac{\omega}{v} \,\Delta l = 2\pi \,\frac{\Delta l}{\lambda} \qquad (6.34)$$

where Δl is the amount by which the length exceeds that required for antiresonance.

If we assume that the length is fixed and that we are interested in the effect of small variations in frequency near antiresonance, we use (6.33) for $\Delta\theta$ and write Eq. (6.31) as

$$Y_s = \frac{1}{Z_0} \left(\alpha l + j \frac{l}{v} \,\Delta\omega \right) \qquad (6.35)$$

At antiresonance, the admittance is simply $Y_s = \alpha l/Z_0$ and the impedance is resistive and much larger than Z_0:

$$Z_{\max} = \frac{Z_0}{\alpha l} \qquad (6.36)$$

Equation (6.35) shows that the input admittance will have a positive susceptance (capacitive) at a frequency above antiresonance and a negative susceptance (inductive) below antiresonance. Thus, an increase in frequency changes the susceptance in a positive direction.

Input Impedance near Resonance. When a low-loss line with a perfectly reflecting termination is near resonance, the incident and reflected voltages nearly cancel at the sending end, while the two components of current add almost directly. This is the reverse of the situation obtained near antiresonance and depicted in Fig. 6.19, and the analysis is identical with the preceding one except that the current and voltage vectors are interchanged. The resulting expression for the input impedance can be anticipated almost immediately from the result given in Eq. (6.31) and is

$$Z_s = Z_0(\alpha l + j\,\Delta\theta) \tag{6.37}$$

If the length is constant and we are interested in the effect of a small variation $\Delta\omega$ from the resonant frequency, we use Eq. (6.33) for $\Delta\theta$ and write

$$Z_s = Z_0\left(\alpha l + j\frac{l}{v}\Delta\omega\right) \tag{6.38}$$

At resonance, the input impedance is the small resistance

$$Z_{\min} = Z_0\alpha l \tag{6.39}$$

Above resonance the impedance has an inductive component; below resonance, a capacitive component. An increase in frequency changes the reactive component in a positive direction.

Example. Given an air-insulated line made of two parallel copper tubes 0.15 in. in radius and spaced 1 in. between centers. The line is driven at one end with a 200-Mc generator and has an adjustable short-circuit termination at the other. Find the input impedance as a function of length, including the effect of resistance losses. Neglect radiation losses from the conductors and from the short-circuited end.

The characteristic impedance is, approximately,

$$Z_0 = 120\log_\epsilon\frac{D}{a} = 120\log_\epsilon\frac{1}{0.15} = 228 \text{ ohms}$$

(The more exact formula, $Z_0 = 120\cosh^{-1}D/2a$, gives 225 ohms.) Except near the resonant and antiresonant points, the lossless formula (6.21) will apply with good accuracy:

$$Z_s = j228\tan\frac{2\pi l}{\lambda} \qquad \text{ohms}$$

For the impedance near the critical points, we use Eqs. (6.31) and (6.37). First we compute the attenuation constant. The resistance per unit length of a parallel wire line composed of copper conductors is (see Chap. 3)

$$R = 8.34 \times 10^{-6} \frac{\sqrt{f}}{a_{cm}} = 8.34 \times 10^{6} \frac{\sqrt{200 \times 10^{6}}}{0.15 \times 2.54}$$

$$= 0.309 \, \text{ohm/meter}$$

Then, neglecting radiation losses and leakage conductance, we have

$$\alpha = \frac{R}{2Z_0} = \frac{0.309}{2 \times 228} = 6.78 \times 10^{-4} \, \text{neper/meter}$$

Assuming a velocity of 3×10^8 meters/sec, the wavelength is

$$\lambda = \frac{v}{f} = \frac{3 \times 10^8}{200 \times 10^6} = 1.5 \, \text{meters}$$

In the neighborhood of the first antiresonant point, the length of the line will be approximately $\lambda/4$, or 0.375 meter. Then, from Eq. (6.31), with $\Delta\theta = 2\pi \, \Delta l/\lambda$, we have

$$Y_s = \frac{1}{228} \left(6.78 \times 10^{-4} \times 0.375 + j2\pi \frac{\Delta l}{\lambda} \right) \qquad \text{mhos}$$

or

$$Z_s = \frac{898,000}{1 + j24,700 \, \Delta l/\lambda} \qquad \text{ohms}$$

where Δl is the amount by which the length exceeds the antiresonant value. At antiresonance, the input impedance is a pure resistance of 898,000 ohms. A phase angle of 45° is introduced, and the magnitude of the impedance is lowered to $898,000/\sqrt{2}$ if the length is changed from the antiresonant value so that $\Delta l/\lambda = 1/24,700$; i.e., if $\Delta l = 0.061$ mm. The antiresonance is seen to be extremely sharp. In practice, a small variable condenser is often connected at the generator end of an antiresonant line for the purpose of providing a fine adjustment.

If the length of the line is adjusted to $\lambda/2 = 0.75$ meter, it will operate in its first resonant mode and, from Eq. (6.39), the sending-end impedance will be

$$Z_{min} = Z_0 \alpha l = 228 \times 6.78 \times 10^{-4} \times 0.75 = 0.116 \, \text{ohm}$$

Finally, suppose that the line is to be operated at its second antiresonant mode, for which $l = 3\lambda/4 = 1.125$ meters. The sending-end impedance at antiresonance will now be

$$Z = \frac{Z_0}{\alpha l} = \frac{228}{6.78 \times 10^{-4} \times 1.125} = 299,000 \, \text{ohms}$$

instead of 898,000 ohms, as for the first mode.

The magnitude of the sending-end impedance is plotted as a function of length in Fig. 6.20, using a logarithmic scale. The peaks are so high that, if the curve were plotted on a linear scale which would permit the highest points to fall on the plot, the graph would consist of a set of sharp spikes of no discernable width located at the odd quarter-wave points. Except near the critical lengths, the graph is virtually identical with the magnitude of the tangent-function variation as calculated for zero losses.

6.9. The Q of Resonant and Antiresonant Lines. Before considering the Q of low-loss transmission lines, we shall review the concepts of Q and band width as applied to lumped resonant circuits.

The Q of Lumped Circuits. The quality factor, or Q, of a lumped inductance coil is often defined as $\omega L/R$, where R is the effective series resistance of the coil. If the coil is connected in series with a lossless condenser, the

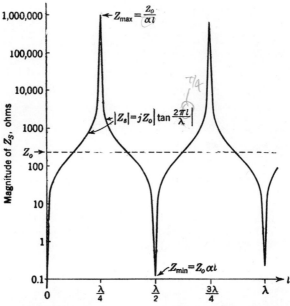

Fɪɢ. 6.20. Variation in the magnitude of the sending-end impedance of a short-circuited line of variable length.

Q is a measure of the sharpness of resonance. For reasonably large values of Q, the impedance near resonance can be shown to be, approximately,

$$Z = R\left(1 + j2Q\,\frac{\Delta\omega}{\omega_0}\right) = R\left(1 + j2Q\,\frac{\Delta f}{f_0}\right) \qquad (6.40)$$

where f_0 is the resonant frequency and Δf is the amount by which the applied frequency is above resonance.[1] This relation is similar in form to Eq. (6.38) for the impedance of a low-loss line near resonance. Inspec-

[1] For a discussion of resonance and Q, see, for example, M.I.T. Staff, "Electric Circuits," pp. 319–331, John Wiley & Sons, Inc., New York, 1940, and R. H. Frazier, "Elementary Electric-circuit Theory," pp. 229–234, McGraw-Hill Book Company, Inc., New York, 1945.

tion of Eq. (6.40) shows that the reactance of the circuit will be equal to
the resistance when the frequency is adjusted either above or below resonance so that $\Delta f/f_0 = \pm 1/2Q$. These two frequencies are

$$
\left.
\begin{aligned}
f_1 &= f_0 \left(1 - \frac{1}{2Q}\right) \\
f_2 &= f_0 \left(1 + \frac{1}{2Q}\right)
\end{aligned}
\right\}
\tag{6.41}
$$

and

If the circuit is driven by a constant-voltage generator, the current at
either of these frequencies will be $1/\sqrt{2}$ of its magnitude at resonance.
The power will then be half the value at resonance, and so these two frequencies are known as the "half-power points." The band width of the
resonant circuit is taken, for convenience, to be the number of cycles per
second between the half-power frequencies; therefore, we have the relation

$$
\frac{BW}{f_0} = \frac{f_2 - f_1}{f_0} = \frac{1}{Q}
\tag{6.42}
$$

where the letters BW signify the band width. The fractional band width,
BW/f_0, is inversely proportional to the Q of the coil.

In communication circuits, the parallel antiresonant circuit is used more
commonly than the series resonant one. If a lumped inductance coil of
reasonably large Q is connected in parallel with a condenser, the admittance of the combination near antiresonance is, approximately,

$$
Y = \frac{1}{Q}\sqrt{\frac{C}{L}}\left(1 + j2Q\,\frac{\Delta\omega}{\omega_0}\right) = \frac{1}{Q}\sqrt{\frac{C}{L}}\left(1 + j2Q\,\frac{\Delta f}{f_0}\right)
\tag{6.43}
$$

Here, as before, $Q = \omega_0 L/R$. If this circuit is supplied with an alternating
current of constant magnitude, the power taken by the circuit will be
half the power at antiresonance when $\Delta f/f_0 = \pm 1/2Q$, and the ratio of
the band width to the resonant frequency is again

$$
\frac{BW}{f_0} = \frac{1}{Q}
$$

The definition $Q = \omega L/R$ is still useful for this case.

Finally, suppose that a lossless inductance L is connected in parallel
with a capacitance C and a resistance R_p (this might be the parallel equivalent of the resistance in series with the inductance, which we have designated as R). When R_p is large, the admittance near antiresonance can
be shown to be, approximately,

$$
Y = \frac{1}{R_p}\left(1 + j2\,\frac{R_p}{\omega_0 L}\,\frac{\Delta\omega}{\omega_0}\right)
\tag{6.44}
$$

We now have a problem in the definition of Q. The original definition was $\omega L/R$, where R was the series resistance of the coil, and, for the first two circuits, this provided a useful result in terms of band width. But the losses in the last circuit are supplied by a parallel, rather than a series, resistance. As a result, the fractional band width is no longer $\omega_0 L/R$, but instead is $R_p/\omega_0 L$. Therefore, we decide to broaden the original definition of Q and base it now on the band width. If we redefine Q as f_0/BW, the first two circuits still have $Q = \omega_0 L/R$. The last one will have $Q = R_p/\omega_0 L$, and Eq. (6.44) will have the same form as Eqs. (6.40) and (6.43). The Q will have the same meaning with respect to band width and sharpness of resonance for all the circuits, and the new definition is therefore broader and more useful than the old one.

The Q of Lines on a Band-width Basis. The band-width definition of Q can be applied directly to resonant or antiresonant lines which are driven at one end and which have a perfectly reflecting termination at the other. Equations (6.35) and (6.38) show that the admittance near antiresonance and the impedance near resonance both involve the following function of frequency:

$$\alpha l + j\frac{l}{v}\Delta\omega$$

The half-power points are obtained when the imaginary part of this is equal to the real part; *i.e.*, when $\Delta\omega = \pm\alpha v$, or $\Delta f = \pm\alpha v/2\pi$. The band width is double the magnitude of this, or

$$BW = \frac{\alpha v}{\pi} \tag{6.45}$$

In general, $\beta = \omega/v$, and so $\omega = \beta v$, from which

$$f = \frac{\beta v}{2\pi} \tag{6.46}$$

The Q at either resonance or antiresonance is, therefore, given by

$$Q = \frac{f}{BW} = \frac{\beta}{2\alpha} \tag{6.47}$$

Since $\beta \approx \omega\sqrt{LC}$ and $\alpha \approx R/2Z_0 + GZ_0/2$, the influence of the line parameters on the Q and band width is not difficult to determine.

Example. Consider the air-insulated line used as an example in the previous section, which was composed of two parallel copper tubes 0.15 in. in radius and spaced 1 in. between centers. The line was short-circuited at one end and was driven with a 200-Mc generator at the other. Neglecting leakage conductance and radiation, the attenuation constant was computed to be

$$\alpha = 6.78 \times 10^{-4} \text{ neper/meter}$$

Assuming a velocity equal to that of light in free space, we have

$$\beta = \frac{\omega}{v} = \frac{2\pi \times 200 \times 10^6}{3 \times 10^8} = 4.19 \text{ rad/meter}$$

Then, using Eq. (6.47), we have

$$Q = \frac{\beta}{2\alpha} = \frac{4.19}{2 \times 6.78 \times 10^{-4}} = 3,090$$

The actual Q may be somewhat lower than this because of losses that we have not taken into consideration, but it is considerably higher than can be obtained with lumped circuit elements. The band width between half-power points will be

$$BW = \frac{f}{Q} = \frac{200 \times 10^6}{3,090} = 64,700 \text{ cps}$$

The resonance is, therefore, extremely sharp. The Q and band width will be the same for the various resonant and antiresonant frequencies indicated in Fig. 6.20 until finally, when the length is made too great, the assumption of a very small over-all loss will be violated and the relations that we have derived on this basis will no longer hold.

The Energy Definition of Q. The definition of Q on the basis of the band width between half-power points, so that $Q = f_0/BW$, requires a knowledge of the input impedance as a function of frequency. For more complicated systems, a still more generalized definition of Q is desirable. A useful definition, which reduces to the previous ones in the simpler cases, is based on energy. It can be applied not only to lumped circuits and transmission lines but also to resonant cavities and to resonant mechanical systems. The definition is

$$Q = 2\pi \frac{\text{stored energy}}{\text{energy dissipated per cycle}} \qquad (6.48)$$

The energy dissipated per cycle, multiplied by the number of cycles per second, is the energy lost per second, which is the dissipated power. Multiplying numerator and denominator by the frequency f, we obtain an equivalent expression for Q:

$$Q = \omega \frac{\text{stored energy}}{\text{average dissipated power}} \qquad (6.49)$$

The stored energy is generally computed by neglecting losses completely. For convenience, an instant is chosen for the computation when the stored energy is either all in the magnetic field (current at its peak, voltage zero) or all in the electric field (current zero, voltage at its peak). The voltage and current standing-wave patterns as obtained for the lossless case are then used to find the I^2R and E^2G losses of the line.

As an example of the use of the energy definition, consider a low-loss quarter-wave line which is short-circuited at the receiving end, the Q of which, on a band-width basis, has been shown to be $\beta/2\alpha$. At an instant when the voltage is zero throughout the length of the line, the current is everywhere at its crest value and is distributed as a sine function:

$$i = \sqrt{2} I_R \sin \frac{\pi x}{2l}$$

where x is measured from the sending end and I_R is the rms value of the receiving-end current. The energy stored in the magnetic field in a length dx is $(L\,dx)i^2/2$, and so, for the whole line,

$$\text{Stored energy} = \int_0^l L I_R^2 \sin^2 \frac{\pi x}{2l}\,dx$$

$$= \tfrac{1}{2} LlI_R^2 \qquad \text{joules}$$

The same result would be obtained from a computation of the energy stored in the electric field at an instant when the current was zero and the voltage maximum. Next, we compute the average power dissipated by the line. The rms current at any point is given by

$$I = I_R \sin \frac{\pi x}{2l}$$

and the I^2R losses, integrated over the length of the line, are

$$P_1 = \int_0^l R I_R^2 \sin^2 \frac{\pi x}{2l}\,dx = \tfrac{1}{2} RlI_R^2 \qquad \text{watts}$$

The rms voltage at any point is

$$E = I_R Z_0 \cos \frac{\pi x}{2l}$$

and the loss in the insulation is

$$P_2 = \int_0^l E^2 G\,dx = \tfrac{1}{2} G l Z_0^2 I_R^2 = \frac{GlL}{2C} I_R^2$$

According to the definition (6.49), the Q of the line is

$$Q = \omega \frac{\text{stored energy}}{P_1 + P_2} = \frac{\omega LC}{RC + LG}$$

The band-width definition would give us an identical expression:

$$Q = \frac{\beta}{2\alpha} = \frac{\omega \sqrt{LC}}{2\left(\dfrac{R}{2}\sqrt{\dfrac{C}{L}} + \dfrac{G}{2}\sqrt{\dfrac{L}{C}}\right)} = \frac{\omega LC}{RC + LG}$$

Whereas a Q of 200 is considered very good for a lumped inductance coil, Q values of several thousand can be obtained at the shorter wavelengths with transmission lines.

PROBLEMS

1. A certain low-loss line has a characteristic impedance of 400 ohms. Determine the standing-wave ratio for the following receiving-end impedances:

a. $Z_R = 70 + j0$ ohms.

b. $Z_R = 800 + j0$ ohms.

c. $Z_R = 650 - j475$ ohms.

2. A transmission line with a characteristic impedance of 70 ohms is terminated at its receiving end in a resistance of 0.50 ohm. The wavelength is 1.5 meters, and the receiving-current is 2.0 amp rms.

a. Sketch the standing-wave patterns of voltage and current *vs.* distance from the load for $0 < d < 1.5$ meters.

b. What is the standing-wave ratio?

c. Determine the line impedance at $d = \lambda/4$, neglecting line losses.

3. A low-loss transmission line with a characteristic impedance of 52 ohms is terminated in a receiving-end impedance of $75 - j61$ ohms. The wavelength is 3 meters, and the maximum voltage on the line is 100 volts rms.

a. Sketch the standing-wave patterns of voltage and current *vs. d* for 1 wavelength. Determine the standing-wave ratio.

b. Compute the power transmitted by the line. Also, determine $|E_{min}|$, $|I_{max}|$, $|I_{min}|$, and $|I_R|$.

4. A low-loss line with a characteristic impedance of 475 ohms is to handle a power of 250 watts. Determine the maximum rms voltage across the line for the following standing-wave ratios: (*a*) $\rho = 1$, (*b*) $\rho = 2$, (*c*) $\rho = 10$.

5. Prove Eq. (6.13): $|E_{min}| = |I_{min}| Z_0$.

6. The impedance that terminates a line can be determined from the standing-wave ratio, ρ, the characteristic impedance of the line, Z_0, and the distance from the load to the first voltage minimum, d_{min}. From Eq. (6.18), show that

$$Z_R = Z_0 \left(\frac{1 - j\rho \tan 2\pi d_{min}/\lambda}{\rho - j \tan 2\pi d_{min}/\lambda} \right)$$

7. A low-loss line with a characteristic impedance of 70 ohms is terminated in a short circuit and is operating at a wavelength of 50 cm. Using one of the transmission-line charts on an admittance basis, determine the shortest lengths that will provide the following sending-end admittances:

a. $Y_s = -j7.15 \times 10^{-3}$ mho.

b. $Y_s = j7.15 \times 10^{-3}$ mho.

8. Show that the average power transmitted by a low-loss line can be expressed as

$$P = P^+(1 - K^2)$$

where P^+ is the power associated with the incident wave, and K is the magnitude of the reflection coefficient.

9. A quarter-wave low-loss line is terminated in a resistance of 140 ohms. The characteristic impedance is 70 ohms and the sending-end voltage is 100 volts rms. Determine the magnitude of the receiving-end voltage.

10. A low-loss line of adjustable length is terminated in an open circuit. It is driven by a generator which has an internal voltage E_g and an internal impedance equal to the characteristic impedance of the line. Show that the voltage across the open end of the line is equal to the internal voltage of the generator, regardless of the length of the line. Neglect line losses entirely.

11. A line of adjustable length is terminated in a short circuit. At the sending end, it is driven by a generator which has an internal voltage E_g and an internal impedance equal to the characteristic impedance of the line. Show that, if line losses are negligible, the receiving-end current has a magnitude equal to E_g/Z_0, regardless of the length of the line.

12. Determine the sending-end impedance of each of the following low-loss lines, each of which has a characteristic impedance of 400 ohms:

a. Total length $= 5\lambda/4$. An 800-ohm resistance is connected across the receiving end; another 800-ohm resistance is connected across the line at $d = \lambda/2$.

b. Total length $= \lambda$. A 400-ohm resistance is connected across the receiving end; another 400-ohm resistance is connected across the line at $d = \lambda/4$.

c. Total length $= \lambda$. A 200-ohm resistance is connected across the receiving end, and an 800-ohm resistance is connected across the line at $d = \lambda/4$.

13. A low-loss line with a characteristic impedance of 77 ohms is terminated in a load impedance of $50 - j50$ ohms. A short-circuited stub ($Z_0 = 77$ ohms) is to be connected in parallel with the load and is to be adjusted in length so that the net terminating impedance of the line is a pure resistance. How long should the stub be, and what is the value of the net terminating impedance? The wavelength is 30 cm. The use of admittances is suggested.

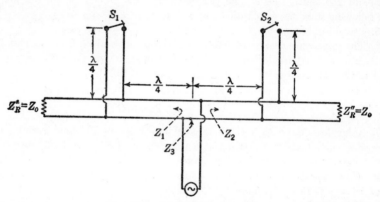

Fig. P14. A switching system.

14. Figure P14 shows a transmission-line system with two loads and one generator. Neglect line losses.

a. With the switch S_1 closed and with S_2 open, determine Z_1, Z_2, and Z_3. Where will the power flow?

b. With S_1 open and S_2 closed, how will the answers of part a be changed?

c. If both switches are open, what will happen in so far as power flow is concerned?

15. A certain air-insulated two-wire line has a characteristic impedance of 400 ohms. Its length is 0.75 meter, and it is terminated in a receiving-end impedance of $Z_R = -j400$ ohms. The frequency is 150 Mc, and the sending-end voltage is 50 volts rms.

a. Compute the sending-end impedance.

b. Sketch the standing-wave patterns of voltage and current.

c. Determine $|E_{max}|$, $|I_{max}|$, $|E_0|$. and $|I_R|$.

16. A low-loss line with a characteristic impedance of 400 ohms is short-circuited at the receiving end. The line is driven at the other end by a generator that has an internal impedance of 400 ohms and an internal voltage given by

$$e_g = 141 \sin \omega t + 141 \sin 2\omega t \qquad \text{instantaneous volts}$$

The length of the line is $\lambda/2$ at the angular frequency ω. Determine and sketch the rms voltage along the length of the line.

17. A certain transmission-line stub is open-circuited at one end and short-circuited at the other. The length of the stub is $\lambda/4$ at the fundamental frequency. A generator is to be connected, as in Fig. 6.16, at such a point that a third harmonic in the generator voltage will be short-circuited. At what fraction of a wavelength from the closed end should the generator be connected?

18. A line connected as shown in Fig. 6.18 is to be used as an antiresonant circuit. If $Z_0 = 200$ ohms, if the length is 30 cm, and if $C = 2.5 \times 10^{-12}$ farad, find the first two antiresonant frequencies. Assume a velocity equal to that of light in free space.

19. A line is to be used as an antiresonant circuit as in Fig. 6.18. If first antiresonant frequency is to be 250 Mc and if $C = 2.5 \times 10^{-12}$ farad, determine required length. $Z_0 = 200$ ohms. Assume a velocity equal to that of light in free space.

20. A certain coaxial line is to be short-circuited at the receiving end and is to be driven by a generator at the other end. The line has copper conductors, is air-insulated, and has a length of 20 cm. The diameter of the inner conductor is 1.15 cm, and the inside diameter of the outer tube is 3.80 cm.

a. Compute the attenuation constant at the first antiresonant frequency. Neglect leakage conductance.

b. Compute the sending-end impedance at the above frequency. Neglect the impedance of the short-circuiting plug at the receiving end.

c. If the length of the line is increased to 40 cm and the frequency remains the same as above, what will be the sending-end impedance?

21. Compute the Q of the line of Prob. 20 at its first antiresonant frequency (length = 20 cm).

22. One of the larger flexible coaxial cables of standard type has $Z_0 = 52$ ohms and $\alpha = 9.05 \times 10^{-3}$ neper/meter at 500 Mc. The phase velocity is 0.670 times the speed of light in free space. A short-circuited quarter-wave section of this cable is to be used as an antiresonant circuit at 500 Mc. Compute the Q, the band width in cycles per second between half-power points, and the sending-end impedance at antiresonance.

23. Show that a line short-circuited at both ends and used in its first mode (see Fig. 6.15a) has a Q given by $\omega L/R$ if the energy losses in the insulating medium can be ignored.

24. An air-insulated coaxial line is to be short-circuited at both ends and used in its first resonant mode. The conductors are silver-plated (resistivity = 1.63×10^{-8} ohm-meter and $\mu = \mu_0 = 4\pi \times 10^{-7}$ henry/meter). The radius of the inner conductor is 0.540 cm and the inside radius of the outer tube is 1.270 cm.

a. Compute the resistance per meter at 600 Mc.

b. Compute the Q, using the results of Prob. 23.

CHAPTER 7

RADIO-FREQUENCY LINES—MEASUREMENTS AND IMPEDANCE MATCHING

7.1. Radio-frequency Measurements. Throughout a large part of the r-f spectrum, the techniques used in measurement do not differ greatly from those used at audio frequencies. Frequency can be measured by comparison with an oscillator of known frequency, or by resonating a circuit of known properties. Thermocouple milliammeters are available which are useful up to the region of hundreds of megacycles. Vacuum-tube voltmeters of commercial design are available for about the same range, and the convenience of these instruments makes them useful in many types of measurements. Radio-frequency bridges can be used for impedance measurements over a wide range of frequencies; in fact, the equivalent of a bridge can be used in the microwave region. At moderate radio frequencies, power is generally determined from a knowledge of the impedance and the voltage or current.

As the frequency is increased, greater care and skill are necessary for accurate measurements. Stray coupling, the loading effects and reflections caused by the instruments, and the presence of standing waves on leads of moderate length, all become more troublesome at the higher frequencies. This brings about a change in method in which transmission-line techniques replace those of the lumped circuit. These techniques have reached a high degree of development in the microwave region because of the stimulus given by radar. The word "microwave" is generally used to denote radio waves in the region of wavelength below perhaps $\frac{1}{2}$ or $\frac{1}{3}$ meter. Here the dimensions of practical apparatus are comparable with the wavelength, and almost every circuit is an electrically long transmission line.

In this chapter, most of the emphasis will be placed on methods of measurement which utilize transmission-line techniques. Many of these methods depend on the use of sections of line that are comparable with a wavelength in extent, and they are, therefore, of use whenever this length is not inconveniently great.

7.2. The Measurement of Standing Waves. A measurement of the standing-wave ratio can be made with an instrument which gives a relative indication of voltage without introducing serious reflections at its point of coupling with the line, and which can be moved along the line over a distance greater than a half wavelength. At the lower radio frequencies, high-impedance instruments are frequently connected directly to the line conductors, but at shorter wavelengths the instrument is usually coupled loosely by maintaining its input terminals at a constant distance from the conductors; the capacitive coupling thus obtained will produce an input voltage which is proportional to the line voltage.

Various commercially available vacuum-tube voltmeters are useful for voltage measurements up to frequencies of hundreds of megacycles, but the high capacity that exists between one input terminal and ground will unbalance a two-wire line at the higher frequencies. Voltage measurements on parallel-wire lines are often made with a thermocouple milliammeter connected at the end of a quarter-wave stub, as shown in Fig. 7.1. As shown by Eq. (6.19), the impedance at the input to the quarter-wave stub will be Z_0^2/R_m, where R_m is the meter impedance, assumed to be a resistance. If R_m is much smaller than Z_0, the input impedance will be quite large, and

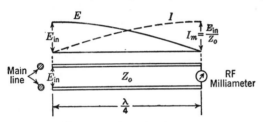

FIG. 7.1.   An r-f voltmeter.

the device will not introduce much reflection at the point of coupling to the line. A standing wave will be produced on the quarter-wave stub with the meter at the position of maximum current, and, as shown by Eq. (6.12), the meter current will be E_{in}/Z_0, which is proportional to the line voltage. This scheme can be extended to still higher frequencies by replacing the thermocouple instrument with a crystal rectifier and d-c microammeter.

Voltage measurements on coaxial lines are generally made with the aid of a section of line that has a longitudinal slot cut for a distance of at least a half wavelength in the outer conductor. A probe projects a short distance into the coaxial line and can be moved lengthwise. The arrangement is shown in a rather schematic way in Fig. 7.2. The probe intercepts a portion of the electric field that exists between inner and outer conductors, and so the voltage between probe and outer conductor is proportional to the line voltage. Figure 7.3 shows a second type of probe which is intended to provide a signal proportional to the magnetic field and therefore proportional to the line current. The electric probe shown in Fig. 7.2 is almost always preferred, however, for it is difficult to construct a magnetic probe that does not respond at all to the electric field. The probe and detector assembly are mounted on a movable carriage which maintains the probe in the middle of the slot and at a constant distance from the center conductor. Slotted coaxial lines of this type are available commercially. The characteristic impedance of the slotted section must, of course, match that of the main line so that the standing-wave pattern is not altered by reflections at the junction.

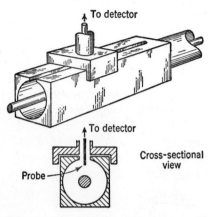

FIG. 7.2. A slotted coaxial line and traveling probe.

Various kinds of detectors and indicators can be used in connection with the slotted line.[1] A simple and satisfactory device is the crystal rectifier and microammeter combination shown in Fig. 7.4. The first illustration shows an untuned probe, which is convenient in broad-band measurements. The second arrangement has a short-circuited stub of variable length which permits tuning out the susceptance that is introduced into the line by the probe. The untuned probe has a second rectifier element which provides a d-c return path for the meter; without this, a d-c potential would build up between the probe and the outer conductor sufficient to bias off the series rectifier. With the tuned probe, the short-circuited stub provides the necessary d-c return path. The r-f energy is short-circuited by a capacitor which is connected just beyond the rectifying elements. The microammeter reads the rectified direct current. For greater sensitivity, the r-f source is sometimes modulated and an amplifier tuned to the modulation frequency is used to amplify the output of the detector.

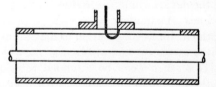

FIG. 7.3. Lengthwise section showing a magnetic-type probe.

A second type of detector is the *bolometer*. This is a small resistive

---

[1] For additional information on the measurement of standing waves, see C. G. Montgomery, "Technique of Microwave Measurements," Chap. 8, McGraw-Hill Book Company, Inc., New York, 1947, and Radio Research Laboratory Staff, "Very High Frequency Techniques," Chap. 2, McGraw-Hill Book Company, Inc., New York, 1947.

element which is arranged to absorb the r-f power and which changes its resistance because of the increased temperature. Bolometers were originally used in the form of blackened platinum strips to measure small

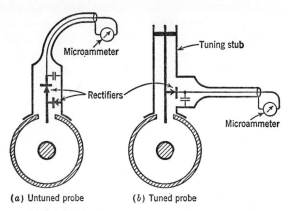

(a) Untuned probe   (b) Tuned probe

Fig. 7.4.   Traveling probes using crystal detectors, shown schematically.

amounts of radiant heat. The form generally used in standing-wave measurements is a thin platinum wire, which is given the special name *barretter*. The usual arrangement is shown in Fig. 7.5. If the r-f source is modulated, the resistance of the bolometer element will vary periodically at the modulation frequency. Then, if a d-c voltage is impressed on the bolometer from an outside source, the resulting current will pulsate at the modulation frequency, and this signal can be applied to a tuned amplifier. If the change in temperature is small, the bolometer response follows a square law accurately.

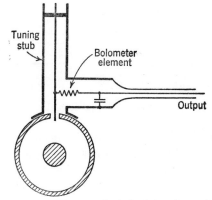

Fig. 7.5.   The use of a bolometer element as a detector.

A crystal is approximately a square-law device for small currents but, to obtain good accuracy, a detector employing a crystal should be calibrated. This can be done by short-circuiting the main coaxial line beyond the slotted section and determining the meter reading as the probe is moved along the line. The standing-wave pattern for the short-circuit termination is known to be sinusoidal, and comparison of the meter readings with a sine wave provides the required calibration, as indicated in Fig. 7.6.

7.3. The Measurement of Wavelength. At the shorter wavelengths, a measurement of wavelength generally replaces the measurement of frequency. The distance between two successive minima on the standing-wave pattern is the half wavelength. If a traveling detector is used to measure the standing-wave pattern, the wavelength can be determined with reasonable accuracy if the standing-wave ratio is high enough to provide sharp minima.

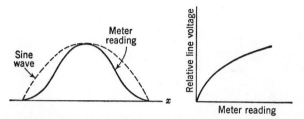

FIG. 7.6. The calibration of a crystal detector.

Resonant air-insulated lines are frequently used to measure wavelength. The line is arranged with a movable short circuit at one end and is loosely coupled to the generator at the other. When the line is adjusted to a resonant length, the energy coupled from the generator is able to build up an

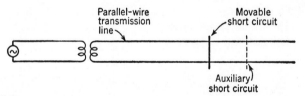

FIG. 7.7. Lecher-wire system for measuring wavelength.

oscillation of large amplitude on the line. In operation, the short circuit is moved along the line, and a measurement is made of the distance between two successive positions that produce resonance. This distance is the half wavelength. A parallel-wire line used for this purpose is called a Lecher-wire system and is illustrated in Fig. 7.7. Resonance can be determined by the reading of a current indicator connected in the short-circuiting slider, or it can be observed by the reaction on the generator. The sharpness of resonance of the open-wire system may be impaired by some of the energy passing the short circuit and entering the supposedly unused portion of the line. This can be prevented by placing one or more additional short circuits behind the main one, preferably with a spacing of about a quarter wavelength.   A half-wavelength spacing between short circuits is ineffective, as this places the auxiliary short circuit at a voltage node.

At short wavelengths, the foregoing principle is utilized with a coaxial line, and an instrument built especially for the purpose is called a *coaxial wavemeter*. It consists of an air-insulated coaxial line which is closed at one end and has a movable short circuit at the other. The source of energy is connected to the wavemeter by means of a small coupling loop of the type shown in Fig. 7.3, which is located at the closed end and is oriented so that it will be linked by the magnetic flux in the annular space. As was discussed in Sec. 6.7, a line short-circuited at both ends is resonant when it is an integral number of half wavelengths long.   When the coaxial wavemeter has been adjusted to one of these lengths, the current in the coupling loop is able to build up an oscillation of large amplitude within the wavemeter.   The distance that the short-circuiting plunger must be moved between two successive resonances is the half wavelength.   The wavemeter has a high Q and resonance is sharp; consequently, considerable accuracy can be obtained if the device is well constructed.

7.4. The Measurement of Impedance with a Transmission Line.   There are a number of methods of using transmission lines to measure impedances at short wavelengths.   The impedance that terminates a line can be determined from the standing-wave ratio and the distance between the load and a voltage maximum or minimum.   The lossless formula for line impedance, Eq. (6.18), can be rearranged to express the receiving-end impedance in terms of these quantities (see Prob. 6 of Chap. 6).   The calculation can also be carried out with the transmission-line chart.   The value of the standing-wave ratio locates the intercept of the appropriate K circle with the real axis.   The chart is entered on this K circle at the voltage maximum or minimum, and the circular locus is followed backward on the wavelength scale until the position representing the load is reached. At this point the normalized load impedance is read.   The distance measurement is usually made to a voltage minimum rather than to a maximum, for the minimum is the more sharply defined of the two points.

When a slotted section is used to make standing-wave measurements on a coaxial cable, the position of a minimum can be located accurately within the slotted section, but the number of wavelengths to the load is not known with accuracy.   This difficulty is overcome by first locating a measuring point within the slotted section which, "by proxy," represents the load.   To do this, a short circuit is placed across the load; the standing-wave pattern thus produced has minima at integral numbers of half wavelengths from the receiving end, and at each of these points the receiving-end impedance is repeated.   The position of one of these minima is located within the slotted section; and this is the measuring point where the line impedance is equal to the receiving-end impedance. The short circuit across the load is now removed, and the line impedance at the measuring point is determined by the method described in the first paragraph of this section.

Example. An unknown impedance is connected to a 50-ohm coaxial cable, and a slotted section is used to measure the standing wave on the line. As shown in Fig. 7.8, a measuring point is located within the slotted section by short-circuiting the receiving end and determining a position of zero voltage. With the unknown

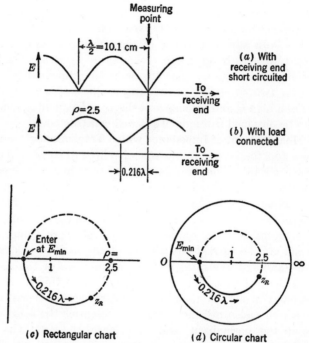

Fig. 7.8. Determination of an unknown receiving-end impedance.

impedance connected to the receiving end, the standing-wave ratio is found to be 2.5. A voltage minimum is located 0.216λ on the generator side of the measuring point.

The appropriate impedance locus is the K circle which intersects the real axis at $\rho = 2.5$, as indicated in Fig. 7.8. The chart is entered on this K circle at the point representing the voltage minimum, i.e., at the point of minimum impedance, which is marked 0.25λ on the distance scale. The locus is followed away from the generator to the position corresponding to the measuring point, which is 0.216λ away and is, therefore, located at $(0.250 - 0.216)\lambda = 0.034\lambda$ on the distance scale. The impedance at this point is equal to the receiving-end impedance and is $Z_R = (2.02 - j0.89)Z_0 = 101 - j44.5$ ohms.

7.5. The Measurement of Power. In the frequency range where a vacuum-tube voltmeter can be connected directly to the line without serious reflections or unbalance, the power carried by a line can be determined from the voltage and the line impedance, as indicated in Sec. 6.3. Another useful and simple method is to use the power to heat the filament of an incandescent lamp, and then determine the amount of low-frequency power required to obtain the same amount of light.

In the microwave region, powers of the order of 1 watt and above are often measured by a calorimetric method: a fluid, generally water, is used as a load to absorb the power and to convert it into heat, and the resulting temperature rise of the fluid is used to determine the power absorbed. But with a power smaller than 1 watt, the temperature rise of a calorimetric fluid becomes too small to measure accurately. The bolometer, first mentioned in Sec. 7.2 as a standing-wave detector, is often employed at low power levels. This device is a resistive element which is used as a load to absorb the power from the line and convert it into heat. The resistance of the bolometer element is changed by its increase in temperature, and this gives a measure of the power dissipated. Two types of bolometers are commonly used: the *barretter* and the *thermistor*.[1] The barretter is a short length of very thin wire, generally made of platinum because of its resistance to corrosion, high melting point, and desirable mechanical properties. The resistance of a barretter increases with temperature. The thermistor, on the other hand, is a small resistive element made of a mixture of metallic oxides which is semi-conducting and which has a negative temperature coefficient of resistance. This device is generally made in the form of a small bead into which two fine wires are embedded. The thermistor has the advantage that its resistance can be varied over a wider range than that of the barretter, and it is less susceptible to being overloaded and burned out.

Figure 7.9 shows a bolometer mount suitable for microwave frequencies, and a bridge for measuring the r-f power absorbed by the bolometer element. The center conductor of the line is supported by a quarter-wave stub, which also provides the d-c return for the bridge. The center conductor of the line is tapered in diameter. The taper gradually alters the characteristic impedance from that of the main line to a higher value equal to the bolometer resistance, and so matches the bolometer to the line without reflection, as is explained in more detail in Sec. 7.11. One end of the bolometer is connected to the inner conductor; the other is connected through a capacitor to the outer conductor. Thus the bolometer forms a load at the end of the line. The capacitor is an r-f short circuit but allows

---

[1] For a thorough treatment of power measurements in the microwave region, see Montgomery, *op. cit.*, Chap. 3.

the d-c bridge to measure the bolometer resistance. In operation, the bridge is first balanced without r-f power, and the d-c power input to the bolometer element is determined. The application of r-f power changes the bolometer resistance and unbalances the bridge. The bridge is brought into balance again by increasing the resistance R in series with the battery, thus reducing the direct current in the bolometer to the point where the total bolometer power is the same as before. Then the bolometer resistance returns to its previous value, and the bridge is again in balance. The required reduction in d-c bolometer power is equal to the r-f power.

7.6. The Directional Coupler. A directional coupler is a device that couples a measuring instrument or an auxiliary line only to the wave traveling in one direction along the main line and ignores entirely a wave traveling in the other direction. The directional coupler has a number of

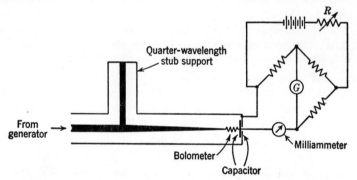

Fig. 7.9. A bolometer mount and bridge for measuring power.

interesting uses, but its chief application has been in monitoring the power transmitted along a line which may connect, for example, a transmitter to an antenna. A measurement of voltage at a single point will not suffice as an indication of power unless the standing-wave ratio is quite small, for a shift in the standing-wave pattern will cause a false change in indication. The uncertainty rapidly becomes greater with an increased standing-wave ratio. A directional coupler, on the other hand, retains the advantages of simplicity and small size, but, by responding only to the forward-traveling wave, it gives an indication that is independent of its position along the standing-wave pattern. The reflected power should, of course, be subtracted from the incident power to obtain the correct power absorbed by the load, but the error introduced by simply metering the incident power does not increase very rapidly with the standing-wave ratio. To illustrate the difference, suppose that the standing-wave ratio is 2, so that $|E_{max}| = 2|E_{min}|$, which corresponds to a reflection coefficient

$K = |E^-|/|E^+| = \frac{1}{3}$. Then the power absorbed by the load is

$$P_L = \frac{|E^+|^2}{Z_0} - \frac{|E^-|^2}{Z_0} = \frac{|E^+|^2}{Z_0}\left(1 - \frac{1}{3^2}\right) = \frac{8}{9}P^+$$

where P^+ is the incident power.   The directional coupler thus will indicate a power equal to $9P_L/8$.   On the other hand, a voltage-measuring device calibrated for a flat line will, if connected at a voltage maximum, indicate a power equal to

$$P' = \frac{|E_{\max}|^2}{Z_0} = \frac{(|E^+| + |E^-|)^2}{Z_0} = \frac{|E^+|^2}{Z_0}\left(1 + \frac{1}{3}\right)^2 = \frac{16}{9}P^+ = 2P_L$$

If connected at a voltage minimum, it will indicate a power equal to $P_L/2$. Thus, the directional coupler gives an indication which is more nearly a measure of the load power than can be obtained with a simple one-point voltage measurement.

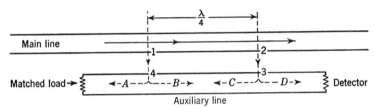

FIG. 7.10.   Schematic diagram of one type of directional coupler.

A directional device can be built by coupling an auxiliary line to the main line through two identical coupling units located a quarter wavelength apart or, alternatively, by coupling at one point through both the electric and magnetic fields.   The first of these methods is illustrated schematically in Fig. 7.10.   Consider a wave traveling to the right on the main line. At point 1 a small portion of its energy is coupled to the auxiliary line and gives rise to the two oppositely traveling waves A and B.   At point 2, a quarter wavelength away, an identical coupling unit gives rise to the two waves C and D.   But wave C, in traveling the path 1-2-3-4, has gone a half wavelength and arrives at point 4 with a phase opposite to that of wave A; hence, A and C cancel and there is no leftward-traveling wave to the left of point 4.   In contrast to this, wave B travels a quarter wavelength along the path 1-4-3, and wave D travels the same distance on the path 1-2-3; hence, these two waves reinforce each other and the detector at the right receives their energy.   From symmetry, we see that, if we reverse the direction of the wave in the main line, the two waves traveling toward the detector will cancel because of the half-cycle lag in one of them. The waves traveling toward the matched load at the left end will reinforce each other, but the load will absorb them without reflection and so they cannot reach the detector. Therefore, the detector will receive energy only from a wave traveling to the right in the main line. One method of obtaining the required coupling between coaxial lines is to run them parallel with each other and join the outer conductors through two openings placed a quarter wavelength apart. Another method is to build the auxiliary line within the center conductor of the main line, the coupling again being obtained with two holes joining the lines.

A second method of obtaining a directional device is to couple the auxiliary line both electrically (proportional to voltage) and magnetically (proportional to current) at one point. A device which has been used for this purpose is the resistive-loop coupler, shown in Fig. 7.11. If the resistance R were infinite, the probe would be of the electric type and would provide a signal proportional to the line voltage. If the resistance were zero, the resulting loop would constitute a magnetic probe and give a signal proportional to the current in the line. A value of resistance between these two extremes causes the probe to respond to both voltage and current. The amount of magnetic coupling can be adjusted by rotating the probe, which changes the amount of magnetic flux that links the loop. The operation of the device depends upon the fact that a backward-traveling wave has the phase of its current reversed, and this reversed signal can be used to cancel the one obtained from the backward-traveling voltage. The output voltage of the coupler is

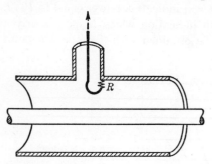

FIG. 7.11. The resistive-loop type of directional coupler.

$$E_c = C_1 E + C_2 I \tag{7.1}$$

where E and I are the voltage and current of the main line, and C_1 and C_2 are constants. But $E = E^+ + E^-$ and $I = E^+/Z_0 - E^-/Z_0$, and so Eq. (7.1) can be written as

$$E_c = C_1(E^+ + E^-) + C_2 \left(\frac{E^+}{Z_0} - \frac{E^-}{Z_0} \right) \tag{7.2}$$

If we adjust the coupling so that $C_1 = C_2/Z_0$, the two terms containing E^- will cancel, leaving only

$$E_c = 2C_1 E^+ \tag{7.3}$$

and the coupler responds only to waves traveling in one direction.

7.7. Impedance Matching. Consider a generator with an internal voltage E and a complex internal impedance $Z_1 = R_1 + jX_1$, as shown in Fig. 7.12. This might be the Thévenin equivalent of any complex linear network, in which case E is the open-circuit terminal voltage of the network and Z_1 is the impedance of the network as viewed from the terminals. Suppose that a variable impedance $Z_2 = R_2 + jX_2$ is connected as a load. The power absorbed by the load is equal to I^2R_2, which can be expressed as

$$P_2 = \frac{E^2 R_2}{(R_1 + R_2)^2 + (X_1 + X_2)^2} \tag{7.4}$$

If we assume that R_1 and X_1 are fixed and that R_2 and X_2 are variable, the conditions for a maximum value of P_2 are obtained by satisfying simultaneously the two relations

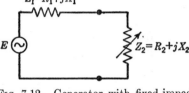

$$\frac{\partial P_2}{\partial R_2} = 0 \text{ and } \frac{\partial P_2}{\partial X_2} = 0 \quad (7.5)$$

The first of these, when applied to Eq. (7.4), gives the result

$$R_1^2 - R_2^2 + (X_1 + X_2)^2 = 0 \quad (7.6)$$

while the second gives

FIG. 7.12. Generator with fixed impedance operating into a variable load.

$$X_2 = -X_1 \tag{7.7}$$

The latter result shows that, for maximum power transfer, the reactance of the load should be equal and opposite to that of the generator. Furthermore, this result, when substituted into Eq. (7.6), gives the second condition:

$$R_2 = R_1 \tag{7.8}$$

Therefore, maximum power in the load is obtained when the load impedance is the complex conjugate of the generator impedance. This condition is sometimes referred to as a "conjugate match."

Suppose that a generator is connected to a load through a lossless transmission line and lossless matching devices. None of the power is lost in the transmission system, and so, if the output of the generator is made a maximum by a conjugate match at its terminals, the power flow at all parts of the system must be at a maximum. Then, if the system is opened at any point, the impedances looking in opposite directions must be the conjugates of each other. This can be made the basis for computing the matching elements to produce maximum power transfer.

In transmission-line systems, there are a number of important reasons for matching impedances, quite apart from power-transfer considerations.

A line terminated in its characteristic impedance has a standing-wave ratio of unity and transmits a given power with a smaller peak voltage; consequently, there is less danger of flashover at large values of power. Also, the efficiency of transmission is greater when there is no reflected wave. Finally, a "flat" line is nonresonant; *i.e.*, its input impedance remains at the value Z_0 when the frequency changes. Variations in frequency will not change the loading of the generator and its power output. Also, a nonresonant line will not tend to pull the generator frequency away from its normal value. In contrast with this, a line many wavelengths long is extremely sensitive to frequency if the reflected wave is large, for the phase of the reflected wave as it arrives at the sending end changes rapidly with frequency. If the line is n wavelengths long, we can write $l = n\lambda = nv/f$, or

$$n = \frac{l}{v} f \qquad (7.9)$$

If the frequency changes by the amount Δf, the number of wavelengths will change by

$$\Delta n = \frac{l}{v} \Delta f$$

which, upon elimination of l/v by means of Eq. (7.9), becomes

$$\Delta n = n \frac{\Delta f}{f} \qquad (7.10)$$

If the line is many wavelengths long, Δn may be a large fraction of a wavelength even if the fractional change in frequency is small. For example, if the line is 250 wavelengths long, a change in frequency of one part in a thousand will produce $\Delta n = 250 \times 0.001 = 0.25$; *i.e.*, the electrical length of the line changes by a full quarter wavelength. On a transmission-line chart, this represents a shift in sending-end impedance halfway around the impedance locus. Therefore, the impedance will change a large amount unless the standing-wave ratio is near unity.

For the foregoing reasons, a matching device is inserted near the load so that the line is terminated in its characteristic impedance at this point. If the generator has been designed to match the characteristic impedance of the line, this one matching device will also produce maximum power transfer to the load. But if the internal impedance of the generator is considerably different from Z_0, the operation of matching at the load may make the impedance match at the generator terminals either better or worse, depending on the generator impedance. Then an additional matching device is needed at the generator end of the line to provide maximum power transfer. Figure 7.13 shows a completely matched low-loss transmission system. At every point the impedances looking in opposite directions are conjugates. The characteristic impedance Z_0, being real, is of course its own conjugate. When the matching devices must be adjusted by trial, the one near the load is first adjusted to provide a flat line and, only after this has been accomplished, the matching unit near the generator is adjusted for maximum power flow.

A very short high-frequency line may be operated unmatched, or it may have a single tuning element to provide a conjugate match and maximum power transfer without concern for the standing-wave ratio.

At audio frequencies the iron-cored transformer is almost universally used as an impedance-matching device. The air-cored transformer is often used at radio frequencies, particularly when the two circuits must be insulated from each other because of the unbalance of one of them with respect to ground. The fractional band width to be passed is generally small at radio frequencies, and the air-cored transformer can be tuned to

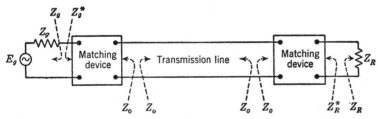

FIG. 7.13.   Illustrating a completely matched transmission system.

resonance with shunt capacitance. When a balance-to-unbalance transition is not necessary, four-terminal networks composed of reactive elements are often used as impedance-matching devices at radio frequencies. These are discussed in Sec. 12.6.

As the wavelength becomes shorter, the lumped elements of the air-cored transformer and of the impedance-matching network become less useful because of stray coupling, distributed effects that are almost unpredictable, and increased losses. At the same time, a quarter-wavelength section of transmission line becomes short enough to be practical as a circuit element. The remainder of this chapter will be devoted to a consideration of matching devices based on transmission-line sections: the quarter-wave transformer, the single-, double-, and triple-stub tuners, and the tapered line.

7.8. The Quarter-wave Transformer. It was shown in Sec. 6.5 that a quarter-wave section of low-loss line inverts the normalized terminating impedance; i.e., $z_s = 1/z_R$, or, in actual ohms,

$$Z_s = \frac{Z_0^{\,2}}{Z_R} \qquad (7.11)$$

If the load impedance is a pure resistance, the input impedance of the quarter-wave line will be a pure resistance also:

$$Z_s = \frac{Z_0{}^2}{R_R} + j0 \qquad (7.12)$$

If a given resistive load does not have the desired magnitude for a particular application, it can be fed with power through a quarter-wave transmission line which has a characteristic impedance selected so as to make its sending-end resistance equal to the desired value. A section of line used in this application is generally called a quarter-wave transformer. The required characteristic impedance of the section can be obtained from Eq. (7.12) and is the geometric mean between the terminating resistance and the desired sending-end resistance:

$$Z_0 = \sqrt{R_R R_s}$$
$$= \sqrt{\text{(load resistance) (desired sending-end resistance)}} \qquad (7.13)$$

An application of this is shown in Fig. 7.14, where a dipole antenna with a resistive input impedance R_R is matched to the main transmission line (characteristic impedance Z_0') by means of a quarter-wave transformer having a characteristic impedance $Z_0 = \sqrt{R_R Z_0'}$. The formulas developed in Chap. 3 can be used to design the matching section for the proper value of Z_0. Matching sections for a parallel-wire line may be made of conductors having a different diameter or different spacing from those of the main line. In coaxial cable, conducting sleeves which fit closely to either the inner or the outer conductor are commonly used, as shown in Fig. 7.15. Obviously, with this arrangement, the characteristic impedance can only be made smaller, not larger, than the normal value. Another method is to insert a "slug" of dielectric material in the annular space and so alter the characteristic impedance for a distance of a quarter wavelength. The phase velocity and wavelength are smaller in the dielectric-filled region, and the length of the slug should be $\lambda/4$ at the reduced velocity.

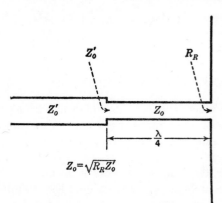

FIG. 7.14. A quarter-wave transformer which matches a dipole antenna to a transmission line.

A load that contains reactance can be matched to a line by including a short section of the main line between the load and the quarter-wave transformer.   The computation, using the rectangular transmission-line chart, is illustrated in Fig. 7.16.   The chart is entered at the normalized load impedance, and the impedance locus is followed around to a point where the line impedance is purely resistive.   This is the position for the "receiving end" of the quarter-wave transformer, and the distance from

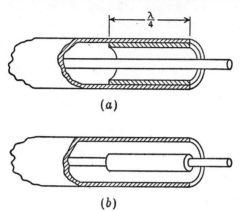

FIG. 7.15.   Quarter-wave sleeves used as impedance transformers in coaxial cable

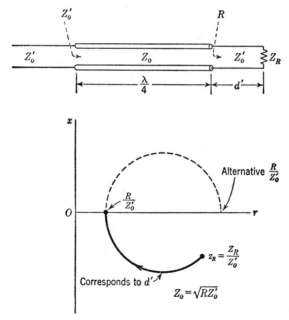

FIG. 7.16.   Matching a complex load by means of a quarter-wave transformer in cascade with a short section of the main line.

the load, d', can be read as the difference between the initial and final readings on the distance scale.   The resistance at the receiving end of the quarter-wave section is read at the intersection of the locus with the r axis, and is inserted into Eq. (7.13) to find the required Z_0 for the quarter-wave section.   As indicated in Fig. 7.16, two different intersections will yield possible solutions.   One of these will make $Z_0 < Z_0'$ and the other will make $Z_0 > Z_0'$.   The choice can be made on the basis of practicability.

The simple quarter-wave transformer suffers from the disadvantage that it is sensitive to changes in frequency, for at a new wavelength the section will no longer be $\lambda/4$ in length.   In some applications the frequency

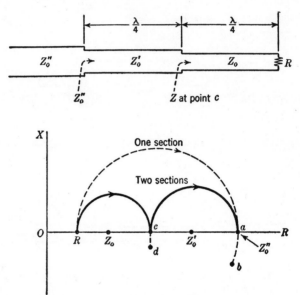

Fig. 7.17.   A two-section transformer.

is confined to a narrow band, and the frequency sensitivity is not a very great disadvantage.   In other applications, broad-band operation may be required.   The transition from one impedance level to another is less sensitive to changes in frequency if it is made in two or more stages.   Figure 7.17 shows a two-section transformer which matches a load of resistance R to a line of characteristic impedance Z_0''.   The reason for the reduced frequency sensitivity can be visualized from the impedance loci of Fig. 7.17, which are plotted in the fashion of a rectangular transmission-line chart except that the impedances are not normalized but are plotted in ohms.   When so plotted, the impedance locus swings about the characteristic impedance of the section that is under consideration at the moment, for this has a status that corresponds to the point $1 + j0$ on a normalized chart. With one section $\lambda/4$ in length, the locus starts at R and ends at Z_0''. However, an increase in frequency will cause the section to be greater than $\lambda/4$ in length, and the end of the locus will move to the point b, causing a mismatch at the input to the transformer. With two sections, however, the locus consists of two arcs, with the point c representing the impedance at the junction between sections. An increase in frequency will increase the lengths of arc for both loci. The end of the first arc will move down to point d; the second arc will start at point d and, because of its increased length, will still end approximately at Z_0''.

It can be shown that, for best results, the characteristic impedances should be chosen so that

$$\left(\frac{Z_0''}{Z_0'}\right)^2 = \frac{Z_0'}{Z_0} = \left(\frac{Z_0}{R}\right)^2 \quad (7.14)$$

Incidentally, this choice makes the impedance at the junction between sections equal to $\sqrt{RZ_0''}$, which is the value that would be chosen for the characteristic impedance of a single matching section. Additional sections reduce the frequency sensitivity still further.[1]

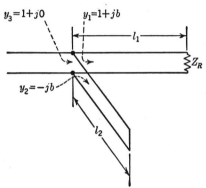

Fig. 7.18. The single-stub tuner.

7.9. The Single-stub Tuner. A single-stub tuner consists of an open- or short-circuited stub line of adjustable length which is shunted across the main line, as shown in Fig. 7.18. The distance from the load, l_1, must be variable. On a parallel-wire line this can be accomplished by sliding the stub along the line. On a coaxial cable a "line stretcher," which is a telescoping section of line, can be used between the stub and the load.

Computations for a single-stub tuner are most easily made with the aid of a transmission-line chart. We shall illustrate the calculation on the rectangular chart because it simplifies the operation of adding the admittances, but it will be evident that the same operations can be performed on the circular chart. Admittances, rather than impedances, will be used because of the parallel connection between the stub and the line. As was explained in Sec. 5.6, admittances are handled in the same manner as impedances on the chart.

---

[1] For further information on multiple-section transformers, see J. C. Slater, "Microwave Transmission," pp. 57–62, McGraw-Hill Book Company, Inc., New York, 1942, and G. L. Ragan, "Microwave Transmission Circuits," Sec. 6.2, McGraw-Hill Book Company, Inc., New York, 1948.

As indicated in Fig. 7.18, the length l_1 is selected so that the normalized admittance has a real part equal to unity at the point where the stub is connected. Looking toward the load, the normalized admittance at this point can then be expressed as $y_1 = 1 + jb$, in which b is an undesired, but unavoidable, susceptance. The length of the stub is adjusted so that its normalized input admittance is $y_2 = -jb$. The sum of the two admittances, as seen by the main line at this point, is $y_3 = y_1 + y_2 = 1 + j0$; *i.e.*, the admittance is $1/Z_0$ mhos and the impedance is Z_0 ohms. At this point the main line is terminated in its characteristic impedance, which is the desired result.

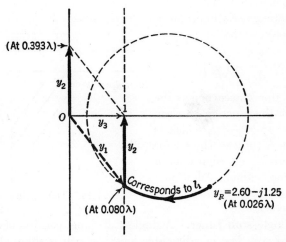

Fig. 7.19. Computation of a single-stub tuner on the rectangular chart.

The computation using the chart is straightforward and is perhaps shown best by an example. Consider a line with $\lambda = 1.50$ meters, $Z_0 = 400$ ohms, and $Z_R = 125 + j60$ ohms. The normalized receiving-end impedance is

$$z_R = \frac{125 + j60}{400} = 0.313 + j0.150$$

The normalized receiving-end admittance is the reciprocal of this and can be computed by the familiar methods of complex algebra or can be found by entering the chart at z_R and swinging around the locus an amount equivalent to $\lambda/4$. In any case, the result is

$$y_R = \frac{1}{z_R} = 2.60 - j1.25$$

As shown in Fig. 7.19, the chart is entered at this value of y_R (at 0.026λ on the distance scale). The admittance locus is followed toward the generator until it intersects the vertical line through unit conductance (at 0.080λ on the distance scale).   Here the stub is to be connected, and its distance from the load is

$$l_1 = (0.080 - 0.026)\lambda = 0.054 \times 1.50 = 0.081 \text{ meter}$$

The admittance at this point is $y_1 = 1 - j1.26$, as shown by the dashed vector extending from the origin.   To the tip of this is added the admittance of the stub, y_2, which must be of sufficient length to make the vector sum equal to $1 + j0$.   On the rectangular chart, where the admittance scale is the same throughout, the vector y_2 can be transferred, without change in length, so that its origin coincides with the origin of the admittance scale.   The distance scale of the chart is arranged so that the required length of a short-circuited stub can be read directly at the tip of the vector. This is true because a short-circuited stub of very short length has a very large negative susceptance (inductive) which coincides with the very small reading of the distance scale far down the vertical axis.   Increased length of the stub brings its admittance up the vertical axis, and, at a length of 0.25λ, its input admittance is zero.   Greater length produces a positive susceptance (capacitive).   In the present problem the tip of the vector is located at $j1.26$.   At this point we read a length of 0.393λ, and so, for a short-circuited stub,

$$l_2 = 0.393 \times 1.50 = 0.590 \text{ meter}$$

The same admittance could be obtained by an open-ended stub with a length either smaller or larger than this by 0.25λ; *i.e.*, an appropriate open-ended stub would have a length of $0.590 - 1.50/4 = 0.215$ meter. Short-circuited stubs are, however, generally used.

It will be observed in Fig. 7.19 that a second solution can be obtained by increasing l_1 until a point on the locus is reached directly above, rather than below, unit conductance.   The required stub will then have a negative susceptance.   Furthermore,   any additional number of solutions can be obtained from the previous two by adding multiples of $\lambda/2$ to l_1.   The shortest value of l_1 reduces the frequency sensitivity of the termination to a minimum, and, other considerations being equal, would be the solution chosen.

7.10. Double-stub and Triple-stub Tuners.   The adjustment of a single-stub tuner requires a change in the distance from the load and, with coaxial cable, this may be inconvenient.   An alternative scheme is to use two stubs which are fixed in position but adjustable in length.   Spacings commonly used between stubs are $\lambda/4$, $3\lambda/8$, and $5\lambda/8$.   A half-wavelength spacing is not used, as this has the effect of merely placing the stubs in parallel. An odd number of eighth wavelengths is generally preferred, as this matches

a wider range of impedances and also leads to more rapid convergence
when the adjustment must be made by trial and error.[1]

Figure 7.20 shows a double-stub tuner and illustrates the computation
for a stub spacing of $3\lambda/8$. The computation is made with the aid of an
auxiliary circle of unit radius which is centered at $1 - j1$ on the diagram.
If one starts at any point on this circle and follows an admittance locus

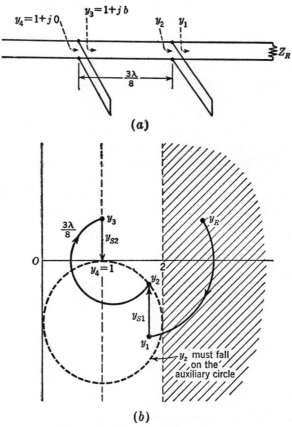

(a)

(b)

Fig. 7.20.   The double-stub tuner.

$3\lambda/8$ toward the generator, the end point of this path will lie on the vertical
line that passes through unity. Now consider Fig. 7.20a and follow the
admittances in reverse order. We wish to have $y_4 = 1 + j0$. The stub
connected here can supply any required susceptance, and so y_3 must
merely have a real part equal to unity; *i.e.*, it may lie anywhere on the
vertical line that goes through unity on the chart. The admittance y_2

---

[1] The effects of various spacings are discussed by Ragan, *op. cit.*, Sec. 8.6.

is $3\lambda/8$ nearer the load than y_3 and must therefore lie on the auxiliary circle. The right-hand stub can supply any desired suceptance, and so y_1 can lie anywhere in the unshaded region. The shaded area, where the normalized conductance is greater than 2, is a forbidden region for the admittance y_1; if y_1 falls within this region the tuner cannot supply the desired impedance transformation.

A calculation proceeds as follows: The chart is entered with the normalized load admittance, y_R, as shown in Fig. 7.20b, and the locus is followed toward the generator until the point is reached corresponding to the position of the right-hand stub. If the admittance here (y_1) has a real part less than 2, a solution is possible. The stub admittance (y_{s1}) is selected to make y_2 fall on the auxiliary circle. From y_2, a new locus is followed $3\lambda/8$ toward the generator, the result being y_3, which falls on the vertical line through unity. The admittance of the left stub (y_{s2}) is selected to close the gap between y_3 and the point $1 + j0$. The result is $y_4 = 1 + j0$, and so the main line is terminated in its characteristic impedance at the point where the left stub is connected.

It will be seen from Fig. 7.20b that a second solution is possible, for, if the vector y_{s1} were drawn downward from y_1, the admittance y_2 could be made to fall on the lower part of the auxiliary circle. This will, of course, change y_{s2}. The disadvantage of the second solution is that the standing-wave ratio between stubs is higher than before.

The double-stub tuner with a spacing of $3\lambda/8$ cannot take care of all possible matching problems, for the admittance y_1 must have a real part less than 2. It can be shown that a spacing of $\lambda/4$ leads to an auxiliary circle that extends only to a conductance of unity, as shown in Fig. 7.21. Hence, a tuner with this spacing is even more limited in its ability. A spacing of $5\lambda/8$ has a circle like that for $3\lambda/8$ but located above the horizontal axis, and so its limit is a conductance of two. These limitations can be circumvented by making the distance to the load adjustable, perhaps by having an additional quarter-wave section of line that can be inserted between the tuner and the load when necessary.

Another method of overcoming the foregoing limitations is to add a third stub. It can be shown that a triple-stub tuner with a quarter-wavelength spacing between stubs is capable of matching all loads except, of course, purely reactive ones. In practice, the two outer stubs are frequently ganged together mechanically to facilitate adjustment.

7.11. The Tapered Line. A tapered line is one in which the characteristic impedance is continuously varied along its length. The usual purpose is to provide a transition from one impedance level to another, and, if the change of impedance per wavelength is not too large, the tapered transition does not have the frequency sensitivity that is characteristic of devices which utilize reflection effects and standing waves. The multiple quarter-wave-section transformer described in Sec. 7.8 can be regarded as an approach toward a smooth transition, and this device has a smaller fre-

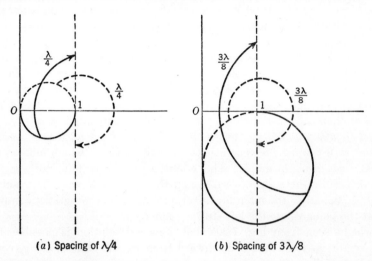

(a) Spacing of λ/4   (b) Spacing of 3λ/8

FIG. 7.21.   The auxiliary circles for double-stub tuners with spacings of λ/4 and 3λ/8.

quency sensitivity than the simple one-section quarter-wave transformer. A tapered section of coaxial line is shown in Fig. 7.22 (see also Fig. 7.9).

So far, our transmission-line analysis has assumed a line of uniform characteristics, and this will now be modified.   Equations (2.23) still hold,

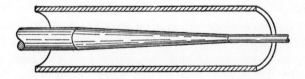

FIG. 7.22.   A tapered section of coaxial cable.

for they referred to conditions at a single point on the line.   If losses are neglected, these equations become

$$\frac{dE}{dx} = -j\omega L I \qquad (7.15)$$

and

$$\frac{dI}{dx} = -j\omega C E \qquad (7.16)$$

Here L is the series inductance per unit length of line at the point in question, and C is the shunt capacitance per unit length at the same point. The uniform line had constant values of L and C; on the tapered line, L and C vary continuously with x. Now we proceed to eliminate the current between the foregoing equations. Taking the derivative of (7.15) with respect to x, we have

$$\frac{d^2E}{dx^2} = -j\omega L \frac{dI}{dx} - j\omega \frac{dL}{dx} I \qquad (7.17)$$

The last term is different from zero because of the variation of L along the line. Substituting for I from (7.15) and for dI/dx from (7.16), we obtain the differential equation for voltage:

$$\frac{d^2E}{dx^2} - \left(\frac{1}{L}\frac{dL}{dx}\right)\frac{dE}{dx} + (\omega^2 LC)E = 0 \qquad (7.18)$$

If the voltage is eliminated between Eqs. (7.15) and (7.16), the result is the differential equation for current:

$$\frac{d^2I}{dx^2} - \left(\frac{1}{C}\frac{dC}{dx}\right)\frac{dI}{dx} + (\omega^2 LC)I = 0 \qquad (7.19)$$

If L and C are constant along the line, the middle terms vanish and we obtain the usual equations for the uniform lossless line. In the general case, the coefficients are functions of x, and a general solution cannot be obtained. But there is a special case that is reasonably simple, namely, the line with an exponential taper. The coefficients of the middle terms can be written as $d(\log L)/dx$ and $d(\log C)/dx$, and these will be constants if L and C vary exponentially with x. Also, the coefficient of the last term, $\omega^2 LC$, will be constant if L and C vary oppositely. Then the differential equations have constant coefficients and can be solved without difficulty.

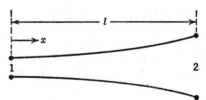

Fig. 7.23. Schematic diagram of a tapered line.

Refer to Fig. 7.23 and write the inductance and capacitance per unit length at any point x as

$$\left. \begin{array}{l} L = L_1 \epsilon^{\eta x} \\[2mm] C = C_1 \epsilon^{-\eta x} \end{array} \right\} \qquad (7.20)$$

and

where L_1 and C_1 are the values at the narrow end. The constant η indicates the rate of taper. At the moment we do not have to specify which of the two ends of the line is the receiving end, for our analysis will apply equally well either way.

The product $LC = L_1C_1$ is constant, but we have

$$\sqrt{\frac{L}{C}} = \sqrt{\frac{L_1}{C_1}}\,\epsilon^{\eta x} \qquad (7.21)$$

If the line were uniform, the quantity $\sqrt{L/C}$ would be the characteristic impedance, but this expression will be modified by the presence of the taper. However, if the taper per wavelength is small, the line will behave as though it were uniform over short distances, and we can reason by analogy with the uniform line. The phase velocity will be constant and very nearly equal to $1/\sqrt{LC}$. The characteristic impedance will be nearly $\sqrt{L/C}$ at each point, and this increases exponentially with x. The result is a transformation of impedance from one end to the other. When terminated without reflection, the line behaves like an ideal transformer which can be used in either step-up or step-down fashion. We would expect physically that the voltage would vary as $\epsilon^{\eta x/2}$ and the current would vary as $\epsilon^{-\eta x/2}$, with the quotient (impedance) being proportional to $\epsilon^{\eta x}$ and the product (power) being constant. The impedance transformation should be independent of frequency so long as the frequency is high enough to keep the taper per wavelength small. In practice, space limitations generally lead to tapers which are not small per wavelength, for which the foregoing conclusions are not reliable. We shall analyze this case with the aid of Eqs. (7.15) to (7.19).

In practice, a linear physical taper is much easier to construct than the taper indicated by Eq. (7.20). If the change in impedance level is not too great, the two will be nearly the same and our analysis will apply to either.

Upon substitution of Eqs. (7.20) into (7.18), we obtain a voltage equation with constant coefficients:

$$\frac{d^2E}{dx^2} - \eta\,\frac{dE}{dx} + \omega^2 LCE = 0 \qquad (7.22)$$

The solution of this differential equation is

$$E = A_1\epsilon^{\eta x/2}\epsilon^{-j\beta x} + A_2\epsilon^{\eta x/2}\epsilon^{j\beta x} \qquad (7.23)$$

where A_1 and A_2 are constants with the dimensions of voltage, and

$$\beta = \sqrt{\omega^2 LC - \frac{\eta^2}{4}} \qquad (7.24)$$

The validity of the foregoing solution can be checked by substituting it into the differential equation, which thereupon will reduce to an identity.

The solution (7.23) consists of two waves. The first, with the phase factor $\epsilon^{-j\beta x}$, travels toward the wide end (Fig. 7.23). The second, with the factor $\epsilon^{j\beta x}$, travels toward the narrow end. Because of the taper, each of the voltage waves is larger at the high-impedance end by the factor $\epsilon^{\eta l/2}$. We have not yet specified which is the load end; hence, the question of which is the incident wave and which is the reflected one is still of no concern.

The phase constant given by Eq. (7.24) is of particular interest, for it will be real only if the frequency is above a certain critical value. Below this frequency, β will be imaginary. Then the factors $\epsilon^{\pm j\beta x}$ in Eq. (7.23) will have real exponents, and the effect will be that of an attenuation constant! The behavior is like that of a high-pass filter, which passes energy freely above a certain frequency but attenuates at lower frequencies. The critical point is given the name *cutoff frequency*, but this is perhaps misleading, for the line does not cut off sharply and energy will reach the load below the critical frequency unless the line is quite long. The critical frequency is obtained when $\omega^2 LC = \eta^2/4$, and so

$$\omega_c = \frac{\eta}{2\sqrt{LC}} \tag{7.25}$$

The phase constant (7.24) can now be written as

$$\beta = \omega\sqrt{LC}\,\sqrt{1 - \frac{\eta^2}{4\omega^2 LC}}$$

or

$$\beta = \omega\sqrt{LC}\,\sqrt{1 - \left(\frac{\omega_c}{\omega}\right)^2} \tag{7.26}$$

This will be lower than the uniform-line value, $\omega\sqrt{LC}$, by 5 per cent when $\omega = 3.21\omega_c$, and will be lower by only 1 per cent when $\omega = 7.10\omega_c$. The phase velocity ω/β and the wavelength $2\pi/\beta$ will be similarly near to the uniform-line values. Practical tapered lines are generally operated at a frequency well above the critical value, where β and λ are nearly the same as for a uniform line.

To find the solution for current, substitute the expression for voltage (7.23) into Eq. (7.15). Using the relation $L = L_1\epsilon^{\eta x}$, the result can be written as

$$I = B_1\epsilon^{-\eta x/2}\epsilon^{-j\beta x} - B_2\epsilon^{-\eta x/2}\epsilon^{j\beta x} \tag{7.27}$$

where

$$B_1 = \frac{\beta + j\eta/2}{\omega L_1}\,A_1$$

and

$$B_2 = \frac{\beta - j\eta/2}{\omega L_1}\,A_2 \tag{7.28}$$

We identify the two terms of (7.27) as current waves that travel toward the high-impedance and low-impedance ends, respectively. Each of the current waves is smaller at the high-impedance end by the factor $\epsilon^{-\eta l/2}$.

The characteristic impedances seen by the two oppositely traveling waves can be obtained from the foregoing expression for current and the equation for voltage (7.23). Unlike the uniform line, the impedances in the two directions are different because of the taper. The impedance seen by the wave that travels toward the high-impedance end will be denoted by Z_{up}, and is obtained by dividing the first term of (7.23) by the first term of (7.27). Then

$$Z_{up} = \frac{\omega L_1 \epsilon^{\eta x}}{\beta + j\eta/2} = \frac{\omega L}{\beta + j\eta/2}$$

This is placed in more convenient form by substituting for β from (7.26) and for η from (7.25), and then rationalizing. The result is

$$Z_{up} = \sqrt{\frac{L}{C}} \left[\sqrt{1 - \left(\frac{\omega_c}{\omega}\right)^2} - j\frac{\omega_c}{\omega} \right]$$

$$= \sqrt{\frac{L_1}{C_1}} \left[\sqrt{1 - \left(\frac{\omega_c}{\omega}\right)^2} - j\frac{\omega_c}{\omega} \right] \epsilon^{\eta x} \qquad (7.29)$$

Similarly, the impedance seen by the wave traveling toward the low-impedance end can be shown to be

$$Z_{down} = \sqrt{\frac{L}{C}} \left[\sqrt{1 - \left(\frac{\omega_c}{\omega}\right)^2} + j\frac{\omega_c}{\omega} \right]$$

$$= \sqrt{\frac{L_1}{C_1}} \left[\sqrt{1 - \left(\frac{\omega_c}{\omega}\right)^2} + j\frac{\omega_c}{\omega} \right] \epsilon^{\eta x} \qquad (7.30)$$

The resistive components of these impedances are very nearly equal to $\sqrt{L/C}$ when the frequency is several times the critical value, but the reactive term is only inversely proportional to ω/ω_c and remains appreciable up to much higher values of this ratio. For example, at $\omega = 7.10\omega_c$ we have $Z_{up} = (0.99 - j0.14)\sqrt{L/C}$. The resistive component is only 1 per cent lower than $\sqrt{L/C}$, but a 14 per cent reactive component remains.

In order to terminate the line without reflection at the high-impedance end, a load impedance equal to Z_{up} should be used [Eq. (7.29) with $x = l$]. The input impedance at the low-impedance end will then be equal to Z_{up} at that end [Eq. (7.29) with $x = 0$]. This will provide a downward transformation of impedance by a factor of $\epsilon^{-\eta l}$. If the load is connected instead at the low-impedance end, it should be equal to Z_{down} at $x = 0$ for no reflection. At the other end, the input impedance will be equal to Z_{down} with $x = l$, and the load impedance will be transformed upward by a factor of $\epsilon^{\eta l}$. For values of ω/ω_c not too near cutoff, a purely resistive load can be given the required reactive component by connecting a series capacitor to match Z_{up}, or by connecting a shunt inductance to match Z_{down}. More complex terminating networks can be derived for a better match over a wide range.[1]

The requirement of a complex load impedance for no reflection is inconvenient and leads, of course, to an input impedance with a reactive component. The difficulty can be avoided by operating the line far above its critical frequency. Then we have, very nearly,

$$Z_{up} \approx Z_{down} \approx \sqrt{\frac{L}{C}} = \sqrt{\frac{L_1}{C_1}}\,\epsilon^{\eta x}$$

Now a purely resistive load can be used, and the impedance transformation $\epsilon^{\eta l}$ is achieved independently of the frequency so long as $\omega \gg \omega_c$. This requires that the taper per wavelength be kept small. We shall denote the wavelength on a uniform line by $\lambda_0 = 2\pi/\omega\sqrt{LC}$. The amount of taper per wavelength can then be expressed with the aid of Eq. (7.25) as

$$\eta\lambda_0 = \left(2\omega_c\sqrt{LC}\right)\left(\frac{2\pi}{\omega\sqrt{LC}}\right) = 4\pi\frac{\omega_c}{\omega} \tag{7.31}$$

If the reactive component of Z_{up} or Z_{down} is to be kept to 2 per cent, we must have $\omega = 50\omega_c$. Then the taper per wavelength is

$$\eta\lambda_0 = \frac{4\pi}{50} = 0.252$$

In each wavelength, the impedance is transformed by a factor of $\epsilon^{0.252} = 1.287$.

In many applications, the space requirements do not permit tapers as long as those indicated by the foregoing considerations, and a resistive load is often used at the end of a tapered line which is operated at a frequency only several times as large as the critical value. The resulting reflection caused by mismatch at the load causes the input impedance to have a reactive component and to vary with frequency. The analogy for a uniform line would be the shift of sending-end impedance around a circular locus on a transmission-line chart. The variation in impedance can be investigated by dividing the voltage (7.23) by the current (7.27), thus obtaining

$$Z = \frac{E}{I} = \frac{A_1 + A_2\epsilon^{-j2\beta x}}{B_1 - B_2\epsilon^{-j2\beta x}}\,\epsilon^{\eta x} \tag{7.32}$$

where B_1 and B_2 are expressed in terms of A_1 and A_2 through Eqs. (7.28).

[1] See Harold A. Wheeler, Transmission Lines with Exponential Taper, *Proc. IRE*, Vol. 27, pp. 65–71, January, 1939.

The analysis of the impedance variation will not be completed in detail except for one simple and important case.[1] The half-wave exponential line, like its uniform counterpart, has an impedance-repeating property, modified only by the transformation ratio $\epsilon^{\eta x}$. This can be seen immediately from Eq. (7.32), for an increase in βx of π rad changes the argument $2\beta x$ by 2π rad, and the impedance remains the same as before except for the transformation ratio $\epsilon^{\eta x}$. The reason is that the reflection arrives at the sending end with the proper phase with respect to the outgoing wave. Therefore, a line which is an integral number of half wavelengths long, terminated in a load resistance R, will have an input impedance equal either to $R\epsilon^{\eta l}$ or to $R\epsilon^{-\eta l}$, depending on whether the line is used in step-up or step-down fashion. This is precisely the desired impedance transformation, but it is sensitive to changes in frequency. For a change in frequency will alter λ, and the line will no longer be an integral number of half wavelengths long. The reflected wave arrives with its worst phase when the line is an odd number of quarter wavelengths long. The variation with frequency can be minimized by making the reflected wave small, i.e., by making R as nearly equal to Z_{up} or Z_{down} as possible. Therefore, (a) the load resistance should be made equal to $\sqrt{L/C}$ at the end where it is connected and (b) the line should be operated as far above its critical frequency as other factors permit; i.e., the line should be made as many half wavelengths long as is practicable. Finally, if the line can be made long enough so that it is operating very far above its critical frequency, we have $Z_{up} \approx Z_{down} \approx \sqrt{L/C}$, and there is no appreciable reflected wave. Then the exact length does not matter.

PROBLEMS

1. A low-loss transmission line is terminated in its characteristic impedance of 400 ohms. Across this line is to be connected an r-f voltmeter of the type shown in Fig. 7.1, which consists of a quarter-wave line with $Z_0 = 400$ ohms, terminated in a meter which has 10 ohms resistance.

  a. Compute the input impedance of the r-f voltmeter, and determine the standing-wave ratio that it will cause on the main line.

  b. The meter requires a current of 50 ma for full-scale deflection, and its deflection is proportional to the square of the current. What input voltage will produce full-scale deflection? What voltage will produce one-fourth of full-scale deflection?

2. A line with $Z_0 = 400$ ohms is terminated in an unknown load impedance. The standing-wave ratio is 5.7, and a voltage minimum occurs at the load. What is the load impedance?

3. A line with $Z_0 = 427$ ohms is terminated in an unknown receiving-end im-

[1] For further information, see Chas. R. Burrows, The Exponential Transmission Line, *Bell System Techn. J.*, Vol. XVII, pp. 555–573, October, 1938. The effect of reflections, as computed for a linear physical taper by a more general method, can be found in "Microwave Transmission Design Data," by Theodore Moreno, pp. 53–55, McGraw-Hill Book Company, Inc., New York, 1948.

pedance. The standing-wave ratio is 3.2, and the first voltage minimum occurs 0.79 meter from the load. The distance between two successive minima on the standing-wave pattern is 2.52 meters. What is the load impedance?

4. A low-loss transmission line with $Z_0 = 60$ ohms was used to measure four different load impedances. In each case, the standing-wave ratio was 3.0. For load (a), a voltage maximum was found exactly at the load. For load (b), the first voltage maximum was $\lambda/8$ back from the load. For load (c), a voltage minimum was found exactly at the load. With load (d) connected, a voltage minimum was found $\lambda/8$ from the load. Determine the four load impedances.

5. A slotted coaxial line with $Z_0 = 52$ ohms was used to measure a load impedance. First, the receiving end of the line was short-circuited. The voltage minima were found to be 20.5 cm apart. One of the minima was marked for use as a measuring point. Next, the unknown impedance was connected to the receiving end of the line. The standing-wave ratio was found to be 2.75, and a voltage minimum was found to be 8.6 cm from the measuring point, on the side toward the load. Determine the unknown load impedance.

6. A directional coupler connected in a transmission line shows an incident power of 20 watts. The standing-wave ratio is 3.0. What power is being absorbed by the load?

7. A low-loss line of variable length is connected between a generator and a load. The line has $Z_0 = 70$ ohms. The load impedance is $Z_R = 210 + j0$ ohms. The generator has an internal voltage of 100 volts rms and an internal impedance of 70 ohms pure resistance. Plot the power received by the load as a function of line length, for $0 < l < \lambda/2$.

8. A slotted section of coaxial line is to be used to determine the velocity of propagation in a flexible coaxial cable which has a solid dielectric. One end of the slotted section is connected to a generator; the other is connected to the flexible cable. The far end of the flexible cable is open-circuited. The characteristic impedance of the slotted section matches that of the flexible line.

a. Voltage nodes are found 30 cm apart in the slotted section. What is the wavelength in the slotted section, and what is the generator frequency?

b. When a 5-cm section is cut off the open end of the flexible cable, the voltage minima in the slotted section move 7.35 cm toward the generator. What is the wavelength in the flexible cable? What is the velocity of propagation?

9. An air-insulated two-wire line with a characteristic impedance of 400 ohms operates into a load resistance of 105 ohms. The frequency is 100 Mc. A quarter-wave transformer is to be used to match the load to the line. If the transformer is to be constructed of tubes having a diameter of 0.250 in., what should be the center-to-center spacing?

10. A low-loss line with a characteristic impedance of 340 ohms is to supply power to a load impedance of $210 + j280$ ohms. A quarter-wave transformer is to be used as shown in Fig. 7.16 to match the load to the line. The wavelength is 5.0 meters.

a. If the characteristic impedance of the quarter-wave section is to be smaller than 340 ohms, what should be the distance d' (see Fig. 7.16)? What value should be used for the characteristic impedance of the quarter-wave section? What should the wire spacing be if the wire diameter is 0.102 in.?

b. Repeat part *a,* but make the characteristic impedance of the quarter-wave section greater than 340 ohms.

11. A certain air-insulated coaxial cable has the following dimensions: Diameter of inner conductor = 0.250 in., inside diameter of outer tube = 0.875 in. The frequency is 300 Mc, and the line is to be terminated in a resistance of 30 ohms. A quarter-wave transformer of the type shown in Fig. 7.15*b* is to be used to match the load to the line. What should be the diameter of the inner conductor for the quarter-wave section, and what should be its length?

12. A certain air-insulated coaxial cable has a characteristic impedance of 75 ohms and operates at a frequency of 1,200 Mc. It is terminated in a resistance of 30 ohms. A quarter-wave "slug" of dielectric material is to be inserted in the line so as to fill the annular space completely. This is to act as a quarter-wave transformer to match the load to the line. What should be the relative dielectric constant of the material? How many centimeters long should the slug be?

13. A two-section impedance transformer of the type shown in Fig. 7.17 is to be used to match a 300-ohm line to a 70-ohm resistive load. Determine the proper characteristic impedances of the two sections.

14. A quarter-wave transformer was designed to match a 400-ohm line to a 100-ohm resistive load at a wavelength of 3.0 meters. The wavelength changes to 3.5 meters. Determine the resulting standing-wave ratio on the 400-ohm line.

15. A single-stub tuner of the type shown in Fig. 7.18 is to match a 400-ohm line to a load of $800 + j300$ ohms. The wavelength is 1.50 meters. Determine the distance from the load to the tuning stub and the proper length for the stub. Locate the stub as near the load as possible.

16. A single-stub tuner is to match a 400-ohm line to a load of $200 - j100$ ohms. The wavelength is 3.00 meters. Determine the shortest distance from the load to the tuning stub and the proper length for the stub.

17. A double-stub tuner of the type shown in Fig. 7.20 is to match a 60-ohm line to a load of $28 + j0$ ohms. The distance from the load to the first stub is $\lambda/18$. The spacing between stubs is $3\lambda/8$. Determine the proper lengths for the two stubs (express as fractions of the wavelength).

18. Repeat Prob. 17, except use a distance of $\lambda/6$ between the load and the first stub.

19. A 70-ohm line is connected to a resistive load of 22.0 ohms. A double-stub tuner is to be inserted to match the load to the line. Within the first wavelength from the load, where should the tuner *not* be inserted? The stub spacing is $3\lambda/8$.

20. A tapered line is to be used as an impedance transformer between a 75-ohm coaxial cable and a resistive load of 150 ohms. The inside radius of the outer tube is to be 1.50 cm, and the inner conductor is to be tapered. The line is air-insulated.

a. Specify the radius of the inner conductor at both ends of the tapered section.

b. If the tapered section is to be operated at 50 times its cutoff frequency, what should be its length in wavelengths? What will be Z_{up} at both ends? How will the phase constant compare with $\omega\sqrt{LC}$?

c. If the length of the tapered section is made a half wavelength, at how many times its cutoff frequency will it operate? What will be Z_{up} at both ends? How will the phase constant compare with $\omega\sqrt{LC}$?

CHAPTER 8

SPECIAL CONSIDERATIONS FOR TELEPHONE AND TELEGRAPH LINES

8.1. Types of Telephone and Telegraph Lines. Telephone and telegraph lines may be either open-wire or cable. Some telegraph systems use the earth as one side of the circuit, while others use a complete metallic path. Modern telephone systems use complete metallic circuits. The conductors of open-wire lines are supported on poles by separate insulators, as illustrated in Figs. 8.9 and 8.10. A telephone cable, on the other hand, is made up of a large number of insulated pairs of wires which occupy the same protective sheath. The two wires of each pair are twisted together, and, in long-distance cables, each set of two pairs is twisted into a group known as a *quad*. A cable can be run overhead on poles, or it can be buried underground. A cable line has an abnormally low inductance and a high capacitance because of the proximity of the conductors and the presence of the solid insulation. The characteristics of the cable are greatly improved by the use of inductive loading, as was discussed in Sec. 2.9. Inductive loading is always used on long voice–frequency telephone cables.

8.2. Frequencies Used in Telephone and Telegraph Communication. The essential frequencies for the transmission of voice messages lie between approximately 250 and 2,700 cps, and the ordinary telephone equipment is designed for this range. The transmission of radio programs over telephone lines requires a frequency range from perhaps 50 to above 7,000 cps for good quality, and certain long lines and their associated equipment are designed for this service.

A number of telephone conversations can be transmitted simultaneously over one pair of wires by use of the *carrier* system. In this system, a message is conveyed over the line by means of a high-frequency "carrier" wave. By using a number of carrier waves of different frequencies and separating them at the end of the line by filters, a number of channels of communication can be obtained simultaneously on one line. A signal is impressed on a carrier voltage by the process of amplitude modulation; *i.e.*, the amplitude of the carrier voltage is made proportional at each instant to the instantaneous value of the signal. As was mentioned in

Sec. 2.10, the resulting amplitude-modulated wave can be resolved into a number of components of different frequencies. One of these components has the same frequency as the carrier; the others have frequencies equal to that of the carrier plus or minus the frequencies present in the original signal. For example, the band of voice frequencies in the range of 250 to 2,700 cps can be translated upward in the frequency scale by causing them to modulate a 20,000-cps carrier. The resulting modulated wave will have a component at 20,000 cps. The other components, or "side bands," will extend on one side of the carrier from $20,000 - 2,700 = 17,300$ cps to $20,000 - 250 = 19,750$ cps, and on the other side of the carrier will extend from $20,000 + 250 = 20,250$ cps to $20,000 + 2,700 = 22,700$ cps. The modulated signal is demodulated at the receiving end, and the original signal is recovered. To conserve band width, the side bands on one side of the carrier are suppressed before transmission, as is also the carrier-frequency component itself. A local oscillator supplies the missing carrier frequency at the receiving end, and the demodulating equipment is arranged to reproduce the original signal from the combination of the local oscillator voltage and the one set of side bands. The carrier frequencies commonly used are in the range of roughly 5,000 to 140,000 cps.

The coaxial cable has been used to extend the range of carrier communication into the megacycle range, making possible the transmission of a large number of telephone channels over one line. The large attenuation at the higher frequencies requires the installation of amplifiers at short intervals along the line.

Telegraph signals are obtained either by interrupting a direct current or by reversing the direction of the current. The essential frequencies to be transmitted range upward to perhaps 25 cps. The teletypewriter operates at a higher speed and requires a wider frequency range than a manually operated system. Telegraph transmission is often superposed on telephone lines. This can be done in any one of several ways. The "simplex" system uses the two wires of a telephone circuit as one conductor for the telegraph, the return being made through ground. The telegraph current flows simultaneously in the same direction along both lines and, so long as the system is balanced, there is no interference. A second method, called "compositing," makes use of the fact that the essential frequencies for the telegraph signal lie below the lowest essential telephone frequencies. The high-frequency components of the telegraph signal, which are caused by the sudden break or make of the circuit, are first removed by means of shunt capacitance and series inductance, and the resulting low-frequency signal can be applied to a telephone line without interference. With a ground return for the telegraph, the two wires of a telephone circuit can be used to provide two telegraph channels.

The carrier system can also be applied to telegraph communication. The "voice-frequency carrier" uses carrier frequencies in the range of 400 to 2,500 cps. A line used in this way cannot, of course, be used simultaneously for voice transmission. Open-wire telephone lines can be used to carry telegraph signals simultaneously by using a telegraph carrier frequency that is somewhat higher than the voice frequencies.

8.3. The Phantom Circuit. The two pairs of wires which provide two separate telephone channels can be used to provide a third circuit by the so-called *phantom* connection. The two principal, or "side," circuits are provided with center-tapped transformers, as shown in Fig. 8.1. The phantom circuit is connected to the center taps of the transformers, and its currents are shown by the dotted arrows. The two wires of one side circuit

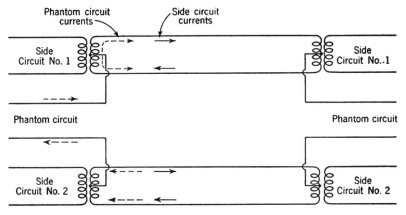

FIG. 8.1. Derivation of a phantom circuit from two side circuits.

are used as one conductor for the phantom circuit, and if the circuits are carefully balanced, there is no interference between the three channels of communication. Phantom circuits are used on both open-wire and cable circuits. In a cable, the four wires which form a "quad" are used for the phantom circuit.

When a transformer is used mainly for purposes of isolation or to provide a center tap, as those shown in Fig. 8.1, it is known in telephone parlance as a *repeating coil*. The name *transformer* is then reserved for those applications where the transformation of voltage, current, or impedance is the primary function of the device.

8.4. Telephone Amplifiers and Repeaters. Before the development of the vacuum-tube amplifier, telephone communication over a considerable distance could be achieved only by the use of large-gauge conductors and heavy inductive loading to reduce the attenuation. The results were, by

present standards, relatively unsatisfactory. The vacuum-tube amplifier
has made it possible to remove the loading from open-wire lines and to
improve the quality of transmission. It has also made practical the use of
long-distance cable lines, whose attenuation would otherwise be too great.

It is essential to space the amplifiers closely enough along a line to
prevent the signal from dropping into the noise level, from which it could
not be recovered. A typical loaded cable has an attenuation of perhaps
0.3 db/mile at 1,000 cps. In a distance of 500 miles, this amounts to a
total attenuation of 150 db, which corresponds to a power ratio of 10^{15}.

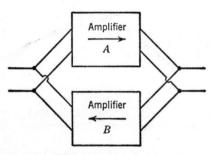

FIG. 8.2. An unsatisfactory attempt to
provide two-way amplification.

Without amplifiers, a power output
at the receiving end of 1 μw would
require an input of 1,000 megawatts
at the sending end! With a practical amount of power entering the
sending end, the power emerging
from the receiving end would be so
small that no amount of amplification could separate it from the random noise that is inherent in all
circuits. Suppose, however, that
amplifiers are spaced along the line
at 50-mile intervals. The attenu-
ation between adjacent amplifiers would be 15 db, which corresponds
to a power ratio of $10^{1.5} = 31.6$. A reasonable input at one end of a
50-mile section will produce an output signal that is well above the
noise level. Furthermore, a gain of 15 db is not at all difficult to
obtain in a vacuum-tube amplifier without danger of instability.

A vacuum-tube amplifier is a one-way device and, in order to amplify
signals in both directions in a two-wire telephone circuit, a special arrangement must be used. Figure 8.2 shows an arrangement which cannot be
used. Each of the boxes represents an amplifier which operates in the
direction of the arrow. Any small disturbance at the input of unit A will
be amplified and delivered at its output with increased amplitude, where it
will be impressed on the input terminals of unit B. The second amplifier
will amplify the disturbance still more and will impress it on the terminals
of unit A again. The disturbance will thus continue to circulate through
the two amplifiers, building up in strength until the amplifiers saturate
because of nonlinearity. The result will be a sustained oscillation which
renders the system inoperable. Evidently, the problem is to provide an
arrangement in which the output of one amplifier is impressed on the line
but not on the input of the other amplifier. This can be done by means of
a three-winding transformer which is known as a *hybrid coil*.

The principle of the hybrid coil, as used in a two-way amplifier, is illustrated in Figs. 8.3 and 8.4. For simplicity, the three coils are assumed to have the same number of turns. Two of the coils are center-tapped, as shown at a and b. Four external circuits are provided, those numbered 1, 2, and 3, and the connection between points a and b. Circuits 2 and 3 terminate in equal impedances, Z. Now assume that a voltage E is applied to circuit 1. This will result in the voltages shown in Fig. 8.3, and a current equal to E/Z will circulate clockwise through circuits 2 and 3. There will be zero voltage between points a and b, which is the desired

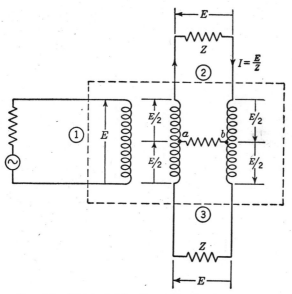

FIG. 8.3. The hybrid coil energized from circuit 1.

result here. Half of the input power will be dissipated in the terminating impedance Z of circuit 2, and the other half will be dissipated in the terminating impedance Z of circuit 3.

Second, assume that a voltage E is applied to circuit 2, as shown in Fig. 8.4. If the terminating impedance of circuit 1 is equal to $Z/2$, and if the impedance connected between points a and b is also equal to $Z/2$, the voltages and currents will be as shown in the figure. The input impedance of circuit 2 is equal to the value Z, which is the desired result. Half of the input power will be delivered to the impedance that terminates circuit 1, the other half will go to the impedance connected between points a and b.

The application of two hybrid coils and two amplifiers to a two-way amplifier is shown in Fig. 8.5. The resulting device is known as a *repeater*. If the impedance relations of both hybrid coils are maintained as was shown in Fig. 8.4, both line 1 and line 2 will be terminated in an impedance equal to Z_0. A signal coming in on line 1 satisfies the conditions of Fig. 8.4. Half of its energy will be dissipated in the output circuit of amplifier B, where it is lost as heat. The other half of the signal energy will be transferred to the input circuit of amplifier A, which corresponds to the impedance connected between points a and b in Fig. 8.4. This part of the signal is amplified and emerges into the second hybrid coil. Here it

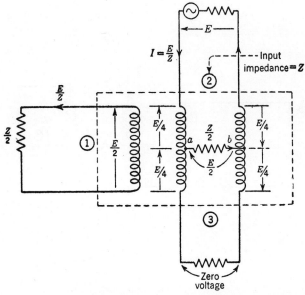

Fig. 8.4. The hybrid coil energized from circuit 2.

satisfies the conditions of Fig 8.3: half of the power will be dissipated in the balancing impedance Z_0, and the other half will be sent down line 2. No signal will appear across the input to amplifier B, thus solving the problem of a circulating signal and possible oscillations.

If the balancing impedances designated as Z_0 in Fig. 8.5 do not match the line impedance at all frequencies, the symmetry of the hybrid-coil connection is destroyed, and some of the signal emerging from one amplifier will appear across the input of the other. The amount of unbalance that can be tolerated depends upon the gain of the amplifiers. For this reason, a very high gain is not attempted in a single repeater.

In the longer cable circuits, one pair of wires is often used to carry a conversation in one direction only, and two pairs of wires are required for two-way conversation. This, of course, eliminates the problem of two-way amplification, and a single one-way amplifier can be used for each direction. Also, in the transmission of radio programs over telephone lines, the problem of two-way amplification does not arise. Program amplifiers must, of course, be designed for a wider frequency band than is required for ordinary voice circuits.

8.5. Noise and Crosstalk. A telephone line operates at a low level of power and is particularly susceptible to the introduction of noise from other electrical circuits in the vicinity, particularly from power lines which may run parallel with the telephone lines. The difficulty is most troublesome when the power lines carry harmonic currents of considerable magnitude, for these frequencies fall in a region where both the telephone equipment and the human ear are more sensitive. A similar difficulty

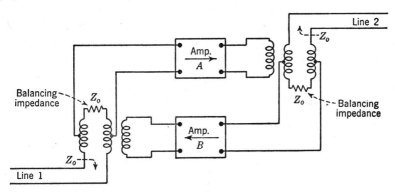

FIG. 8.5. A two-way repeater circuit for a two-wire telephone system.

is the coupling of a signal from one telephone line into an adjacent one. This is known as *crosstalk*. If the crosstalk comes simultaneously from a number of other telephone circuits, the result is an unintelligible scramble of voice sounds called *babble*.

The coupling between a telephone line and another electrical circuit generally comes about in either or both of two ways: through inductive coupling, in which the magnetic flux of one circuit links the other, or through capacitive coupling, in which the electric field of one circuit causes a voltage in the second circuit. These types of coupling are illustrated separately in Figs. 8.6 and 8.7. It is also possible to have conductive coupling between two sets of lines through faulty insulation, but this does not generally happen on well-kept systems. Figure 8.6 shows schematically the induction of voltage from line A into line B through magnetic coupling. This is equivalent to placing a generator in series with line B. The disturbance that appears at the end of line B nearest the original source of signal is

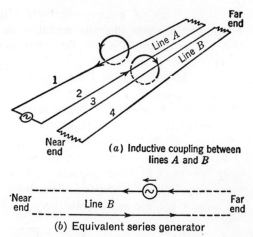

(a) Inductive coupling between
lines A and B

(b) Equivalent series generator

FIG. 8.6.   The induction of voltage in line B by inductive coupling with line A.

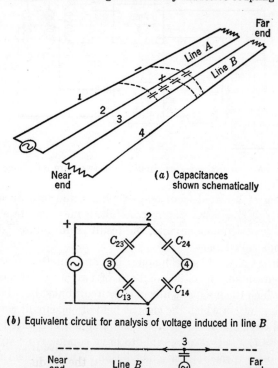

(a) Capacitances
shown schematically

(b) Equivalent circuit for analysis of voltage induced in line B

(c) Equivalent Thévenin generator in line B

FIG. 8.7.   The induction of voltage in line B by capacity coupling with line A.

called "near-end" crosstalk, and that which appears at the end farthest
from the signal source is called "far-end" crosstalk. With inductive
coupling, the current in any one line is in the same direction on both the
near and the far sides of the point of coupling. Figure 8.7 shows the
induction of voltage from line A into line B through capacity coupling.
The equivalent circuit at the point of coupling is shown in Fig. 8.7b and is
seen to be a bridge circuit. Since C_{23} is greater than C_{14}, the bridge is
unbalanced, and a voltage appears across line B. As indicated in Fig. 8.7c,
the currents resulting from capacity coupling are in opposite directions on
opposite sides of the point of coupling.

A wave that travels down a long telephone line is attenuated considerably
by the time it reaches the receiving end. When the sending end of one
circuit is in close proximity to the receiving end of another circuit, there is
greater likelihood of difficulty with troublesome crosstalk in the low-power

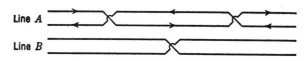

Line A

Line B

FIG. 8.8.   Transposed lines.

circuit.   Circuits with different power levels should be kept apart as much
as is feasible.

The principal method of reducing noise and crosstalk to a low value on
open-wire lines is by *transposing* the conductors of each circuit. In
transposition, the two conductors of each circuit exchange positions at
intervals along the line so as to balance out the voltages induced from
adjacent circuits. This is illustrated in Fig. 8.8. Obviously, the two
circuits should not both be transposed at the same points throughout their
length, as this would not balance the induced voltages. There should be a
number of transpositions per wavelength for best effectiveness. The proper
transposition, combined with resistance symmetry between the two wires
of a circuit (*e.g.*, no faulty joints), permits the circuits to be run rather
close together on the pole crossarms. Figure 8.9 shows a standard arrangement of wires on an open-wire voice-frequency line. The side circuits are
indicated by the double-headed arrows, and the phantom groups are
shown enclosed by dashed lines. Another standard arrangement, in which
high-frequency carrier currents are superimposed on voice-frequency signals,
is shown in Fig. 8.10. Here the spacing between the conductors of a circuit
is smaller than the spacing between circuits, because of the greater difficulty
in avoiding crosstalk between circuits at the carrier frequencies.

The crosstalk problem is somewhat more difficult in telephone cables because of the proximity of the many circuits and the lack of perfect mechanical symmetry which leads to capacity unbalances. The difficulty is alleviated somewhat by the fact that the two wires of a circuit are

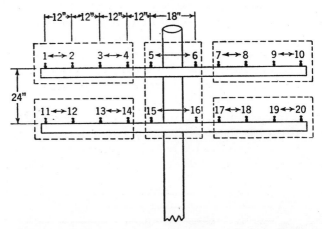

Fig. 8.9. Arrangement of conductors on an open-wire telephone line carrying voice-frequency signals. The phantom groups are shown enclosed by dashed lines.

separated by only a thin layer of insulation, for, with this small separation, the main portion of the magnetic flux is confined to a very small region near the conductors. The two wires are transposed by twisting them together.

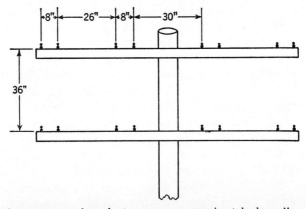

Fig. 8.10. Arrangement of conductors on an open-wire telephone line that carries voice-frequency signals and superimposed high-frequency carrier signals. Phantom circuits not used.

In addition, the two pairs of wires which form a quad are twisted to provide transposition for the phantom circuit. As a result, the crosstalk caused by magnetic flux is negligible at voice frequencies, and the main coupling is

produced by capacity unbalances.   A satisfactory remedy is the connection
of small capacitors to correct the unbalance.

   In the longer cables which employ a number of repeaters, the near-end
crosstalk can be virtually eliminated by using separated circuits for the
two directions of transmission.   This is illustrated in Fig. 8.11.   Crosstalk
which is induced from circuit A' into circuit B', such as that shown at
point P, cannot travel to the near end of circuit B because of the one-way
action of the amplifier between.   The physical separation between the
circuits for the two directions also minimizes the possible difficulty associated with having the high-level output of an amplifier in proximity
with the low-power-level lines coming in from the other direction.

   In cables which operate at carrier frequencies, both magnetic and
electric coupling are important.   Separately shielded cables are used for
the two directions of transmission, thus eliminating near-end crosstalk as

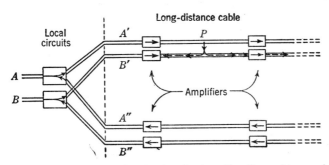

FIG. 8.11.   The use of separate circuits for the two directions of transmission on a
long cable line.

explained in the preceding paragraph, and minimizing the effects caused by
the proximity of lines with different power levels.

   When phantom circuits are used, there is another possibility for crosstalk
which has not been considered.   This is the presence of an unbalance in
resistance, such as would be caused by a faulty joint.   As illustrated in
Fig. 8.1, the independence of the phantom and the side circuits depends
on the phantom currents dividing equally between the two wires of the
side circuit.   If these two paths have different resistances, the phantom
currents will divide unequally, and the difference between them has the
same effect as a true side-circuit current.   The result will be a coupling
between the phantom circuit and the side circuit.   The remedy is, of
course, to keep both sides of the side circuit as nearly balanced as possible.

   The same measures which reduce crosstalk between telephone lines are
also effective in reducing the noise which may be induced into the system
from nearby power lines.   However, since the power level in a power
line is so much greater than that of a telephone line, serious noise may be

introduced into even a well-balanced telephone system. The obvious solution is to keep the two systems well separated, but this is not always feasible. Other effective methods of noise reduction are available, including the transposition of power-line conductors, and the control of the harmonic content of the power-line current and voltage by the use of appropriate transformer connections and the installation of filters. The measures taken to avoid inductive interference between power and communication lines are given the name *inductive coordination*.

CHAPTER 9

SPECIAL CONSIDERATIONS FOR POWER LINES

9.1. Losses and Efficiency. In a communication system, the impedance of the load is generally matched to that of the source in order to obtain the greatest power from the available equipment. The efficiency of power transfer is then only 50 per cent, but the total amount of power is comparatively small, and so the price paid for the higher output is not too great. On the other hand, a power system must handle large amounts of energy, and consequently it is essential to have a high efficiency of transmission to avoid both excessive heating and the uneconomical wastage of large amounts of energy. Therefore, a power system is operated with a load impedance that is much greater than the internal impedance of the source.

The loss in the insulators of a power transmission line is generally much smaller than the I^2R loss of the conductors. The transmission losses can be diminished by reducing the current, and this, for a given amount of power, means that the voltage must be increased correspondingly. Longer transmission lines, where the losses become increasingly serious, are operated at higher voltages than short lines.

9.2. Long and Short Lines. Since the wavelength on an air-insulated line at 60 cps is of the order of 3,000 miles, the longest power transmission lines are only a fraction of a wavelength long. On the longer lines, however, the distribution of the parameters must be taken into account in calculation. Simpler approximate methods are appropriate for short lines.

A power line is said to be "short" if its distributed capacitance can be ignored completely. This can generally be done for lengths up to perhaps 25 or 30 miles. On lines of such short length, the current taken by the line capacitance—called *charging current* in this connection—is negligible. The properties of the line can then be computed by considering only its series resistance and inductance. For a balanced three-phase system, the computations are made for a single phase, using the line current and the line-to-neutral voltage. This is illustrated in Fig. 9.1. The equivalent circuit for the line consists of a lumped resistance in series with a lumped inductance, and the calculations proceed according to elementary circuit theory. The resistance in the equivalent circuit is equal to Rl, where R is the resistance per mile of one conductor and l is the length of the line in miles. The inductance in the equivalent circuit is equal to Ll, where L is the inductance per mile which is associated with one phase of the three-phase line.

As the length of line increases, the effect of the distributed capacitance increases rapidly, for not only does the charging current increase, but also it flows through a larger series impedance. The computations for long lines can be made from the basic transmission-line relations developed in Chap. 4. A convenient method is to use the π form of the equivalent network, which was developed in Sec. 4.8 and which is illustrated in Fig. 9.2. This network is used on a per-phase basis, using the line current and the line-to-neutral voltage. Since there is only a single frequency, the elements of the equivalent network are constant for a given line. The shunt conductance of the line, G, is generally neglected. Then the characteristic impedance is given by

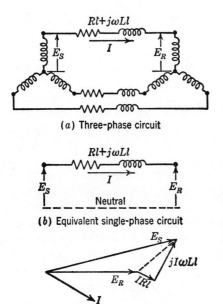

(a) Three-phase circuit

(b) Equivalent single-phase circuit

(c) Vector diagram

FIG. 9.1. Equivalent circuits and vector diagram for a short power line.

$$Z_0 = \sqrt{\frac{R + j\omega L}{j\omega C}} \quad \text{ohms}$$

and the propagation constant is

$$\gamma = \sqrt{(R + j\omega L)(j\omega C)} \quad \text{per mile.}$$

9.3. The Inductance and Capacitance of Three-phase Lines. We shall assume that the line consists of three conductors placed symmetrically at the corners of an equilateral triangle, as indicated in Fig. 9.3. The effect of the earth will be neglected. Balanced three-phase currents will be assumed.

The inductance which should be associated with one phase of the three-phase line can be found by using superposition and the results obtained for a two-wire line in Sec. 3.8. In a balanced three-phase system, we have $I_a + I_b + I_c = 0$. Hence, the current in line a is $I_a = -I_b - I_c$. Now we consider the effect of the current I_b in line b and the equal and opposite component of current

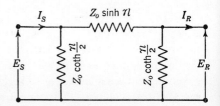

FIG. 9.2. The equivalent π network for a long transmission line.

$I_a' = -I_b$ that flows in line a, as shown in Fig. 9.4a. The flux caused by these currents is the same as that caused by a two-wire line and, from Eq. (3.54), the external flux linkages of this circuit are

$$\psi = \frac{\mu I_a'}{\pi} \log_\epsilon \frac{D}{a}$$

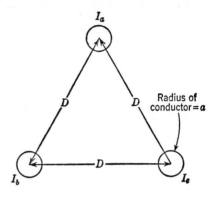

webers/meter length

where μ is the permeability of the insulating medium in henrys per meter, D is the wire spacing, and a is the radius of the wire. Only half of this flux surrounds conductor a, and so, for phase a and the component of current I_a', we have the external flux linkages

$$\psi_a' = \frac{\mu I_a'}{2\pi} \log_\epsilon \frac{D}{a}$$

webers/meter (9.1)

FIG. 9.3. A three-phase line with equilateral spacing.

In a similar manner we consider the effect of the current I_c in conductor c and the corresponding component of current in conductor a, $I_a'' = -I_c$.

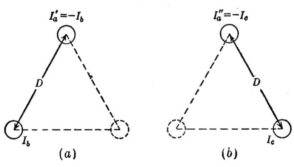

FIG. 9.4. For determining the inductance associated with phase a. $I_a = I_a' + I_a''$.

The external flux linkages of phase a caused by this current are

$$\psi_a'' = \frac{\mu I_a''}{2\pi} \log_\epsilon \frac{D}{a} \qquad \text{webers/meter} \tag{9.2}$$

The total flux linkages of conductor a are the sum of Eqs. (9.1) and (9.2), and so we have

$$\psi_a = \psi_a' + \psi_a'' = \frac{\mu(I_a' + I_a'')}{2\pi} \log_\epsilon \frac{D}{a}$$

$$= \frac{\mu I_a}{2\pi} \log_\epsilon \frac{D}{a} \qquad \text{webers/meter} \tag{9.3}$$

The inductance is equal to the flux linkages per ampere. Dividing Eq. (9.3) by I_a and substituting the permeability of free space ($\mu = \mu_0 = 4\pi \times 10^{-7}$ henry/meter), we have for the inductance of one phase, caused by the flux external to the conductor,

$$L_{\text{ext}} = 2 \times 10^{-7} \log_\epsilon \frac{D}{a} \quad \text{henrys/meter length} \quad (9.4)$$

The inductance caused by the flux within the conductor will depend upon the construction of the conductor; for example, a hollow conductor will have a smaller internal flux than a simple stranded one. Also, the skin effect may have some influence even at 60 cps. Assuming the special case of a solid cylindrical conductor with a relative permeability of unity and negligible skin effect, the flux within the conductor adds an inductance of 0.5×10^{-7} henry/meter (see Eq. 3.58). Then the total inductance of one phase will be

$$L = \left(2 \log_\epsilon \frac{D}{a} + \frac{1}{2}\right) \times 10^{-7} \quad \text{henry/meter} \quad (9.5)$$

This can be converted to a mile basis by multiplying by 1,609 meters/mile. Comparison of Eqs. (9.5) and (3.61) shows that the inductance associated with one wire of an equilateral three-phase line is just half the inductance of a two-wire line that has the same conductor size and spacing.

We can determine the expression for the capacitance associated with one phase by using the fact that, in air, a hypothetical lossless line would have a phase velocity $v = 1/\sqrt{LC} = 3 \times 10^8$ meters/sec. Here, L represents the inductance caused by the external flux only. Then we have

$$C = \frac{1}{9 \times 10^{16} L_{\text{ext}}} \quad \text{farads/meter}$$

Substituting from Eq. (9.4), we obtain the expression

$$C = \frac{10^{-9}}{18 \log_\epsilon D/a} \quad \text{farads/meter} \quad (9.6)$$

An equilateral spacing is not generally a convenient one to use in practice. However, a line with unsymmetrical spacing can be balanced by properly transposing the conductors at regular intervals. An equivalent equilateral spacing can be found for the resulting balanced line, and the foregoing formulas will apply to the equivalent equilateral line. In transposition, the conductors exchange positions at regular intervals along the line, so that each conductor occupies every position for equal distances. One cycle of a transposition arrangement is shown in Fig. 9.5. The equivalent equilateral spacing can be shown to be the geometric mean of the three spacings:

$$D = \sqrt[3]{D_{ab}D_{bc}D_{ca}} \qquad (9.7)$$

where D_{ab}, D_{bc}, and D_{ca} are the distances between individual conductors as indicated by the subscripts.[1]

The conductors most generally used for power transmission are either stranded copper, stranded aluminum with a steel core for reinforcement, or hollow copper. Tables of resistance and reactance are available from the manufacturers. The internal inductance indicated in Eq. (9.5) for a solid cylinder will not, of course, be strictly accurate for these more complex

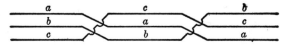

FIG. 9.5. Transposition arrangement to balance an unsymmetrical three-phase line.

conductors. In the tables, the effect of the internal flux upon inductance is indicated by a quantity called the *geometric mean radius* of the conductor. This takes into account the shape of the conductor and the effect of stranding. Its meaning can be understood by considering the special case for which Eq. (9.5) was derived: a solid cylindrical conductor with a relative permeability of unity and negligible skin effect. Equation (9.5) can be arranged as follows:

$$L = 2 \left(\log_\epsilon \frac{D}{a} + \frac{1}{4} \log_\epsilon \epsilon \right) \times 10^{-7}$$

$$= 2 \left(\log_\epsilon \frac{D}{a} + \log_\epsilon \epsilon^{\frac{1}{4}} \right) \times 10^{-7}$$

or

$$L = 2 \times 10^{-7} \log_\epsilon \left(\frac{D}{a\epsilon^{-\frac{1}{4}}} \right) \qquad \text{henry/meter} \qquad (9.8)$$

This is of the same form as Eq. (9.4), which took into account only the flux external to the conductor. The quantity $a\epsilon^{-1/4}$ can be considered as the radius of a new hypothetical conductor which is hollow and therefore has no internal flux, but which provides the same total inductance as the

[1] See L. F. Woodruff, "Principles of Electric Power Transmission," 2d ed., John Wiley & Sons, Inc., New York, 1938; J. G. Tarboux, "Introduction to Electric Power Systems," International Textbook Company, Scranton, Pa., 1944; and M. P. Weinbach "Electric Power Transmission," The Macmillan Company, New York, 1948.

actual conductor. This is shown schematically in Fig. 9.6. If both conductors carry the same current, they will have the same flux beyond the radius a; this is the entire external flux for the solid conductor. In addition, the solid conductor has internal flux linkages which add to its inductance. The hollow conductor has no internal flux, but its external flux between the radii $a\epsilon^{-\frac{1}{4}}$ and a gives it an extra inductance exactly equal to the internal inductance of the solid cylinder. Thus, the total inductance is

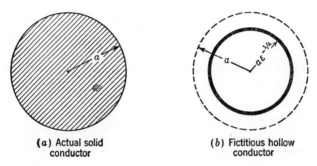

(a) Actual solid conductor   (b) Fictitious hollow conductor

FIG. 9.6. Geometric mean radius of a solid cylinder.

the same for either conductor. The formula for the total inductance of a single conductor in a three-phase configuration can be expressed as

$$L = 2 \times 10^{-7} \log_\epsilon \frac{D}{GMR} \quad \text{henrys/meter} \tag{9.9}$$

where D is the equivalent equilateral spacing of the line, and GMR is the geometric mean radius of the conductor. As shown by Eq. (9.8), a solid cylinder has $GMR = a\epsilon^{-\frac{1}{4}}$; for other conductors, the manufacturer's specifications should be consulted.

9.4. Example: Long Line. Consider a three-phase line which is 225 miles in length and which has an equivalent equilateral spacing of 24 ft. The conductors are made of standed copper with a diameter of 0.811 in. From the manufacturer's data, each conductor has a 60-cycle resistance of 0.120 ohm/mile (this is about 2 per cent higher than the d-c value), and has a geometric mean radius of 0.0256 ft. We shall compute the characteristics of the line, per phase, and shall determine the equivalent π network.

The inductance will be obtained from Eq. (9.9) and will be converted to a mile basis. Then, per conductor,

$$L = 1{,}609 \times 2 \times 10^{-7} \log_\epsilon \left(\frac{24}{0.0256} \right)$$

$$= 2.20 \times 10^{-3} \text{ henry/mile}$$

The capacitance is obtained from Eq. (9.6), using the actual conductor radius of 0.4055 in. = 0.0338 ft. Then, on a mile basis,

$$C = \frac{1,609 \times 10^{-9}}{18 \log_e (24/0.0388)}$$

$$= 0.0136 \times 10^{-6} \text{ farad/mile}$$

The series impedance of one phase per mile is

$$Z = R + j\omega L = 0.120 + j377 \times 2.20 \times 10^{-3}$$

$$= 0.120 + j0.830 \text{ ohms/mile}$$

The shunt admittance associated with one phase, neglecting the leakage conductance, is

$$Y = j\omega C = j377 \times 0.0136 \times 10^{-6}$$

$$= j5.12 \times 10^{-6} \text{ mhos/mile}$$

The characteristic impedance per phase is

$$Z_0 = \sqrt{\frac{Z}{Y}} = \sqrt{\frac{0.120 + j0.830}{j5.12 \times 10^{-6}}}$$

$$= 401 - j29 \text{ ohms}$$

The propagation constant is

$$\gamma = \sqrt{ZY} = \sqrt{(0.120 + j0.830)\,(j5.12 \times 10^{-6})}$$

$$= 1.48 \times 10^{-4} + j2.06 \times 10^{-3} \text{ per mile}$$

Thus, the attenuation constant is 1.48×10^{-4} neper/mile, and the phase constant is 2.06×10^{-3} rad/mile. The phase velocity is $v = \omega/\beta = 183,000$ miles/sec, and the wavelength is $\lambda = v/f = 3,050$ miles.

To determine the equivalent π network for the 225-mile line, we refer to Fig. 9.2. We have

$$\alpha l = 1.48 \times 10^{-4} \times 225 = 0.0333 \text{ neper}$$

and

$$\beta l = 2.06 \times 10^{-3} \times 225 = 0.464 \text{ rad} = 26.6°$$

Using the hyperbolic identity (4.37), we compute

$$\sinh \gamma l = \sinh \alpha l \cos \beta l + j \cosh \alpha l \sin \beta l$$

$$= \sinh 0.0333 \cos 26.6° + j \cosh 0.0333 \sin 26.6°$$

$$= 0.0298 + j0.450$$

The series branch of the equivalent network has the impedance

$$Z_0 \sinh \gamma l = (401 - j29)(0.0298 + j0.450)$$
$$= 25.0 + j179.8 \text{ ohms}$$

For the shunt branches of the equivalent network, we use the hyperbolic identities (4.37) and compute

$$\sinh \frac{\gamma l}{2} = 0.0163 + j0.231$$

and

$$\cosh \frac{\gamma l}{2} = 0.976 + j0.004$$

Then we have

$$Z_0 \coth \frac{\gamma l}{2} = (401 - j29)\left(\frac{0.976 + j0.004}{0.0163 + j0.231}\right)$$
$$= 4 - j1{,}699 \text{ ohms}$$
$$\approx -j1{,}699 \text{ ohms}$$

The shunt branch is very nearly a pure capacitive reactance. The resulting equivalent network is shown in Fig. 9.7.

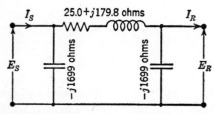

Since the line is less than a tenth of a wavelength long, it should behave nearly the same as though its elements were lumped. The series branch of the equivalent circuit should not be greatly different from the total series impedance of the line, and the shunt branch should be nearly the same as would be obtained from the total shunt capacitance of the line. This will provide a rough check on the results. The total series impedance of the line is

Fig. 9.7. Equivalent π network for the example.

$$Zl = (0.120 + j0.830) \times 225 = 27.0 + j187 \text{ ohms}$$

The total shunt admittance of the line is $Yl = j5.12 \times 10^{-6} \times 225 = j1.152 \times 10^{-3}$ mho. If we associate half of this with each end of the line, the reactance of each half will be $-j1{,}736$ ohms. These impedances are not greatly different from those of the equivalent circuit. If the line were a greater portion of a wavelength long, the elements of the equivalent network would not be so nearly equal to the total series impedance and shunt admittance of the line.

9.5. Circle Diagrams. In this section we shall develop a convenient graphical method for computing the terminal relations of a power line under the conditions of a varying load. The result will be a circle diagram which is plotted in terms of power. Circle diagrams can be drawn for both the sending and the receiving ends, but we shall discuss here only the one for the receiving end. Further information can be obtained from the references given in Sec. 9.3.

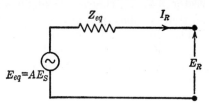

FIG. 9.8. The Thévenin equivalent of a transmission line as viewed from the receiving end.

The derivation will be based on Thévenin's theorem. This theorem states that, when viewed from its output terminals, any linear network behaves like a single voltage generator connected in series with a single impedance (see Sec. 10.7). This is illustrated in Fig. 9.8. The equivalent voltage E_{eq} is equal to the open-circuit voltage of the network; for the transmission system this will be proportional to the sending-end voltage, E_s. It is denoted by $E_{eq} = AE_s$ in Fig. 9.8, where A is a complex number. The equivalent impedance Z_{eq} is equal to the impedance that would be measured between the output terminals if the internal sources of emf were short-circuited. Both E_{eq} and Z_{eq} can readily be computed from the equivalent π network of Fig. 9.2. They will be regarded as known quantities in the derivation that follows.

From the Thévenin equivalent circuit of Fig. 9.8, we can write the following voltage equation:

$$AE_s - I_R Z_{eq} = E_R$$

Solving for the receiving-end current, we obtain the relation

$$I_R = -\frac{E_R}{Z_{eq}} + \frac{AE_s}{Z_{eq}} \qquad (9.10)$$

The vector diagram corresponding to this is shown in Fig. 9.9, with the receiving-end voltage taken along the real axis. The vector current I_R is equal to the sum of the two vectors $-E_R/Z_{eq}$ and AE_s/Z_{eq}. The angle between the voltages at the two ends of the line is the same as the angle between AE_s/Z_{eq} and AE_R/Z_{eq}. The first of these appears on the vector diagram; the angle of the second differs from that of the vector E_R/Z_{eq} by the angle of the complex number A, and its direction is shown by the dashed line in the illustration. The angle between E_s and E_R is denoted by θ_E.

Now we shall assume that both E_R and E_s are fixed in magnitude, but that the angle θ_E between them is variable. In the diagram, the vectors $-E_R/Z_{eq}$ and AE_s/Z_{eq} are now fixed in magnitude. The first of these is also fixed in angle by the fact that E_R is along the reference axis, but the vector AE_s/Z_{eq} changes its angle with a change in θ_E. As θ_E varies, the locus of possible values of I_R is a circle, as shown in Fig. 9.9.

The three components of the vector triangle in Fig. 9.9 are currents. If each of these is multiplied by the magnitude of the receiving-end voltage, the locus of the tip of I_R is transformed into a locus of receiving-end power,

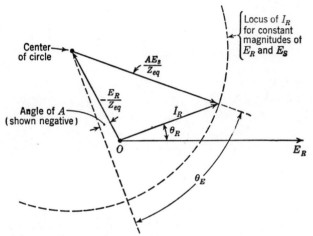

FIG. 9.9. The circular locus of the receiving-end current.

with the horizontal component representing real power in watts (P_R) and the vertical component representing reactive volt-amperes (Q_R). The resulting diagram is shown in Fig. 9.10. A negative value of Q_R corresponds with a lagging, or inductive, current.

For given values of E_s and E_R there is a maximum amount of power which can be transferred to the receiving end. This is given by the maximum horizontal coordinate of the power circle. A larger value of E_s will increase the radius of the circle; a larger value of E_R will move the center of the circle away from the origin and also increase its radius.

If E_s and E_R are expressed in line-to-neutral volts, the coordinates of Fig. 9.10 will represent watts per phase and reactive volt-amperes per phase, respectively. But if E_s and E_R are expressed in line-to-line volts, which are $\sqrt{3}$ times as large as the phase voltages, the power will be larger by a factor of 3 and represent total three-phase watts and volt-amperes.

The Thévenin equivalent shown in Fig. 9.8 produces the correct results in so far as the terminal relations of the network are concerned, but the

power loss in the equivalent impedance of the Thévenin circuit is not the
same as the power loss in the actual network which it replaces. The
Thévenin circuit is designed to produce the correct exterior relations, not to
reproduce interior phenomena. If one wishes to compute the power loss
of a long transmission line, he can do so from the π network of Fig. 9.7,
but not from the two-terminal Thévenin equivalent of Fig. 9.8.

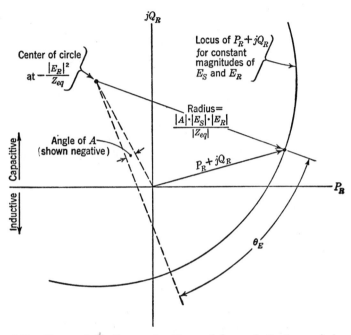

Fig. 9.10. Power circle diagram for the receiving end of a transmission line.

Example. Construct the power-circle diagram for the receiving end
of the line specified in Sec. 9.4, for which the equivalent π network is
given in Fig. 9.7. Use $|E_s| = 250$ kv line to line, $|E_R| = 210$ kv line
to line.

First, we shall determine from Fig. 9.7 the elements of the Thévenin
equivalent as viewed from the receiving end. To find E_{eq}, we imagine that
the receiving end is open-circuited. Then the receiving-end voltage is

$$E_{eq} = \frac{-j1,699}{25 - j1,519} E_s = (1.12\underline{/-0.9^\circ})E_s$$

Since $E_{eq} = AE_s$, we have

$$A = 1.12\underline{/-0.9^\circ}$$

The impedance Z_{eq} is equal to the impedance that would be measured between the receiving-end terminals when E_s is short-circuited. This is

$$Z_{eq} = \frac{-j1,699(25 + j179.8)}{-j1.699 + 25 + j179.8} = 31.3 + j200 \text{ ohms}$$

In the receiving-end power-circle diagram, the center of the circle is at

$$-\frac{|E_R|^2}{Z_{eq}} = \frac{(210 \times 10^3)^2}{31.3 + j200}$$

$$= (-34 + j215) \times 10^6 \text{ total three-phase volt-amperes}$$

The radius of the circle is

$$\frac{|A| \cdot |E_s| \cdot |E_R|}{|Z_{eq}|} = \frac{1.12 \times 250 \times 210 \times 10^6}{202}$$

$$= 291 \times 10^6 \text{ total three-phase volt-amperes}$$

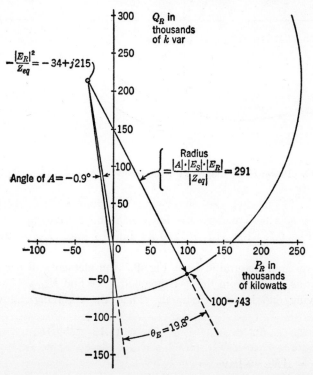

FIG. 9.11. Example of a power-circle diagram for the receiving end of a transmission line.

The resulting diagram is shown in Fig. 9.11. If an output of 100,000 kw (100 megawatts) is desired, the corresponding reactive power for these particular voltages is 43,000 kva in the inductive direction. If the actual reactive kva requirements of the load are different from this value, a synchronous condenser may be used to supply the difference. The phase angle between E_s and E_R under the assumed conditions is $\theta_E = 19.8°$.

PROBLEMS

1. A three-phase power line is 10 miles long and operates at 60 cps. The equivalent equilateral spacing of the conductors is 3 ft. The conductors are No. 4/0 stranded copper, with a diameter of 0.552 in. and a geometric mean radius of 0.0175 ft. The resistance is 0.278 ohm/mile for one conductor.

a. Compute the inductance of the line and the total impedance per phase.

b. The line is to supply 2,400 kw at 80 per cent power factor (current lagging). The receiving-end voltage is to be 13,200 volts rms line to line. Compute the required sending-end voltage (line to line). Determine the power loss in the line and the efficiency of transmission.

2. The three conductors of a certain three-phase power line are placed so that the centers of the conductors fall on the same straight line. The distance between adjacent conductors is 30 ft. Find the equivalent equilateral spacing.

3. A certain 50-mile three-phase power line has an equivalent equilateral spacing of 8 ft, and operates at 66 kv line to line. The conductors are 250,000-cir-mil stranded copper with an outside diameter of 0.574 in. and a geometric mean radius of 0.0181 ft at 60 cps. The resistance per conductor is 0.235 ohm/mile. Compute the series impedance associated with one phase per mile of line, also the shunt admittance per mile, all at 60 cps.

4. Using the series expansions for the hyperbolic functions, show that, for $\gamma l << 1$, the branches of the equivalent circuit of Fig. 9.2 can be expressed, approximately, as

$$Z_0 \sinh \gamma l \approx Zl \left(1 + \frac{ZYl^2}{6}\right)$$

and

$$Z_0 \coth \frac{\gamma l}{2} \approx \frac{2}{Yl} \left(1 + \frac{ZYl^2}{12}\right)$$

where $Z = R + j\omega L$, the series impedance per mile
$Y = j\omega C$, the shunt admittance per mile
(These approximations are sufficiently accurate for 60-cps lines up to perhaps 30 miles in length. See "Electrical Transmission and Distribution Reference Book," by the Central Station Engineers of the Westinghouse Electric and Manufacturing Company, p. 45.)

5. A three-phase 60-cps power transmission line has an equivalent equilateral spacing of 30 ft. The length of the line is 200 miles. The conductors are made of 600,000-cir-mil hollow copper, with a diameter of 1.558 in., and a geometric mean radius of 0.0615 ft. The resistance per conductor is 0.0960 ohm/mile.

a. Compute the inductance and capacitance of the line, and determine the π equivalent network.

b. Determine the Thévenin equivalent of the line as viewed from the receiving end, and draw the power-circle diagram for the receiving end. The receiving-end voltage is to be 190,000 volts line to line, and the sending-end voltage is to be 230,000 volts line to line.

c. Using the foregoing circle diagram, determine the reactive volt-amperes and the angle between E_R and E_s for a receiving-end power of 150,000 kw. Also, determine the maximum power output and the resulting angle between E_R and E_s.

d. Use the π network and the results of part *c* to find the line loss and the efficiency of transmission at a power output of 150,000 kw. 90.4 %

Part II
FOUR-TERMINAL NETWORKS

CHAPTER 10

A REVIEW OF ELEMENTARY NETWORK ANALYSIS

THE MATERIAL of this chapter is given mainly for review and reference. More detailed treatments can be found in various books on elementary circuit theory.[1]

10.1. Network Definitions. An electrical network consists of an interconnected set of inductances, resistances, and capacitances, together with sources of emf. A network is said to be passive if it contains no sources or sinks of electrical energy within its boundaries, otherwise it is said to be active. If the resistance, inductance, or capacitance of a given element is independent of the current passing through it, that element is said to be linear. Then, at a given frequency, the current through the element will be linearly proportional to the voltage across its terminals. A nonlinear element is one which has a nonlinear relation between voltage and current at a given frequency; this leads to greatly increased difficulties in analysis. We shall be concerned here only with linear circuits.

Figure 10.1 illustrates some of the terms used in the description of network geometry. A junction point between elements in a network, as at point a or point c, is called a node. An element joining two nodes, such as the element ac, is called a branch. Sometimes it is convenient to think of point b as a node and the elements ab and bc as two separate branches. For other purposes it is convenient to consider the group abc as a single branch,

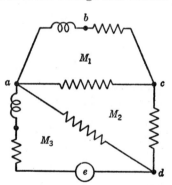

FIG. 10.1. Illustrating network terminology.

in which case point b is not considered a node. Any closed path through the network is called a loop; thus, the paths $abca$ and $abcda$ are loops. The word mesh has its ordinary meaning: it designates the open spaces which appear in the drawing of the network. Thus, the spaces M_1,

---

[1] For example, see "Electric Circuits," by the M.I.T. Staff, John Wiley & Sons, Inc., New York, 1940; "Elementary Electric-circuit Theory," by R. H. Frazier, McGraw-Hill Book Company, Inc., New York, 1945; "Alternating-current Circuit Theory," by Myril B. Reed, Harper & Brothers, New York, 1948.

M_2, and M_3 are the meshes of the network in Fig. 10.1. The shortest closed paths through the network form the contours of the meshes. The location of the meshes depends on the way the circuit is drawn; for example, the elements bordering the meshes M_1 and M_2 would be altered by interchanging the positions of the branches abc and ac in Fig. 10.1. Thus, although a mesh is a convenient concept, it is not a fundamental entity of the circuit. A network which can be drawn on a flat surface without crossovers is sometimes called a "flat" network. The network shown in Fig. 10.2 would cease to be flat if, for example, an additional element were joined between the mid-points of branches Z_a and Z_e. The concept of a mesh loses its simplicity and much of its usefulness in nonflat networks.

10.2. Loop and Node Equations. Kirchhoff's laws are used as the basis for setting up the equations which describe the performance of a network. These laws state that at every instant (1) the algebraic sum of the currents approaching any node is zero and (2) the algebraic sum of the voltages

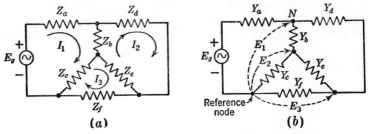

FIG. 10.2. Illustrating mesh currents and node voltages.

around any closed loop is equal to zero. The two laws are applied to the steady-state analysis of linear a-c circuits by summing the complex (or "vector") currents at each node and the complex voltages around each loop (see Sec. 2.1). The currents and voltages are assigned positive directions by means of arrows or polarity marks, as in Fig. 10.2. The instantaneous current alternates in direction, of course, and so is positive in one half cycle (flows in the direction of the arrow) and is negative in the next half cycle (flows opposite to the arrow).

There are two particularly convenient methods of applying Kirchhoff's laws to a given network. One of these uses currents as the unknowns and is called the loop-current or the mesh-current method. The other uses voltages as the unknowns and is known as the node-voltage method.

In the loop-current method, the unknowns are chosen to be currents which are visualized as flowing around closed loops in the network. For flat networks, the chosen paths are generally the contours of the meshes, as illustrated in Fig. 10.2a. If an element forms a mutual boundary between two meshes, as Z_b in the illustration, both mesh currents contribute to the actual current in the element.   Thus, the current flowing downward through Z_b is $I_1 - I_2$.   This automatically causes Kirchhoff's current law to be satisfied at each junction.   For a nonflat network, where the location of the meshes is not obvious, the required number of unknown currents that must be assigned can be determined from the relation

No. of independent loops = No. of branches − No. of of nodes + 1   (10.1)

If the network has two or more separate parts which are coupled only by mutual inductance, the number of independent loops can be found separately for each part, then added to find the total.   In a flat network, the number of independent loops is equal to the number of meshes, or open spaces, in the circuit diagram.   When the unknown currents have been assigned, an equal number of voltage equations can be written around the loops, making sure to include each element at least once in the group of equations.   For flat networks, this is generally done by writing one voltage equation around the contour of each mesh.   For example, in Fig. 10.2a, we would write for the first mesh:

$$E_g - I_1 Z_a - (I_1 - I_2)Z_b - (I_1 - I_3)Z_c = 0$$

or, collecting terms,

$$(Z_a + Z_b + Z_c)I_1 - Z_b I_2 - Z_c I_3 = E_g \qquad (10.2)$$

Similarly, two other independent voltage equations can be written, one around each of the other two meshes.   These equations are

$$-Z_b I_1 + (Z_b + Z_d + Z_e)I_2 - Z_e I_3 = 0 \qquad (10.3)$$

and

$$-Z_c I_1 - Z_e I_2 + (Z_c + Z_e + Z_f)I_3 = 0 \qquad (10.4)$$

The three equations can then be solved for the three unknown currents. Because of the linearity of the circuit, the Z's are independent of current, and the result is a set of linear algebraic equations with complex coefficients.

Observe in Eq. (10.2), which was written for the first mesh, that the coefficient of I_1 is the total impedance around the loop.   The coefficient of I_2 is the mutual impedance between meshes 1 and 2, with a negative sign prefixed because the positive directions of I_1 and I_2 are opposite through the mutual element.   The coefficient of I_3 is the mutual impedance between meshes 1 and 3, again with a negative sign because I_1 and I_3 are shown flowing oppositely through the mutual link.   Similar considerations hold for the equations written for the other two meshes.

A second method of applying Kirchhoff's laws to a network is to assign voltages between nodes as the unknowns, and to write the required equations from Kirchhoff's current law at the junction points. One node can be chosen to be the reference, as indicated in Fig. 10.2b, with the voltages to other nodes measured from this one. Obviously, the number of unknown voltages is one less than the number of nodes. One current equation is written for each of the nodes at which a voltage is assigned. For example, consider node N in Fig. 10.2b. According to Kirchhoff's voltage law, the voltage acting downward across Y_b is $E_1 - E_2$. The current flowing through this element away from node N is equal to $(E_1 - E_2)Y_b$, where Y_b is the admittance of the branch and is equal to $1/Z_b$. Similarly, the currents flowing through the admittances Y_a and Y_d are, respectively, $(E_1 - E_g)Y_a$ and $(E_1 - E_3)Y_d$. According to Kirchhoff's current law, we can write the algebraic sum of the currents flowing away from node N as

$$(E_1 - E_g)Y_a + (E_1 - E_2)Y_b + (E_1 - E_3)Y_d = 0$$

This equation can be arranged more conveniently in the form

$$(Y_a + Y_b + Y_d)E_1 - Y_bE_2 - Y_dE_3 = Y_aE_g \tag{10.5}$$

Similarly, a current equation can be written at each of the other two nodes, excluding the reference node. These equations are

$$-Y_bE_1 + (Y_b + Y_c + Y_e)E_2 - Y_eE_3 = 0 \tag{10.6}$$

and

$$-Y_dE_1 - Y_eE_2 + (Y_d + Y_e + Y_f)E_3 = 0 \tag{10.7}$$

We now have three independent linear equations which can be solved for the three unknown voltages. Once the voltages are known, the current through any element can be computed.

The mesh-current method is the older of the two and is perhaps used more frequently than the node-voltage method, but neither has a fundamental advantage over the other. The circuit of Fig. 10.2 has three unknowns either way, but another network may give an advantage of fewer unknowns to one method or the other, depending on the network configuration. The mesh-current method will be adequate for most of the purposes of this book, and so we shall concentrate on it for the time being.

The voltage equations for a network with n meshes can be written as

$$\left. \begin{array}{l} z_{11}I_1 + z_{12}I_2 + \cdots + z_{1n}I_n = E_1 \\ z_{21}I_1 + z_{22}I_2 + \cdots + z_{2n}I_n = E_2 \\ \cdots\cdots\cdots\cdots\cdots\cdots\cdots\cdots\cdots\cdots \\ z_{n1}I_n + z_{n2}I_2 + \cdots + z_{nn}I_n = E_n \end{array} \right\} \tag{10.8}$$

Here z_{11} represents the total impedance around the contour of mesh 1. The quantity z_{12} is the impedance which is held mutually between meshes 1 and 2, and it is assigned a positive or negative sign depending on whether the two mesh currents flow in the same or in opposite directions through the mutual element. Similarly with the other impedances: two like subscripts denote the total impedance around a mesh; two unlike subscripts denote a mutual impedance between two meshes. The voltages E_1, E_2, $\cdots$, E_n represent impressed emfs in the respective meshes. The positive signs associated with the emfs in Eqs. (10.8) are correct if each emf has its positive direction oriented so that it aids the current flowing in that mesh, otherwise the sign of the emf must be taken as negative. The mutual impedance z_{12} which appears in the first equation is the same impedance that is labeled z_{21} in the second equation. In general, for the type of system that we are considering, we have $z_{jk} = z_{kj}$.

10.3. Driving-point and Transfer Impedances. The solution of a linear system of equations such as (10.8) can be systematized by using determinants. This method does not necessarily shorten the work required in making numerical computations for a given problem, but it is of great aid in determining various important properties of networks in general. For simplicity, we shall now assume that all the impressed emfs are zero except E_1. The theory of determinants shows that the solutions for the current in the first mesh and for the current in the kth mesh can be expressed, respectively, as

$$I_1 = \frac{A_{11}}{D} E_1 \qquad (10.9)$$

and

$$I_k = \frac{A_{1k}}{D} E_1 \qquad (10.10)$$

where D is the determinant formed from the coefficients of the left side of Eq. (10.8) and is

$$D = \begin{vmatrix} z_{11} & z_{12} & \cdots & z_{1n} \\ z_{21} & z_{22} & \cdots & z_{2n} \\ \cdots\cdots\cdots\cdots\cdots \\ z_{n1} & z_{n2} & \cdots & z_{nn} \end{vmatrix} \qquad (10.11)$$

The quantities labeled A are determinants derived from the determinant D and are called cofactors. A cofactor A_{1k} is the determinant that remains if the first row and kth column of the determinant D are canceled out, this determinant to be prefixed with the sign $(-1)^{1+k}$. (The cofactor without the prefixed sign is called the minor; the latter name is often used loosely for either quantity.)   Since we have $z_{jk} = z_{kj}$, the determinant (10.11) is symmetrical about its principal diagonal.

The input impedance seen by the source of emf is often called the *driving-point impedance*.   From Eq. (10.9), this impedance can be expressed in determinant notation as

$$Z_{11} = \frac{E_1}{I_1} = \frac{D}{A_{11}} \tag{10.12}$$

The ratio of the voltage impressed in one mesh to the resulting current in another mesh is called the *transfer impedance* between the two meshes. The transfer impedance from mesh 1 to mesh k can be expressed by means of Eq. (10.10) as

$$Z_{1k} = \frac{E_1}{I_k} = \frac{D}{A_{1k}} \tag{10.13}$$

Sometimes it is simpler to use the reciprocals of the foregoing impedances. The reciprocals are called the driving-point and transfer admittances, respectively.

Unfortunately, there is some possibility for confusion in the use of the name "transfer impedance," for it can be used in two different ways. In the foregoing definition, we regarded the driving-point voltage as the cause and a current in some other part of the network as the effect.   Sometimes, however, it is convenient to regard the current entering the driving-point terminals as the cause and the voltage across some interior part of the network as the effect.   The ratio between these two quantities is a "transfer" quantity and has the dimensions of an impedance, but it is not equal to Eq. (10.13) and in fact has an entirely different meaning.   To avoid confusion, either additional descriptive words should be used, or the context should make the meaning clear.

10.4. The Principle of Superposition.   The general principle of superposition applies to linear systems of all types.   Applied to electrical networks, it can be stated as follows: If a number of electrical sources exist in a linear network, the resulting current or voltage in any part of the system is equal to the sum of the currents or voltages that would be caused by each source acting separately.

The principle can be demonstrated with the system of equations (10.8). Suppose that the voltage E_1 is made up of two components, E_1' and E_1'', so that $E_1 = E_1' + E_1''$.   Also, suppose that all the other emfs are zero except one other in the jth mesh, E_j.   From the theory of determinants, the solution for the current in the kth mesh is given by

$$I_k = \frac{A_{1k}}{D} E_1' + \frac{A_{1k}}{D} E_1'' + \frac{A_{jk}}{D} E_j \tag{10.14}$$

Each of these terms is the current that would be caused by one of the voltages acting alone in the network, and the total current is the sum of the separate contributions.

The principle of superposition is useful not only for computing the total current caused by sources located in different loops, but also for determining the net effect of sources of different frequency which are applied simultaneously in one loop.   If an applied voltage is nonsinusoidal, it can be resolved by Fourier analysis into sinusoidal components with different frequencies.   In a linear system, the effect of each component can be computed separately, and the net effect is found as the sum of the individual effects.

The methods of determinants apply only to linear equations.   If a system is nonlinear, the net current resulting from two or more sources is entirely different from the sum of the currents that would be produced by each source acting separately, and the principle of superposition does not apply.

10.5. The Reciprocity Theorem.   Suppose that all emfs in a network are zero except E_1.   Then the current in the kth mesh will be

$$I_k = \frac{A_{1k}}{D} E_1 \qquad (10.15)$$

On the other hand, if an emf E_k acted in the kth mesh and all the other emfs were zero, the current in the first mesh would be

$$I_1 = \frac{A_{k1}}{D} E_k \qquad (10.16)$$

Now, the cofactor A_{1k} is obtained by canceling out the first row and kth column of the determinant (10.11) and prefixing the sign $(-1)^{1+k}$.   The other cofactor, A_{k1}, is obtained by canceling out the kth row and the first column, then prefixing the sign $(-1)^{k+1}$.   But these two operations produce identical results, for the general relation $z_{jk} = z_{kj}$ makes the elements of any particular column equal to those of the corresponding row.   Hence, the transfer impedance from one mesh to another is the same in either direction.

The result is often expressed as follows: If a voltage E applied in one mesh causes a current I to flow in the kth mesh, then the voltage E applied in the kth mesh would cause a like current I to flow in the first mesh.

10.6. Equivalent Voltage and Current Sources.   The idealized concept of an impedanceless source of constant emf is a familiar one.   Also useful is the concept of a current source which supplies a current of constant amplitude regardless of the voltage into which it operates.   A constant emf in series with an impedance is precisely equivalent at its terminals to a source of current that is shunted with an impedance, as indicated in Fig. 10.3. In circuit analysis, and particularly in using equivalent circuits, it is often convenient to replace one of these representations by the other.

The terminal equivalence of the two sources in Fig. 10.3 is easily shown. The terminal voltage in Fig. 10.3a can be expressed as

$$E_t = E_g - I_t Z_g \qquad (10.17)$$

where I_t is the current flowing at the terminals. On the other hand, the relations satisfied by the circuit of Fig. 10.3b can be described by writing Kirchhoff's current law at the node:

$$I_t = \frac{E_g}{Z_g} - \frac{E_t}{Z_g} \qquad (10.18)$$

But if this equation is multiplied through by Z_g and rearranged, it is identical with (10.17); therefore, the terminal conditions of the two circuits are described by identical equations and the circuits are equivalent.

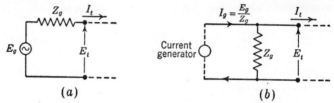

(a)   (b)

FIG. 10.3. Equivalent voltage and current sources.

Although the two circuits produce identical external results, their power relations are quite different in the interior. To illustrate this, observe that the voltage generator will dissipate zero power in its internal impedance when it operates into an open circuit. On the other hand, the current generator will dissipate zero power in its interior when its terminals are short-circuited.

10.7. Thévenin's Theorem. In many practical applications, a linear network which is energized by one or more sources of emf is utilized by making a connection to two output terminals. Thévenin's theorem states that, at its terminals, such a network cannot be distinguished from a single source of emf in series with a single impedance (see Fig. 10.4). The equivalent emf has a value equal to the open-circuit voltage between the terminals; the equivalent impedance is the impedance that would be measured between the terminals if the internal sources of energy were disconnected (emfs short-circuited, current sources open-circuited). If desired, the equivalent voltage source shown in Fig. 10.4b can be replaced by an equivalent current source in the manner outlined in the preceding section.

Although the Thévenin equivalent produces the correct exterior results, its internal power relations may be quite different from those of the network that it replaces. The power dissipated in the equivalent impedance of the Thévenin circuit is not the same as the power dissipated in the resistances of the actual network.

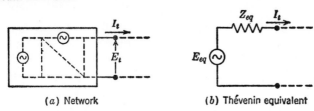

(a) Network   (b) Thévenin equivalent

FIG. 10.4.   Illustrating Thévenin's theorem.

Thévenin's theorem follows directly from the principle of superposition. Suppose that the network is terminated in an impedance Z_t across the terminals ab, as shown in Fig. 10.5. A current I will flow through Z_t. Now imagine that the connection is broken at the point P. Across this break will appear the open-circuit voltage of the network, E_{eq}. A generator having this voltage could be inserted in the break without causing any current to flow (see Fig. 10.5b). Now we apply superposition, and imagine that this zero current is the sum of two equal and opposite components, one the original current I which is caused by the internal generators of the network, and the other caused by the generator inserted in the break, also equal to I. But this latter current is given by

$$I = \frac{E_{eq}}{Z_t + Z_{ab}} \qquad (10.19)$$

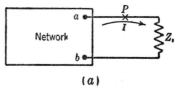

(a)

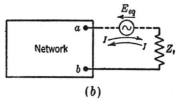

(b)

FIG. 10.5.   Derivation of Thévenin's theorem by superposition.

where Z_{ab} is the impedance presented by the network at the terminals ab. This is equal to the current which flowed in the original circuit of Fig. 10.5a and is expressed in terms of the open-circuit voltage and the network impedance as seen from the terminals. Furthermore, it is precisely the expression that we would compute from the Thévenin equivalent of Fig. 10.4b, with $Z_{eq} = Z_{ab}$. Therefore, the Thévenin equivalent of Fig. 10.4 gives the correct result.

Problem. The *compensation theorem* provides a means of calculating the change of current in any part of a network caused by the introduction of an additional impedance ΔZ in one of the branches. The current that originally flowed in the modified branch will be denoted by I. According to the theorem, the change in current is equal to the current that would be produced by an emf equal to $-I \cdot \Delta Z$ acting in series with the modified branch. Use Thévenin's theorem to show that the compensation theorem gives the correct change of current in the modified branch.

10.8. Maximum Power Transfer. The conditions for maximum power transfer from a source to a load were discussed in Sec. 7.7 under the assumption that the source consisted of a constant emf in series with a fixed impedance, and that the load had an impedance which was adjustable in both magnitude and phase angle. It was shown that maximum power will be absorbed by the load when its impedance is made the complex conjugate of the source impedance. With a conjugate match, the two impedances will have equal resistive components, and the reactances will be equal but of opposite sign. The efficiency of power transfer is only 50 per cent under these conditions, and this would be intolerably wasteful if energy were to be transferred in large quantities. Consequently, power systems are operated with load impedances much greater than the impedance of the source, thus improving the efficiency to a high value. No attempt is made to draw maximum power from the source. But if the power to be handled is comparatively small, as in a communications system, it may be more important to obtain maximum power from the available equipment than to economize on losses, and conditions approaching those of maximum power transfer are generally used.

Sometimes the phase angle of the load is fixed, and only its magnitude can be changed (as by a transformer). Then the greatest power transfer is obtained when the magnitude of the load impedance is made equal to the magnitude of the source impedance. Denote the emf of the source by E, the impedance of the source by $Z_1 = R_1 + jX_1$, and the load impedance by $Z_2 = |Z_2| (\cos \theta + j \sin \theta)$, where θ is the phase angle of Z_2. The power absorbed in the resistance $|Z_2| \cos \theta$ is

$$P_2 = |I|^2 \cdot |Z_2| \cos \theta$$

or

$$P_2 = \frac{E^2 |Z_2| \cos \theta}{(R_1 + Z_2 \cos \theta)^2 + (X_1 + Z_2 \sin \theta)^2} \qquad (10.20)$$

This is maximized by placing $dP_2/d|Z_2| = 0$, and results in the relation

$$|Z_2|^2 = R_1^2 + X_1^2$$

which is the same as

$$|Z_2| = |Z_1| \qquad (10.21)$$

This is the relation that was to be proved.

10.9. The Wye-Delta, or T-π, Transformation.   Two networks are said to be equivalent at a given frequency if electrical tests performed between the terminals at this frequency cannot distinguish between them.   The three-terminal wye and delta networks shown in Fig. 10.6 can be made equivalent to each other by choosing the impedances of one in the proper relationship with the impedances of the other.   To determine the required relations, we assume that terminal voltages E_1 and E_2 are applied as indi-

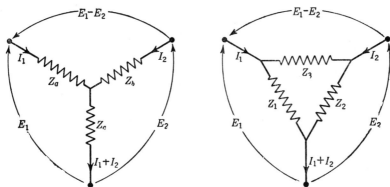

Fig. 10.6.   Equivalent wye and delta networks.

cated in the figure, and then write equations for the currents that enter the terminals.   For the wye network, we have

$$E_1 = (Z_a + Z_c)I_1 + Z_cI_2$$

and

$$E_2 = Z_cI_1 + (Z_b + Z_c)I_2$$

$$(10.22)$$

In the delta network, the current flowing downward through Z_1 is E_1/Z_1, and the current flowing to the right through Z_3 is $(E_1 - E_2)/Z_3$. Therefore, at the upper left node we have

$$\frac{E_1}{Z_1} + \frac{E_1 - E_2}{Z_3} = I_1 \qquad (10.23)$$

Similarly, at the upper right-hand node we have

$$\frac{E_2}{Z_2} - \frac{E_1 - E_2}{Z_3} = I_2 \qquad (10.24)$$

To permit comparison with the equations for the wye network, we solve Eqs. (10.23) and (10.24) for E_1 and E_2.   The results are

$$E_1 = \frac{(Z_1Z_2 + Z_1Z_3)I_1 + Z_1Z_2I_2}{Z_1 + Z_2 + Z_3}$$

and

$$E_2 = \frac{Z_1Z_2I_1 + (Z_2Z_3 + Z_1Z_2)I_2}{Z_1 + Z_2 + Z_3}$$

$$(10.25)$$

A comparison of Eqs. (10.22) and (10.25) shows that the terminal equations for the two networks will be identical if

$$Z_a = \frac{Z_1 Z_3}{Z_1 + Z_2 + Z_3}$$

$$Z_b = \frac{Z_2 Z_3}{Z_1 + Z_2 + Z_3}$$

and

$$Z_c = \frac{Z_1 Z_2}{Z_1 + Z_2 + Z_3}$$

(10.26)

When the impedances of the delta branches are known, the foregoing equations can be used to find the equivalent wye.

FIG. 10.7.   The wye and delta redrawn as the T and π, respectively.

The equations for the reverse transformation can be found by solving Eqs. (10.22) for I_1 and I_2 and comparing the results with Eqs. (10.23) and (10.24).   The comparison shows that equivalence is obtained when

$$Z_1 = \frac{N}{Z_b}$$

$$Z_2 = \frac{N}{Z_a}$$

$$Z_3 = \frac{N}{Z_c}$$

(10.27)

where $N = Z_a Z_b + Z_b Z_c + Z_c Z_a$.

If the foregoing relations are satisfied, the two networks will have identical terminal relationships at the given frequency, and they will behave in the same way if placed in any circuit operating at this frequency.

The main usefulness of an equivalent circuit is in the simplification of calculations, and there is generally no attempt to construct the equivalent physically.   In fact, it would often be impossible to construct the equivalent out of passive elements, for sometimes it requires negative resistances. Also, the branches of the equivalent may be functions of frequency that cannot be obtained with passive elements.   For example, it may be impossible to build a passive wye network that will duplicate the performance of a given physical delta at all frequencies. (A noteworthy exception is a network composed entirely of elements of the same type, in which case an equivalent that is good at all frequencies can be constructed

FIG. 10.8.   Two networks that can be made equivalent at all frequencies (Example 1).

from passive elements.) Although the foregoing peculiarities limit the actual construction of the equivalent circuits, they are no great hindrance when one is merely making calculations. At any one frequency, the impedances of the equivalent circuit will be fixed, and then the equivalent is an especially convenient aid to computation.

For some purposes, it is more convenient to draw the wye and delta networks in the forms shown in Fig. 10.7. Then they are known as the T and π networks, and the relations expressed by Eqs. (10.26) and (10.27) are often called the T-π transformation.

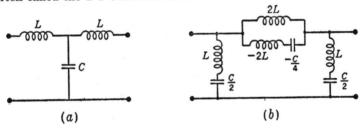

FIG. 10.9.   This π equivalent can be realized physically with passive elements only at a fixed frequency (Example 2).

Example 1.   The T network of Fig. 10.8a has $Z_a = j\omega L_a$, $Z_b = j\omega L_b$, and $Z_c = j\omega L_c$. The application of Eqs. (10.27) shows that the π network of Fig. 10.8b will be precisely equivalent to the T at all frequencies if

$$L_1 = \frac{P}{L_b}, \qquad L_2 = \frac{P}{L_a}, \qquad L_3 = \frac{P}{L_c}$$

where $P = L_aL_b + L_bL_c + L_cL_a$.

Example 2.   The T network of Fig. 10.9a has $Z_a = Z_b = j\omega L$ and $Z_c = -j/\omega C$. The use of Eqs. (10.27) shows that the equivalent π should have

$$Z_1 = Z_2 = j\omega L - \frac{j}{\omega C/2}$$

and

$$Z_3 = j(2\omega L - \omega^3 L^2 C)$$

The second term of Z_3 is a capacitive reactance whose magnitude varies as the cube of the frequency, and this cannot be obtained physically with only passive elements. But if the frequency is fixed, the reactances will be constant and can be duplicated with ordinary inductance and capacitance. The proper variation with frequency can be obtained with the network of Fig. 10.9b, which involves a negative inductance and a negative capacitance.

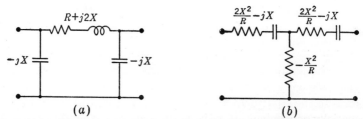

(a)   (b)

FIG. 10.10.   This equivalent T requires a negative resistance (Example 3).

A negative inductance provides a negative (capacitive) reactance whose magnitude is directly proportional to frequency, and a negative capacitance provides a positive (inductive) reactance whose magnitude varies inversely with frequency. Therefore, we conclude that this equivalent π can be realized physically with passive elements only at a fixed frequency.

Example 3.   Figure 10.10a shows a π network that operates at a fixed frequency. The use of Eqs. (10.26) shows that the T equivalent is that shown in Fig. 10.10b. The shunt branch of the T is a negative real impedance, *i.e.*, a negative resistance, and so the equivalent T cannot be realized physically with passive elements even at a fixed frequency.

CHAPTER 11

THE CHARACTERISTICS OF PASSIVE FOUR-TERMINAL NETWORKS

11.1. Introduction. The analysis and synthesis of both power and communication networks are based to a considerable extent on the properties of the passive network which has two entrance terminals and two exit terminals. Such a network is variously called a transmission network, a four-terminal network, a quadripole, a coupling network, or a two-terminal pair. An ordinary two-winding transformer is a four-terminal network, and so is a transmission line. Another important four-terminal network is the filter, which is designed to pass freely only certain bands of frequency. An attenuation equalizer is a network which provides an attenuation that varies in some desired manner with frequency, in order to correct undesirable attenuation *vs.* frequency characteristics in other parts of a system. A phase equalizer has a similar task in producing a desired phase characteristic. An attenuator is a four-terminal network that provides a constant amount of attenuation throughout the whole range of frequencies. An impedance-matching network matches the impedance of a source to that of a load in order to minimize reflections and to produce a good transfer of power.

Each of the foregoing devices is a four-terminal network. Furthermore, a chain of two or more of them connected in cascade forms a larger four-terminal network, the over-all properties of which are related to the properties of the individual networks which compose the whole. In four-terminal network theory, suitable methods of representing the terminal characteristics of a network are first deduced. Next, criteria are developed for connecting a chain of four-terminal networks so that the characteristics of each contribute separately and in a predictable manner to the performance of the whole. This permits different functions to be designed into separate units, a most desirable feature in a complex system. Finally, methods are developed for synthesizing individual networks to perform various functions. This leads into a wide field in which a great deal still remains to be known. Of the various branches of this subject, the filter network is probably the best known and most fully developed, and one of the most useful. We shall consider filter networks later in some detail.

Transmission circuits are almost always operated in one or the other of two limiting relations with respect to ground potential. One of these is referred to as "balanced" operation, the other as "unbalanced" operation In a balanced system, the two sides of the transmission circuit always have equal and opposite potentials with respect to ground. In order to maintain this symmetrical relation, any interconnected apparatus should insert equal impedances in both lines and should present equal impedances to ground. Dipole antennas, ordinary parallel-wire transmission lines, and the transmission circuits shown schematically in Figs. 11.1*b* and *d* are all balanced circuits. An unbalanced circuit, of which a coaxial cable is an

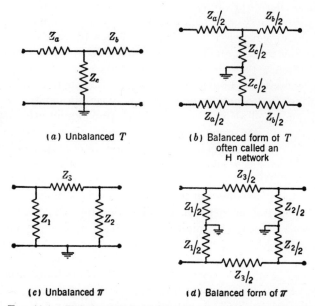

(a) Unbalanced T

(b) Balanced form of T often called an H network

(c) Unbalanced π

(d) Balanced form of π

Fig. 11.1.   Unbalanced and balanced transmission networks.

example, is operated with one side grounded; the path on this side ideally has the same potential all along its length. Figures 11.1*a* and *c* show unbalanced transmission networks. Conversion from a balanced system to an unbalanced one is generally done with a transformer at the lower frequencies.

Generally, the terminals of transmission networks are used in pairs, one pair as the input and one pair as the output. The networks are, therefore, not always as general as a four-terminal network can be, and this is the reason for the various other names that are often used. For example, the network of Fig. 11.1*a* is a true transmission network, but it could get along just as well with only three terminals. In spite of this.

the name "four-terminal network" is often used for transmission networks in general, and we shall use the various names indiscriminately.

When the terminals of the network are used in pairs, the relations between the output and input voltages and currents are the same whether the network is balanced or unbalanced. Since it will not generally make any difference in the computed results, most of our calculations will be referred to the unbalanced type of circuit for convenience.

11.2. Equivalent Networks. A simple method of representing the terminal characteristics of a four-terminal network is by means of a simplified equivalent circuit. In this section we shall show that, at a given frequency, either a T or a π network is a suitable equivalent.

In Fig. 11.2, the box is intended to indicate a concealed network which is linear and passive, but which may be of any degree of complexity. The ends will be numbered 1 and 2, and the terminal voltages and currents will be numbered accordingly.

If we knew the details of the network, we could write a set of equations similar to Eqs. (10.8) to relate the currents in the network, the various impedance elements, and the voltages E_1 and E_2.

FIG. 11.2. A four-terminal network.

The latter voltage will have to be written as $-E_2$ in these equations, for the positive direction of the voltage is shown in Fig. 11.2 as opposing the flow of the current I_2 around the mesh. If we solved the resulting equations for the currents I_1 and I_2, we would obtain relations similar to those in Sec. 10.3:

$$I_1 = \frac{A_{11}}{D} E_1 - \frac{A_{21}}{D} E_2 \qquad (11.1)$$

and

$$I_2 = \frac{A_{12}}{D} E_1 - \frac{A_{22}}{D} E_2 \qquad (11.2)$$

Here, as in Sec. 10.3, the symbol D represents the determinant formed from the impedance coefficients in the equations, and each A represents the cofactor of the row and column indicated by the two subscripts. We also recall that, by the principle of reciprocity, the two cofactors A_{12} and A_{21} are equal. The impedances of the network elements will vary with frequency, and so the determinant and its cofactors will vary with frequency in both magnitude and angle. They will be constants at any one frequency. At the moment, we are not particularly interested in the determinants themselves or in the details of the elements that form them, but they provide a convenient starting point in describing the terminal characteristics of the network.

In the ordinary operation of a communication system, the voltage E_1 will be caused by an impressed emf, but E_2 will be the voltage caused by the current I_2 flowing through a load impedance as yet unspecified. Then $E_2 = I_2 Z_R$, where Z_R is the impedance into which the network operates. In a power system, both E_1 and E_2 may be caused by impressed emfs.

Now, observe in Eqs. (11.1) and (11.2) that the quantities A/D have the dimensions of admittances, and that only three of these are necessary to describe the terminal relations of the network. Next, consider the π network shown in Fig. 11.3. The current I_1 that approaches the left-hand

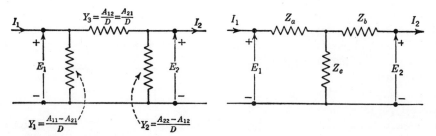

FIG. 11.3. Equivalent π network.   FIG. 11.4. Equivalent T network.

node is equal to the sum of the currents that flow away from the node through the admittances Y_1 and Y_3. The current flowing through Y_1 is

$$Y_1 E_1 = \frac{A_{11} - A_{21}}{D} E_1$$

and that flowing through Y_3 is

$$Y_3(E_1 - E_2) = \frac{A_{21}}{D} (E_1 - E_2)$$

Equating I_1 to the sum of these, we obtain

$$I_1 = \frac{A_{11}}{D} E_1 - \frac{A_{21}}{D} E_2 \qquad (11.3)$$

A similar consideration of the currents at the right-hand node provides the relation

$$I_2 = \frac{A_{12}}{D} E_1 - \frac{A_{22}}{D} E_2 \qquad (11.4)$$

These two equations are precisely the same as Eqs. (11.1) and (11.2); therefore, the π network of Fig. 11.3 is equivalent at its terminals to the network of Fig. 11.2. The admittances Y_1, Y_2, and Y_3 will, in general, be functions of frequency—perhaps complicated functions if the network of Fig. 11.2 is a complicated one. Therefore, it may not be possible to assemble a set of physical elements into the π form so that the π will reproduce the terminal characteristics of the actual network at all frequencies. But, at any one frequency, the elements of the π will be fixed, and then it becomes a convenient equivalent circuit to replace the actual network in calculations. An obvious advantage of the equivalent circuit is the relative ease of visualizing the circuit relations.

For some purposes, it is more convenient to have an equivalent circuit in the form of a T, as shown in Fig. 11.4. The elements of the T can be found from those of the π by means of the π-T, or delta-wye, transformation. This transformation was discussed in Sec. 10.9.

The equivalence between the T or π and the actual network is limited in a way that will not generally be serious in the networks we shall consider. We have visualized the network of Fig. 11.2 as having two input and two output terminals without outside interconnection between the two pairs. If outside interconnections exist, the T and π "equivalents" will not necessarily produce the correct results. For example, the networks of Figs. 11.3 and 11.4 have zero impedance between the two lower terminals, whereas this is not necessarily true for the network of Fig. 11.2. But, so long as the terminals are used in the two specified pairs, no electrical tests performed at the given frequency will be able to distinguish between the actual network and its T and π equivalents.

11.3. Open-circuit and Short-circuit Impedances. The most convenient method of obtaining information about the terminal relations of a four-terminal network is to measure or compute its open-circuit and short-circuit impedances. This provides data from which an equivalent circuit can be determined. The impedances referred to are the following:

1. The impedance measured at end 1 with end 2 open. This will be denoted by the symbol Z_{1o}, where the subscript 1 refers to the terminals at which the impedance is measured, and the subscript o refers to the open circuit at the other end.

2. The impedance measured at end 1 with end 2 short-circuited. This will be denoted by Z_{1s}, where the second subscript refers to the short circuit at the far end.

3. The impedance measured at end 2 with end 1 open-circuited, Z_{2o}.

4. The impedance measured at end 2 with end 1 short-circuited, Z_{2s}.

In much of our subsequent work we shall find it convenient to refer to the T equivalent network. Therefore, we shall now express the elements of the equivalent T in terms of the open-circuit and short-circuit terminal impedances. If we refer to Fig. 11.4, we can write by inspection of the diagram:

$$Z_{1o} = Z_a + Z_c \tag{11.5}$$

$$Z_{2o} = Z_b + Z_c \tag{11.6}$$

$$Z_{1s} = Z_a + \frac{Z_b Z_c}{Z_b + Z_c} \tag{11.7}$$

and

$$Z_{2s} = Z_b + \frac{Z_a Z_c}{Z_a + Z_c} \tag{11.8}$$

Since there are only three independent quantities to be determined, one of these four equations must be superfluous; *i.e.*, there should be a relation between the four open- and short-circuit impedances. Inspection of the equations shows that $Z_{1s}Z_{2o} = Z_{1o}Z_{2s}$; hence we have the relation

$$\frac{Z_{1o}}{Z_{1s}} = \frac{Z_{2o}}{Z_{2s}} \tag{11.9}$$

This relation can be used to check either measured or computed data on the open- and short-circuit impedances.

Now, using the first three equations, (11.5) through (11.7), we can solve for the elements of the T. For Z_c we obtain

$$Z_c = \sqrt{Z_{2o}(Z_{1o} - Z_{1s})} \tag{11.10a}$$

An alternative form of this equation is obtained by using the relation (11.9) to eliminate Z_{1s}. The result is

$$Z_c = \sqrt{Z_{1o}(Z_{2o} - Z_{2s})} \tag{11.10b}$$

For the other elements, we simply use

$$Z_a = Z_{1o} - Z_c \tag{11.11}$$

and

$$Z_b = Z_{2o} - Z_c \tag{11.12}$$

The resulting equivalent circuit is shown in Fig. 11.5. If the open- and short-circuit impedances of a four-terminal network are known from either measurements or computation, the equivalent T can be determined and then used in subsequent calculations.

In extracting the square root indicated in Eq. (11.10), there may sometimes be doubt whether to attach a plus or minus sign to the result. Either one will produce an equivalent circuit that has the proper open- and short-circuit impedances. (One of them may produce a negative resistance in the shunt branch, but this is not uncommon in equivalent circuits, as was discussed in Sec. 10.9.) The difference lies in the fact that the two output terminals of the network can be interchanged without affecting the open- and short-circuit impedances, but the transposition will reverse the phase of the output with respect to the input. The two values of Z_c correspond to these two possibilities. We shall not often be troubled by the sign of Z_c but, in case of doubt, the matter can be settled by choosing the sign of Z_c so that the equivalent circuit produces the correct phase position for the open-circuit output voltage.

For some purposes it is convenient to have the terminal relations written down in equation form. By reference to Fig. 11.5, these equations are

$$Z_{1o}I_1 - \sqrt{Z_{2o}(Z_{1o} - Z_{1s})}I_2 = E_1 \tag{11.13}$$

and

$$-\sqrt{Z_{2o}(Z_{1o} - Z_{1s})}I_1 + Z_{2o}I_2 = -E_2 \tag{11.14}$$

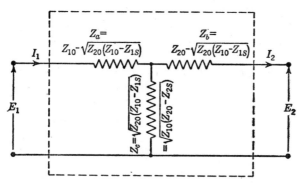

FIG. 11.5. Equivalent T with the elements expressed in terms of the open- and short-circuit impedances of the network.

In all the foregoing, three independent parameters were required to describe the terminal relationships of either the actual network or its equivalent. A further simplification results if the network is symmetrical in such a way that it can be turned end for end without altering its behavior.

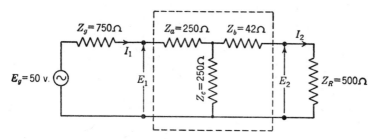

FIG. 11.6. Equivalent network for the example.

Then we have $Z_{1o} = Z_{2o}$ and $Z_{1s} = Z_{2s}$, and only two independent parameters remain. The equivalent T will then have $Z_a = Z_b$.

Example. The open- and short-circuit terminal impedances of a certain four-terminal network are measured with a bridge, with the following results: $Z_{1o} = 500 + j0$ ohms, $Z_{1s} = 285.5 + j0$ ohms, $Z_{2o} = 292 + j0$ ohms, and $Z_{2s} = 167 + j0$ ohms. (1) Check whether the data are self-consistent; (2) determine an equivalent T for this network; (3) compute the current that will flow through a 500-ohm load resistor connected across end 2 if the network is driven at end 1 with a generator that has an internal resistance of 750 ohms and an open-circuit voltage of 50 volts.

To determine the self-consistency of the data, we compute $Z_{1o}/Z_{1s} = 500/$ $285.5 = 1.751$ and compare this with $Z_{2o}/Z_{2s} = 292/167 = 1.749$. This is satisfactory agreement.

To find the equivalent T network, we merely substitute the data into Eqs. (11.10) to (11.12), or else use Fig. 11.5. The resulting network is shown enclosed within the dashed rectangle in Fig. 11.6.

The current that will flow through the load can be found by applying elementary circuit theory to the network of Fig. 11.6, or it can be found by use of Eqs. (11.13) and (11.14). The latter method will be used for purposes of illustration. Evidently, $E_2 = I_2 Z_R$ and $E_1 = E_g - I_1 Z_g$. Using these relations and substituting numerical data, Eqs. (11.13) and (11.14) become

$$500 I_1 - 250 I_2 = 50 - 750 I_1$$

and

$$-250 I_1 + 292 I_2 = -500 I_2$$

Solving these equations simultaneously for I_2, we obtain

$$I_2 = 0.0135 \text{ amp}$$

11.4. Sending-end Impedance and Input-Output Current Ratio. Suppose that the equivalent circuit of Fig. 11.5 is terminated on the right with an impedance Z_R. The sending-end impedance at end 1 can be found either by using the relation $E_2 = I_2 Z_R$ in Eqs. (11.13) and (11.14) and solving for the ratio E_1/I_1, or simply by combining impedances in series and parallel in Fig. 11.5. Using the latter method, we have

$$Z_s = Z_a + \frac{Z_c(Z_b + Z_R)}{Z_c + Z_b + Z_R} \tag{11.15}$$

Substituting the expressions for Z_a, Z_b, and Z_c from Eqs. (11.10) to (11.12) and clearing of fractions, the result is found to be

$$Z_s = Z_{1o} \frac{Z_R + Z_{2s}}{Z_R + Z_{2o}} \tag{11.16}$$

It is interesting to see how this relation applies to the uniform transmission line. The line has $Z_{1s} = Z_{2s} = Z_0 \tanh \gamma l$ and $Z_{1o} = Z_{2o} = Z_0/\tanh \gamma l$. Substitution of these relations into Eq. (11.16) produces a result identical with Eq. (4.42):

$$Z_s = Z_0 \frac{Z_R + Z_0 \tanh \gamma l}{Z_0 + Z_R \tanh \gamma l}$$

Since a current divides between two parallel branches inversely as the impedances, the output-input current ratio of the network is obtained by inspection of Fig. 11.5 as

$$\frac{I_2}{I_1} = \frac{Z_c}{Z_b + Z_c + Z_R}$$

or, substituting the expressions for Z_b and Z_c,

$$\frac{I_2}{I_1} = \frac{\sqrt{Z_{2o}(Z_{1o} - Z_{1s})}}{Z_R + Z_{2o}} = \frac{\sqrt{Z_{1o}(Z_{2o} - Z_{2s})}}{Z_R + Z_{2o}} \tag{11.17}$$

The foregoing equations for Z_s and for the current ratio will be used in the next chapter when we discuss the interconnection of four-terminal networks.

11.5. Various Methods of Expressing Terminal Relations. The terminal relations of a four-terminal network can be specified by giving the elements of its equivalent circuit, as was done, for example, in Fig. 11.6. Sometimes it is convenient to express the terminal relations in algebraic form. This was done in Eqs. (11.13) and (11.14), which are repeated here for convenience:

$$E_1 = Z_{1o}I_1 - \sqrt{Z_{2o}(Z_{1o} - Z_{1s})}I_2 \tag{11.18}$$

and

$$-E_2 = -\sqrt{Z_{2o}(Z_{1o} - Z_{1s})}I_1 + Z_{2o}I_2 \tag{11.19}$$

Here the two terminal voltages are expressed in terms of the two currents I_1 and I_2, the two open-circuit impedances Z_{1o} and Z_{2o}, and a third impedance coefficient $\sqrt{Z_{2o}(Z_{1o} - Z_{1s})}$. The latter is sometimes called the open-circuit transfer impedance because, with end 2 open-circuited, it is the ratio of the output voltage E_2 to the input current I_1.

The relations (11.18) and (11.19) can be placed in various other forms. Each of these forms contains the same information, and the additional ones will be given here merely to acquaint the reader with their appearance. The second form is obtained by solving Eqs. (11.18) and (11.19) for the two currents, which gives the result:

$$I_1 = \frac{E_1}{Z_{1s}} - \frac{1}{Z_{1s}}\sqrt{\frac{Z_{1o} - Z_{1s}}{Z_{2o}}} E_2 \tag{11.20}$$

and

$$I_2 = \frac{1}{Z_{1s}}\sqrt{\frac{Z_{1o} - Z_{1s}}{Z_{2o}}} E_1 - \frac{E_2}{Z_{2s}} \tag{11.21}$$

These equations express the two terminal currents in terms of the two voltages E_1 and E_2, the two short-circuit impedances Z_{1s} and Z_{2s}, and a third impedance coefficient $Z_{1s}\sqrt{Z_{2o}/(Z_{1o} - Z_{1s})}$. The latter is generally called the short-circuit transfer impedance because, with the output terminals short-circuited, it is equal to the ratio of the input voltage to the output current. The foregoing equations are often written in terms of admittances rather than impedances.

The third common form expresses E_1 and I_1 in terms of E_2 and I_2. If the current I_1 is eliminated between Eqs. (11.18) and (11.19), the result can be written as

$$E_1 = \frac{Z_{1o}}{\sqrt{Z_{2o}(Z_{1o} - Z_{1s})}} E_2 + \frac{Z_{1s}Z_{2o}}{\sqrt{Z_{2o}(Z_{1o} - Z_{1s})}} I_2 \qquad (11.22)$$

The second equation is obtained by writing Eq. (11.19) in the form

$$I_1 = \frac{1}{\sqrt{Z_{2o}(Z_{1o} - Z_{1s})}} E_2 + \frac{Z_{2o}}{\sqrt{Z_{2o}(Z_{1o} - Z_{1s})}} I_2 \qquad (11.23)$$

These relations are abbreviated to the form

$$E_1 = AE_2 + BI_2 \qquad (11.24)$$

and

$$I_1 = CE_2 + DI_2 \qquad (11.25)$$

Since only three coefficients can be independent, there must be a relation between the four coefficients A, B, C, and D. Reference to Eqs. (11.22) and (11.23) shows that

$$AD - BC = 1 \qquad (11.26)$$

Equations (11.24) and (11.25) are particularly convenient when the desired values of E_2 and I_2 are specified and it is required to find the necessary value of E_1 and the resulting I_1. This is a common situation in power-system calculations.

11.6. The Transformer.[1] The treatment given in this section assumes that the reader is already somewhat familiar with the conventional theory of the coupled circuit and of the transformer. Some of the possibilities of the equivalent circuit are explored, including the use of an equivalent circuit as an aid in visualizing and computing transients.

The transformer is a highly important four-terminal network, for it serves the purposes of changing the voltage level, transforming the impedance level, and providing d-c insulation between two parts of a circuit. A ferromagnetic core is generally used at frequencies up through the audio range, and this introduces a nonlinearity. Also, the distributed capacitances of the transformer introduce important effects at the higher audio

---

[1] This section may be omitted without destroying the continuity of the text.

frequencies. Each of these effects is difficult to analyze rigorously, the first because of the usual difficulties with nonlinear analysis, and the second because of the complicated geometry of the actual distributed circuit. Both effects will be ignored at first in the analysis. The effect of the nonlinearity can be seen later in a qualitative way and, fortunately, in the usual applications of a well-designed transformer, it is not of major importance. The effect of the capacitances will be added later in an approximate manner.

The equivalent circuit of the transformer is useful in visualizing and calculating transients, as well as in its usual application to steady-state computations. To show this, we shall start the analysis from the differential equations of the circuit, using instantaneous values of voltage and current. We begin with the schematic diagram of Fig. 11.7, which shows the conventions of positive direction for both voltage and current. The relative polarities of the two coils are indicated by dots. If the current

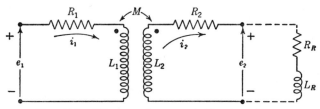

Fig. 11.7. Diagram of a two-winding transformer.

in a coil is increasing into the dotted terminal on that side, the resulting induced voltage in the other coil will make its dot positive with respect to its undotted terminal. Now we consider the instantaneous voltages and currents of the transformer and write the voltage equations around the two loops:

$$R_1 i_1 + L_1 \frac{di_1}{dt} - M \frac{di_2}{dt} = e_1$$

and

$$-M \frac{di_1}{dt} + R_2 i_2 + L_2 \frac{di_2}{dt} = -e_2$$

(11.27)

Here R_1 and L_1 are, respectively, the resistance and self-inductance of the first coil, and R_2 and L_2 are the corresponding quantities for the second coil. The quantity M is the mutual inductance between coils. If the right-hand coil operates into a passive load, the relation between the voltage e_2 and the current i_2 will be determined by the character of the load. One possible type of load is shown with a dashed connection in Fig. 11.7.

Now, let the symbol a represent any number. (One of our choices for a will be the turns ratio N_1/N_2, but it is not restricted to this value and other choices are occasionally more useful.) Multiply and divide the foregoing equations by the number a so that they are written as

$$R_1 i_1 + L_1 \frac{di_1}{dt} - (aM) \frac{d}{dt}\left(\frac{i_2}{a}\right) = e_1$$

and

$$-(aM)\frac{di_1}{dt} + (a^2 R_2)\left(\frac{i_2}{a}\right) + (a^2 L_2)\frac{d}{dt}\left(\frac{i_2}{a}\right) = -ae_2$$

(11.28)

Now, the point is that these are precisely the equations that we would write for the network of Fig. 11.8, which contains only conductively coupled impedance links. This network is, therefore, an equivalent circuit for the transformer. Of course, the equivalence is a direct one only as viewed from the terminals at the left, for the number a is interposed in the

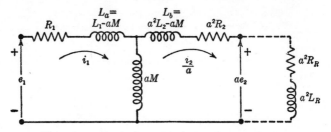

Fig. 11.8.   Equivalent circuit for the transformer.

secondary voltage and current. This will be corrected later by using an ideal transformer in connection with the equivalent circuit but will be ignored for the moment. Figure 11.7 showed a particular kind of load for illustrative purposes. The corresponding load to be connected in the equivalent circuit can be found as follows, again using a differential equation so that the conclusions will hold for transients as well as for the steady state.   For the particular load of Fig. 11.7, we have $e_2 = R_R i_2 + L_R di_2/dt$. Multiplying and dividing this by the number a, the relation between ae_2 and i_2/a is found to be

$$ae_2 = (a^2 R_R)\left(\frac{i_2}{a}\right) + (a^2 L_R)\frac{d}{dt}\left(\frac{i_2}{a}\right)$$

This establishes the effective load values $a^2 R_R$ and $a^2 L_R$ to be used for this particular load in the equivalent circuit of Fig. 11.8.

Now, the equations we have just written are quite general and hold for wave shapes of any form and for transients as well as steady state. Also, the number a is still unrestricted and can have any value. A particularly convenient value is the turns ratio N_1/N_2, for this will ensure that the series elements have positive inductances. It can be shown that $M = k_1L_1N_2/N_1$, where k_1 is the fraction of the first coil's flux that links the second coil.[1] Similar considerations with respect to the second coil show that $M = k_2L_2N_1/N_2$, where k_2 is the fraction of the second coil's flux that links the first coil. (Multiplying these two relations together gives $M = k\sqrt{L_1L_2}$, where $k = \sqrt{k_1k_2}$ is the coefficient of coupling.) The series inductance in the first loop can now be expressed as

$$L_a = L_1 - aM = L_1\left(1 - k_1a\frac{N_2}{N_1}\right) \qquad (11.29)$$

Similarly, the series inductance in the second loop can be written as

$$L_b = a^2L_2 - aM = a^2L_2\left(1 - \frac{k_2}{a}\frac{N_1}{N_2}\right) \qquad (11.30)$$

Each of the fractions k_1 and k_2 can approach unity in value; therefore, the only choice of a which ensures against negative inductances in both places simultaneously is $a = N_1/N_2$. Then we have $L_a = L_1(1 - k_1)$ and $L_b = L_2(1 - k_2)(N_1/N_2)^2$. With this choice of a, the inductances L_a and L_b are called "leakage inductances" because they are associated with the fact that only part of either coil's flux links the other coil. With perfect coupling and no leakage of flux, k_1 and k_2 would both be unity and the leakage inductances would be zero. The flux associated with the shunt-connected inductance (aM) represents the mutual flux linking both windings, and in an iron-cored transformer this path is mostly through the core. This shunt branch, therefore, is the nonlinear element in the circuit. It is clear that, if the current taken by the shunt element is small compared with that flowing straight through the equivalent circuit, the effect of the nonlinearity will be negligible. Thus, the choice $a = N_1/N_2$ has the additional advantage of giving a clear physical meaning to each of the elements, and it permits one to visualize the effect of the transformer on the circuits with which it is associated. The equivalent circuit on this basis is sometimes said to reduce the transformer to an equivalent one-to-one turns ratio.

The choice $a = N_1/N_2$ is a useful and logical one, but it is not the only one that can be made. We can, for example, choose a somewhat larger than N_1/N_2 so that L_a as given by Eq. (11.29) is precisely zero. This choice increases L_b and places all the leakage inductance on the right side of the equivalent circuit. Or, the number a can be chosen somewhat less than N_1/N_2 to make L_b precisely zero, thus placing all the leakage inductance

[1] For example, see "Electric Circuits," by the MIT Staff, p. 385, John Wiley & Sons, Inc., New York, 1940.

on the left. The usefulness of this point of view is discussed in a later paragraph.

A spectacular choice for the value of a is unity, which reduces the equivalent circuit to that shown in Fig. 11.9. Here the voltage and

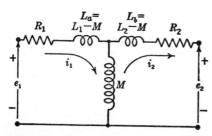

current at the exit terminals are precisely equal to those of the actual transformer, and the circuit is a true equivalent from either end. This equivalent circuit is particularly useful for transformers which are loosely enough coupled so that the inductances $L_1 - M$ and $L_2 - M$ are both positive. For more tightly coupled transformers, particularly when the turns ratio is

FIG. 11.9. Equivalent circuit with the number a chosen equal to unity.

not near unity, one or the other of these inductances will turn out to be negative. This offers no theoretical difficulty for paper calculations, but it makes visualization difficult. Physically, the effect of a negative inductance can be obtained at a single frequency by using a capacitive reactance. But the impedance of a true negative inductance must vary with frequency as $-j\omega L$, while that of a capacitance is $-j/\omega C$, and so the use of a capacitance does not provide a correct equivalent for nonsinusoidal operation or for transients.

The distributed winding capacitances, which have so far been ignored, can now be taken into account in a rough way by showing them as lumped

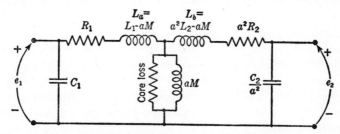

FIG. 11.10. Equivalent circuit with the effects of winding capacitance and of core loss represented approximately.

capacitances across each coil. This is analogous to representing a short-circuited transmission line by a simple parallel LC circuit, an approximation that is reasonably good for frequencies up to and somewhat beyond the first antiresonance. In addition, the core loss can be represented approximately by a resistance connected across the shunt branch in the equivalent circuit. The resulting equivalent circuit is shown in Fig. 11.10. The impedance of the secondary capacitance is transformed by the factor a^2, and so the capacitance itself is transformed by the factor $1/a^2$. The capacitance which exists between one winding and the other is not shown; this may sometimes modify the operation by providing a path other than the mutual inductance by which energy can flow through the transformer. If necessary, this path can be shunted to ground by placing a grounded conducting shield around each winding.

Transformers in the audio-frequency range generally have completely closed cores of ferromagnetic material. If we use the equivalent circuit of Fig. 11.10 with $a = N_1/N_2$, various elements in the circuit become important in different frequency ranges. At low frequencies the leakage inductances and the capacitances have negligible effect, but the current taken by the shunt branch becomes large enough to be important. In the middle range of frequencies, the transformer can often be regarded as ideal to a very good approximation. At higher frequencies, the current taken

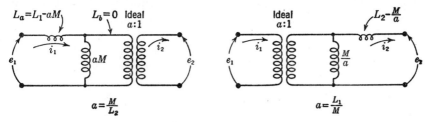

FIG. 11.11. Two examples of true four-terminal equivalents of the transformer.

by the shunt branch can be ignored, but the capacitances and the leakage inductances become important. The principal effect is the resonance that eventually occurs between the secondary capacitance and the leakage inductances. As the frequency rises above resonance, the secondary capacitance rapidly short-circuits the output terminals.

Except when the number a is chosen to be unity, the circuits given previously are equivalent only at the primary terminals, for the output voltage and current of the network must be, respectively, divided and multiplied by a to obtain the true output quantities for the actual transformer. A true four-terminal equivalence can be obtained by using an ideal transformer in combination with the equivalent network. An ideal transformer is one which has zero losses, perfect coupling and therefore zero leakage inductance, and an infinite mutual inductance. The ideal transformer has the well-known properties of transforming voltage and current oppositely by a factor equal to the turns ratio, and of transforming impedance by the square of this ratio. The equivalent circuit of Fig. 11.10 can be divided into two parts by a vertical line placed at any point, and an ideal transformer with a turns ratio of $a:1$ interposed between the two portions. The impedances to the right of the ideal transformer must be multiplied by the factor $1/a^2$ to maintain the equivalence. Then the circuit is equivalent to the actual transformer from both sets of terminals. Two useful examples of this are shown in Fig. 11.11. The resistances and capacitances are omitted from these illustrations for simplicity; if desired, the winding resistances can later be added in series with the terminals. In the first of these equivalent circuits, the ideal transformer has been placed on the right and the quantity a has been given the special value

$$a = \frac{M}{L_2} = k_2 \frac{N_1}{N_2}$$

where k_2 is the fraction of the secondary flux which links the primary. As shown by Eq. (11.30), this choice for a makes L_b equal to zero, and places all the leakage reactance on the left of the shunt branch. If the transformer is operating out of a transmission network which, in its design, can make use of the two inductances L_a and aM, the only remaining effect of the transformer will be that of an ideal one. It is interesting to note that the foregoing value of a is precisely the open-circuit ratio E_1/E_2 when end 2 is energized. This ratio approaches the value N_1/N_2 as k_2 approaches unity.

In the second circuit of Fig. 11.11, the ideal transformer is placed on the left, and the quantity a is given the value

$$a = \frac{L_1}{M} = \frac{1}{k_1} \frac{N_1}{N_2}$$

where k_1 is the fraction of the primary flux which links the secondary. This value of a is the open-circuit ratio E_1/E_2 when end 1 is energized. According to Eq. (11.29), this choice of a makes L_a equal to zero and places all the leakage reactance on the right. Now, if the transformer operates into another network which can utilize the two inductances in performing its function, the remaining effect of the transformer is ideal.

PROBLEMS

1. The following open- and short-circuit impedances were measured on a certain four-terminal network: $Z_{1o} = 2,590 + j0$ ohms, $Z_{1s} = 2,015 + j0$ ohms, $Z_{2o} = 1,300 + j0$ ohms, and $Z_{2s} = 1,010 + j0$ ohms.

   a. Check the self-consistency of the data.

   b. Determine the equivalent T network.

   c. A resistance of 500 ohms is connected across end 2. End 1 is driven with a generator that has an internal emf of 50 volts rms and an internal resistance of 1,000 ohms. Compute the input and output currents of the network.

2. Given the resistance network shown in Fig. P2.

a. Compute the open- and short-circuit impedances of the network, and determine the equivalent T.

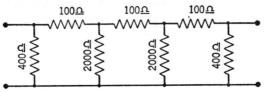

Fig. P2. Resistance network.

b. A resistance of 460 ohms is connected across end 2. The network is driven at end 1 with a generator that has an internal emf of 25 volts rms and an internal resistance of 460 ohms. Compute the input and output currents of the network.

3. A certain reactive network has the following open- and short-circuit impedances: $Z_{1s} = Z_{2s} = j378$ ohms, and $Z_{1o} = Z_{2o} = j425$ ohms. With end 2 open-circuited and a voltage applied to end 1, E_2 is 180° out of phase with E_1.

a. Determine the T equivalent circuit.

b. A resistance of 400 ohms is connected across end 2. If $E_1 = 10/\underline{0°}$ volts, find the magnitude and phase of I_1 and I_2.

c. Compute the sending-end impedance of the network for the conditions of part *b.*

d. Compute the transfer impedance between the sending-end voltage and the load for the conditions of part *b* $(Z_{tr} = E_1/I_2)$.

4. *a.* Find the A, B, C, D constants for the network of Prob. 1 and write the A, B, C, D equations numerically for this network.

b. If $Z_R = 400 + j0$ ohms and $E_2 = 2.00$ volts, what will be the values of E_1 and I_1?

5. The following tests were made on a four-terminal network [refer to Fig. 11.2 and Eqs. (11.24) and (11.25)]. With end 2 open and $E_1 = 10.0$ volts, $I_1 = 0.166$ amp and $E_2 = 6.73$ volts, both quantities in phase with E_1. With end 2 short-circuited and $E_1 = 10.0$ volts, $I_1 = 0.199$ amp and $I_2 = 0.0486$ amp, both quantities in phase with E_1.

a. Find the A, B, C, D constants of the network and write the A, B, C, D equations numerically. Check the test data by determining whether the constants satisfy the proper relationship among themselves.

b. For $I_2 = 0.025$ amp and $Z_R = 300 + j0$ ohms, what must be the values of E_1 and I_1?

6. The following tests were made on a four-terminal network which is known to consist of resistances only: With either end short-circuited, the impedance measured at the other end is 7.5 ohms. A voltage of 10 volts applied to end 1 produces a short-circuit current of 0.167 amp at end 2. Determine the π equivalent network.

7. *a.* Determine the T equivalent for the network of Prob. 6.

b. A resistance of 10 ohms is connected across one end of the network. A generator with E_g = 5.0 volts and Z_g = 20 ohms is connected to the other end. Using the T equivalent, compute the input and output currents of the network.

8. An air-cored transformer has the following constants: N_1 = 120 turns, N_2 = 240 turns, R_1 = 1.26 ohms, R_2 = 2.52 ohms, L_1 = 0.0410 henry, L_2 = 0.164 henry, M = 0.0368 henry. Draw the equivalent circuit, first using a = 1 and then using $a = N_1/N_2$.

9. Repeat Prob. 8, using M = 0.0736 henry. What is the coefficient of coupling for this value of M?

10. Repeat Prob. 8 with M = 0.0368 henry, but choose the quantity a so as to make L_b = 0.

11. A certain transformer has L_1 = 0.20 henry, L_2 = 1.80 henrys, N_2/N_1 = 3, and a coefficient of coupling k = 0.950. Sketch two equivalent circuits like those of Fig. 11.11, giving the values of all elements.

CHAPTER 12

THE IMAGE AND ITERATIVE OPERATION OF FOUR-TERMINAL NETWORKS

12.1. The Image and Iterative Impedances. Transmission networks are generally connected in cascade, as shown in Fig. 12.2, so that the output of one feeds into the input of the next. In this interconnection, it is important to have each network operate out of and into appropriate impedances so that the conditions of maximum power transfer are approximated over the range of frequencies to be transmitted. The terminating impedances should be selected in a consistent way so that the terminal relations of each network can be predicted separately and so that they contribute in a simple manner to the performance of the whole chain. It is important from a power standpoint to be able to change the impedance level in the chain wherever necessary, in order to match impedances. These conditions are met by the so-called image-impedance connection, which is illustrated for a single network in Fig. 12.1. Here the network operates out of an impedance Z_g and into an impedance Z_R, both of which are so chosen that, at either set of terminals, the impedance is the same looking in either direction. The impedance in one direction is said to be the image of that in the other direction. These impedances will be designated by a subscript I for "image," with an additional subscript to identify the terminals as necessary. A communication system is generally operated so that the image impedances are very nearly real over the frequency range to be transmitted. Under this condition, the image termination is the same as the conjugate match which will provide maximum power transfer. Furthermore, by proper design of the networks, the image impedances can be altered as desired from one set of terminals to the other for impedance-matching purposes.

Figure 12.2 shows two dissimilar networks connected in cascade on an image basis. The load and generator impedances are fixed in value. Network A has been designed to have the image impedance $Z_{I1} = Z_g$ at the left. At the connection between the networks, their two image impedances are made equal. Network B is designed to have the image impedance $Z_{I3} = Z_R$ at the right. This method of connection can obviously be extended to a chain of any number of four-terminal networks.

Another method of terminating four-terminal networks is on the so-called

271

iterative basis. Some multisection networks are designed on this basis, for example L-type attenuators. The concept of the iterative connection is also useful for those cases where the image impedance reduces to the iterative one. Figure 12.3 shows a chain of two identical networks terminated at both ends on an iterative basis. The receiving-end impedance has a particular value Z_{K1} which is so chosen that the input impedance of network A', looking toward the right, is precisely equal to Z_{K1}. The network A now operates into this same impedance and, being identical in

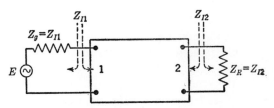

Fig. 12.1. A single four-terminal network terminated on an image basis.

structure with A', also presents the impedance Z_{K1} at its input terminals. The impedance is repeated, or "iterated." In the illustration, the chain is also connected in iterative fashion looking in the other direction, for the generator impedance has the particular value Z_{K2} which makes this impedance repeated when looking leftward from any set of terminals. The iterative connection can be extended to a chain of any number of networks provided, of course, that they all have identical terminal characteristics

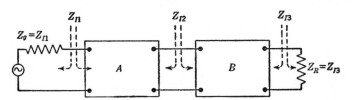

Fig. 12.2. Two networks connected in cascade on an image basis.

If the networks are symmetrical, so that any one of them can be turned end for end without altering its performance, the two iterative impedances Z_{K1} and Z_{K2} are equal. Now, with the proper terminations at both ends, the impedances at any junction are equal in both directions, and so this is also the image connection for these networks. In this case the name "characteristic impedance" is often used. A uniform transmission line is a familiar example of this type of structure.

The image impedances of a network can be expressed in terms of its open- and short-circuit impedances. To do this, we use Eq. (11.16), which expressed the input impedance of a network in terms of the terminating impedance Z_R at the other end. This relation was

$$Z_s = Z_{1o} \frac{Z_R + Z_{2s}}{Z_R + Z_{2o}} \tag{12.1}$$

Now we refer to Fig. 12.1 and use $Z_R = Z_{I2}$ and $Z_s = Z_{I1}$. Then (12.1) becomes

$$Z_{I1} = Z_{1o} \frac{Z_{I2} + Z_{2s}}{Z_{I2} + Z_{2o}} \tag{12.2}$$

Equation (12.1) can also be used to obtain the impedance looking to the left at end 2. For this, we consider Z_g as the load and interchange the subscripts 1 and 2 in the equation. Then, viewed from end 2, we have for the image termination,

$$Z_{I2} = Z_{2o} \frac{Z_{I1} + Z_{1s}}{Z_{I1} + Z_{1o}} \tag{12.3}$$

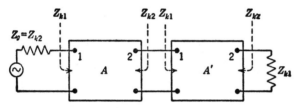

FIG. 12.3. A chain of two networks terminated in its iterative impedances at both ends.

Equations (12.2) and (12.3) can be solved simultaneously for the two image impedances. Upon noting from Eq. (11.9) that $Z_{1o}Z_{2s} = Z_{2o}Z_{1s}$, the results reduce simply to

$$Z_{I1} = \sqrt{Z_{1o}Z_{1s}} \tag{12.4a}$$

and

$$Z_{I2} = \sqrt{Z_{2o}Z_{2s}} \tag{12.4b}$$

Thus, the image impedance at either end is simply the geometric mean between the open- and short-circuit impedances at that end. A similar relation will be recalled for the uniform transmission line, which was symmetrical and had equal image and iterative impedances.

The iterative impedances of a network can be expressed in a similar way. We refer to network A of Fig. 12.3 and again use Eq. (12.1). Here we have $Z_s = Z_{K1}$ when $Z_R = Z_{K1}$. Then

$$Z_{K1} = Z_{1o} \frac{Z_{K1} + Z_{2s}}{Z_{K1} + Z_{2s}} \tag{12.5}$$

Solving for the iterative impedance viewed rightward, we obtain

$$Z_{K1} = \tfrac{1}{2}[(Z_{1o} - Z_{2o}) \pm \sqrt{(Z_{2o} - Z_{1o})^2 + 4Z_{1o}Z_{2s}}] \qquad (12.6)$$

Similarly, using Eq. (12.1) with the subscripts 1 and 2 interchanged, and with Z_s and Z_R now both equal to Z_{K2}, we obtain for the leftward-looking iterative impedance:

$$Z_{K2} = \tfrac{1}{2}[(Z_{2o} - Z_{1o}) \pm \sqrt{(Z_{1o} - Z_{2o})^2 + 4Z_{2o}Z_{1s}}] \qquad (12.7)$$

Because of the relation $Z_{1o}Z_{2s} = Z_{2o}Z_{1s}$, the quantity under the radical is the same in both the foregoing equations. The $\pm$ signs provide two results for each iterative impedance; the sign should be chosen to make the resistive component positive.

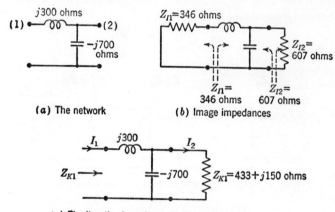

(a) The network

(b) Image impedances

(c) The iterative impedance looking rightward

FIG. 12.4. Network for the example.

For a symmetrical network, which has $Z_{1o} = Z_{2o}$ and $Z_{1s} = Z_{2s}$, the four impedances, iterative and image, all become equal.

Example. Figure 12.4 shows an example of a simple L-type four-terminal network, so named because its two elements are joined to form an inverted L. From inspection of the diagram, we see that the open- and short-circuit terminal impedances are $Z_{1o} = j300 - j700 = -j400$ ohms, $Z_{1s} = j300$ ohms, and $Z_{2o} = -j700$ ohms. With end 1 short-circuited, we combine the two elements in parallel and obtain $Z_{2s} = j525$ ohms. Using the relation $Z_{1o}Z_{2s} = Z_{2o}Z_{1s}$ as a check, we find that each product is equal to 210,000 ohms^2.

Now, from Eqs. (12.4) we find the two image impedances of the network to be

$$Z_{I1} = \sqrt{(-j400)(j300)} = 346 \text{ ohms}$$

and

$$Z_{I2} = \sqrt{(-j700)(j525)} = 607 \text{ ohms}$$

The network is shown terminated in its image impedances in Fig. 12.4b.   On each side, the impedances looking in opposite directions are equal.

The rightward-looking iterative impedance is obtained from Eq. (12.6):

$$Z_{K1} = \tfrac{1}{2}[(-j400 + j700) \pm \sqrt{(j300)^2 + 4(-j400)(j525)}]$$
$$= 433 + j150 \text{ ohms}$$

In Fig. 12.4c, the network is shown terminated in its rightward-looking iterative impedance, thus producing an exactly equal input impedance at the left-hand terminals. Similarly, substitution in Eq. (12.7) shows that the leftward-looking iterative impedance is $Z_{K2} = 433 - j150$ ohms.

A simple and useful illustration of the concept of image impedance is given by the use of a transformer in matching the resistance of a load to that of a source. This is shown in Fig. 12.5, where the transformer is assumed to be ideal and to have a turns ratio $N_1/N_2 = \sqrt{R_g/R_R}$. At either

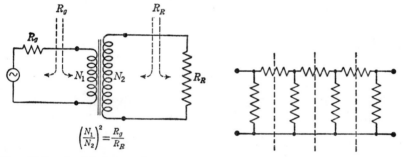

FIG. 12.5. An ideal transformer con-   FIG. 12.6. A ladder network formed of
nected on an image-impedance basis.   two L sections and two T sections.

the input or output terminals, the impedances looking in opposite directions are equal. The transformer, considered as a four-terminal network, is terminated on an image basis. Because the impedances are pure resistances, this is also the condition for maximum power transfer. The ideal transformer differs from most four-terminal networks because its image impedances are not fixed quantities but may have any values so long as they are in the ratio $Z_{I1}/Z_{I2} = (N_1/N_2)^2$. The ideal transformer possesses two rather trivial iterative impedances: the open circuit and the short circuit.

12.2. The Impedances of the L, T, and π Sections.   The L structure and the symmetrical T and π are frequently used in the design of transmission networks, and so we shall develop special formulas for their impedances. One of these structures is sometimes used alone, but frequently a group is connected in cascade to form a ladder structure, as illustrated in Fig. 12.6. The connection is generally made on an image basis.

An L section is shown in Fig. 12.7. By inspection of the diagram, we determine the following short- and open-circuit impedances:

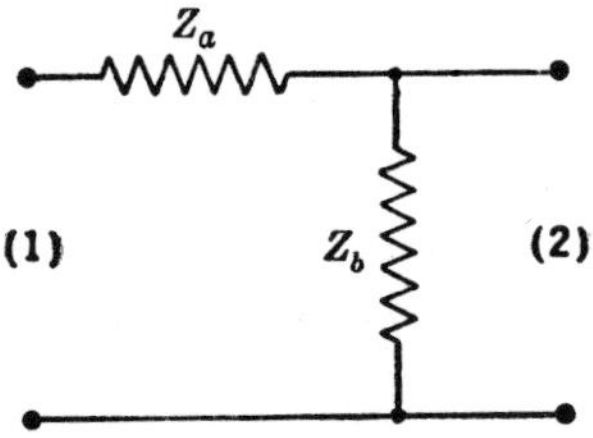

FIG. 12.7. The L section.

$$Z_{1o} = Z_a + Z_b$$

$$Z_{1s} = Z_a$$

$$Z_{2o} = Z_b$$

and

$$Z_{2s} = \frac{Z_a Z_b}{Z_a + Z_b}$$

(12.8)

Substitution of these impedances into Eqs. (12.4) gives the image impedances:

$$Z_{I1} = \sqrt{Z_a(Z_a + Z_b)} = \sqrt{Z_a Z_b}\sqrt{1 + \frac{Z_a}{Z_b}} \qquad (12.9)$$

and

$$Z_{I2} = \frac{\sqrt{Z_a Z_b}}{\sqrt{1 + Z_a/Z_b}} \qquad (12.10)$$

Using Eq. (12.6) and (12.7), we obtain the iterative impedances for each direction:

$$Z_{K1} = \frac{Z_a}{2}\left(1 \pm \sqrt{1 + \frac{4Z_b}{Z_a}}\right) \qquad (12.11a)$$

and

$$Z_{K2} = \frac{Z_a}{2}\left(-1 \pm \sqrt{1 + \frac{4Z_b}{Z_a}}\right) \qquad (12.11b)$$

Now, suppose that we take two identical L structures and join them back to back as in Fig. 12.8a, and then terminate the resulting network in its image impedances at both ends. The structure is now connected on an image basis in the middle and at each end and, upon combination of the two shunt branches, is seen to be a symmetrical T. The image impedance at each end is obtained by using the notation of Fig. 12.8 in Eq. (12.9), i.e., we use $Z_a = Z_1/2$ and $Z_b = 2Z_2$. The result is

$$Z_{I1} = \sqrt{Z_1 Z_2}\sqrt{1 + \frac{Z_1}{4Z_2}}$$

The completed section, being symmetrical, has equal image and iterative impedances, and so the name "characteristic impedance" is appropriate. The symbol Z_T will be used hereafter for this impedance. The image impedance in the middle of the shunt branch is obtained by using Eq. (12.10), and is

$$Z_{I2} = \frac{\sqrt{Z_1 Z_2}}{\sqrt{1 + Z_1/4Z_2}}$$

Next, consider Fig. 12.8b, which shows the same L sections reversed and connected together, thus forming a symmetrical π section. Now, the image impedance at the ends is Z_{I2}, and this, being the characteristic impedance of the π section, will hereafter be denoted by Z_π. The image impedance in the middle of the series arm is the same quantity that we have just given the symbol Z_T. Now we have for the characteristic impedances of the symmetrical T and π sections:

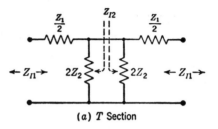

(a) T Section

$$Z_T = \sqrt{Z_1 Z_2}\sqrt{1 + \frac{Z_1}{4Z_2}} \quad (12.12)$$

and

$$Z_\pi = \frac{\sqrt{Z_1 Z_2}}{\sqrt{1 + Z_1/4Z_2}} \quad (12.13)$$

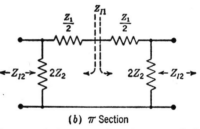

(b) π Section

FIG. 12.8. The formation of symmetrical T and π sections from two L sections.

These two impedances are often called the mid-series and the mid-shunt image impedances, respectively.

Figure 12.9 shows the symmetrical T, the symmetrical π, and the L "half section," and indicates their terminal image impedances. The L may, of course, be faced either way. Its property of transforming from the mid-series image impedance to the mid-shunt one is often useful in filter design.

Example. Consider a T section which, at a certain frequency, has $Z_1 = j600$ ohms and $Z_2 = -j350$ ohms. The characteristic impedance of the section is obtained from Eq. (12.12), and is

$$Z_T = \sqrt{(j600)(-j350)}\sqrt{1 + \frac{j600}{4(-j350)}}$$

$$= \sqrt{210,000}\sqrt{1 - 0.429}$$

$$= 346 \text{ ohms}$$

This T section is the same as would be obtained by joining two L sections that have the constants shown in Fig. 12.4a, and the value that we have just computed for Z_T is identical with the image impedance Z_{I1} that was computed for the L. If the shunt branch of the T were split into two equal parallel branches, the image impedance between the parallel halves (mid-shunt image impedance) would be

$$Z_\pi = \frac{\sqrt{(j600)(-j350)}}{\sqrt{1 - 0.429}} = 607 \text{ ohms}$$

This is the same as the image impedance Z_{I2} for the L of Fig. 12.4.

If a π section were made up with $Z_1 = j600$ ohms and $Z_2 = -j350$ ohms, the characteristic impedance of the section would be $Z_\pi = 607$ ohms, and the mid-series image impedance would be the same as Z_T, or 346 ohms.

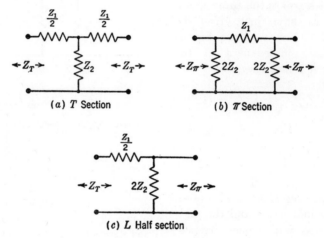

(a) T Section   (b) π Section

(c) L Half section

FIG. 12.9.   T, π, and L sections.

12.3. The Image and Iterative Transfer Constants. In the study of uniform transmission lines we defined the propagation of energy along the line in terms of a propagation constant γ, which was measured per unit of length. The real part of this was denoted by the symbol α and gave the attenuation of the propagated wave. The imaginary part, β, determined the relative phase of the wave at any point. The propagation constant entered the analysis as an exponent; to take a simple case, a line terminated in its characteristic impedance had a voltage that varied along the line as $E_s \epsilon^{-\gamma x}$, where x was the distance from the sending end. A line of unit length, terminated in its characteristic impedance, would have the ratio ϵ^γ between its input and output voltages. Two identical lines of unit length, connected in cascade and properly terminated, would have the over-all voltage ratio $\epsilon^\gamma \cdot \epsilon^\gamma = \epsilon^{\gamma+\gamma} = \epsilon^{2\gamma}$. Observe the additive property of the exponents for the cascade connection.

In a similar manner, we shall define a propagation constant, or, as it is more usually called, a *transfer constant*, for the four-terminal network. This is sometimes called the *transfer function* to emphasize that it is generally a function of frequency. The performance of a network depends, of course, upon its terminations. Either of the two consistent systems of termination, image or iterative, can be made the basis for defining a transfer constant, and deviations from the given ideal can be expressed in terms of reflections. First, we shall discuss a transfer constant based on the important image-type termination, and then we shall discuss another which is based on the iterative connection.

Figure 12.10*a* shows a network which is terminated in its image impedances and is transmitting power from end 1 to end 2. Since the impedance

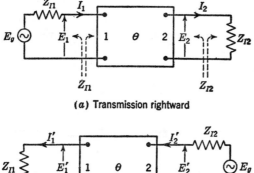

(a) Transmission rightward

(b) Transmission leftward

Fig. 12.10. A network terminated in its image impedances.

level may be different at the two ends, the most useful basis for defining the transfer constant is not on voltage or current alone, but on a volt-ampere basis. The *image transfer constant*, θ, will be defined as

$$\epsilon^{\theta} = \sqrt{\frac{E_1 I_1}{E_2 I_2}} \tag{12.14}$$

where the E's and I's are those obtained with the image termination. Since we have $E_1 = I_1 Z_{I1}$ and $E_2 = I_2 Z_{I2}$, Eq. (12.14) can also be written as

$$\epsilon^{\theta} = \frac{I_1}{I_2} \sqrt{\frac{Z_{I1}}{Z_{I2}}} = \frac{E_1}{E_2} \sqrt{\frac{Z_{I2}}{Z_{I1}}} \tag{12.15}$$

For any symmetrical network, where $Z_{I1} = Z_{I2}$, Eq. (12.15) reduces merely to a voltage or current ratio. Our definition of θ, when applied to a uniform transmission line, is the same as the definition of the propagation constant of the line, except that here we define θ on a per-network basis rather than per unit length. The foregoing equations can be solved for θ, giving

$$\theta = \log_\epsilon \sqrt{\frac{E_1 I_1}{E_2 I_2}} = \log_\epsilon \frac{I_1}{I_2} \sqrt{\frac{Z_{I1}}{Z_{I2}}} \qquad (12.16)$$

In general, the image transfer constant will be a complex number:

$$\theta = \alpha + j\beta \qquad (12.17)$$

where α is the image attenuation constant of the network in nepers, and β is the image phase constant in radians.

It is not hard to show by the reciprocity theorem that the same image transfer constant holds for the transmission of power in either direction. To do this, we consider the two directions of transmission shown in Figs. 12.10a and b, and compare the ratio $E_1 I_1 / E_2 I_2$ for rightward transmission

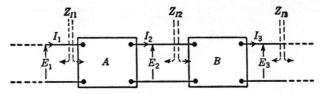

FIG. 12.11. Sections connected in cascade on an image basis.

with the ratio $E_2' I_2' / E_1' I_1'$ for leftward transmission. In Fig. 12.10a we have $I_1 = E_g/2Z_{I1}$, $E_1 = E_g/2$, and $E_2 = I_2 Z_{I2}$. Then

$$\frac{E_1 I_1}{E_2 I_2} = \frac{E_g^2}{4 Z_{I1} Z_{I2} I_2^2}$$

With the generator shifted to the right side, as in Fig. 12.10b, we have $I_2' = E_g/2Z_{I2}$, $E_2' = E_g/2$, and $E_1' = I_1' Z_{I1}$, and so for this direction of transmission, we have the ratio

$$\frac{E_2' I_2'}{E_1' I_1'} = \frac{E_g^2}{4 Z_{I1} Z_{I2} (I_1')^2}$$

But, by the reciprocity theorem, the current I_2 in Fig. 12.10a is equal to the current I_1' in Fig. 12.10b, and so the two volt-ampere ratios are equal.

The exponential definition of θ provides an additive property for the transfer constants of networks connected in cascade on an image basis. In Fig. 12.11, the image transfer constant of network A is defined by

$$\epsilon^{\theta_A} = \sqrt{\frac{E_1 I_1}{E_2 I_2}}$$

and that of network B is defined by

$$\epsilon^{\theta_B} = \sqrt{\frac{E_2 I_2}{E_3 I_3}}$$

The product $\epsilon^{\theta_A} \cdot \epsilon^{\theta_B}$ is

$$\epsilon^{\theta_A + \theta_B} = \sqrt{\frac{E_1 I_1}{E_3 I_3}}$$

which is precisely the definition required for the over-all transfer constant of networks A and B combined in cascade. So long as the image connection is maintained throughout, the over-all image transfer constant of a cascaded group of transmission networks is the sum of their individual transfer constants:

$$\theta = \theta_A + \theta_B + \theta_C + \cdots \qquad (12.18)$$

Breaking this apart into its real and imaginary parts, as indicated by Eq. (12.17), it can be seen that the over-all attenuation constant is the sum of the individual attenuation constants, and likewise with the phase constants:

$$\left. \begin{aligned} \alpha &= \alpha_A + \alpha_B + \alpha_C + \cdots \\ \beta &= \beta_A + \beta_B + \beta_C + \cdots \end{aligned} \right\} \qquad (12.19)$$

The image transfer constant of a network can be expressed rather simply in terms of the open- and short-circuit impedances. This is done most conveniently by computing the square of Eq. (12.15). The output-input current ratio was given by Eq. (11.17) for a network terminated in a receiving-end impedance Z_R. Using $Z_R = Z_{I2}$ for the image termination, the square of that ratio becomes

$$\left(\frac{I_2}{I_1}\right)^2 = \frac{Z_{1o}(Z_{2o} - Z_{2s})}{(Z_{2o} + Z_{I2})^2}$$

Into the square of Eq. (12.15) we substitute this ratio and the relations $Z_{I1} = \sqrt{Z_{1o}Z_{1s}}$, $Z_{I2} = \sqrt{Z_{2o}Z_{2s}}$, and obtain

$$\epsilon^{2\theta} = \frac{\left(1 + \sqrt{Z_{2s}/Z_{2o}}\right)^2}{1 - Z_{2s}/Z_{2o}} \sqrt{\frac{Z_{2o}Z_{1s}}{Z_{1o}Z_{2s}}}$$

Now we use the identity $Z_{1o}Z_{2s} = Z_{2o}Z_{1s}$ and find that the quantity under the radical is unity. Also, we note that the denominator $(1 - Z_{2s}/Z_{2o})$ can be factored into

$$\left(1 - \sqrt{\frac{Z_{2s}}{Z_{2o}}}\right)\left(1 + \sqrt{\frac{Z_{2s}}{Z_{2o}}}\right)$$

the second factor of which can be canceled into the numerator. Therefore, the result simplifies to

$$\epsilon^{2\theta} = \frac{1 + \sqrt{Z_{2s}/Z_{2o}}}{1 - \sqrt{Z_{2s}/Z_{2o}}} = \frac{1 + \sqrt{Z_{1s}/Z_{1o}}}{1 - \sqrt{Z_{1s}/Z_{1o}}} \qquad (12.20)$$

This expresses the image transfer constant in terms of the open- and short-circuit impedances at either end. Although the result is useful as it stands, a more compact expression can be found by solving for $\sqrt{Z_{1s}/Z_{1o}}$. The result is

$$\sqrt{\frac{Z_{1s}}{Z_{1o}}} = \frac{\epsilon^{2\theta} - 1}{\epsilon^{2\theta} + 1}$$

$$= \frac{\epsilon^{\theta} - \epsilon^{-\theta}}{\epsilon^{\theta} + \epsilon^{-\theta}}$$

The latter expression will be recognized as the hyperbolic tangent, and so we have

$$\tanh \theta = \sqrt{\frac{Z_{1s}}{Z_{1o}}} = \sqrt{\frac{Z_{2s}}{Z_{2o}}} \qquad (12.21)$$

An identical expression has been derived previously for the transmission line.

The iterative system of connection, illustrated in Fig. 12.3, can also be used to define a transfer constant for a network. The iterative connection produces the same impedance looking rightward at all terminals, and so the definition of iterative transfer constant that is analogous to Eq. (12.15) is simply

$$\epsilon^{P} = \frac{E_1}{E_2} = \frac{I_1}{I_2} \qquad (12.22)$$

where P is the *iterative transfer constant* for transmission to the right, and the E's and I's are now the ones that are obtained with the *iterative* connection. The reciprocity theorem can be used to show that the iterative transfer constant is the same for either direction of transmission through the network. Because of its exponential definition, the over-all iterative transfer constant of a chain of networks is equal to the sum of their individual iterative transfer constants. For a symmetrical network, the iterative and image impedances are identical, and the transfer constant is then the same on either basis.

Example: Image and Iterative Transfer Constants. The image and iterative impedances of a reactive L network were determined in the example of Sec. 12.1, and the results are shown in Fig. 12.4. We shall now determine the image and iterative transfer constants for this network.

The open- and short-circuit impedances of the network as measured from end 1 were $Z_{1o} = -j400$ ohms and $Z_{1s} = j300$ ohms. For the image transfer constant, we use Eq. (12.20):

$$\epsilon^{2\theta} = \frac{1 + \sqrt{Z_{1s}/Z_{1o}}}{1 - \sqrt{Z_{1s}/Z_{1o}}}$$

$$= \frac{1 + j0.866}{1 - j0.866} = 1.00\underline{/82.0^\circ}$$

Since $\epsilon^{2\theta} = \epsilon^{2\alpha}\epsilon^{j2\beta} = \epsilon^{2\alpha}\underline{/2\beta}$, we have $\epsilon^{2\alpha} = 1.00$ and $2\beta = 82.0^\circ$. Then $\alpha = 0$, and $\beta = 41.0^\circ = 0.715$ rad. The image transfer constant is simply $\theta = j\beta = j0.715$.

To illustrate the meaning of the image transfer constant, refer to Fig. 12.4b Suppose that end 2 is terminated in its image load of 607 ohms, and that end 1 is supplied from a generator that has an internal resistance equal to the image imped-ance at that end, 346 ohms. Then, according to the volt-ampere definition of ϵ^θ given by Eq. (12.14), we have

$$\frac{E_1 I_1}{E_2 I_2} = \epsilon^{2\theta} = 1\underline{/82.0^\circ}$$

The output and input volt-amperes are equal in magnitude. Using Eq. (12.15), we have for the current ratio

$$\frac{I_1}{I_2} = \epsilon^\theta \sqrt{\frac{Z_{I2}}{Z_{I1}}} = (1.00\underline{/41.0^\circ}) \sqrt{\frac{607}{346}} = 1.325\underline{/41.0^\circ}$$

Thus, the output current is smaller than the input current by the factor $1/1.325$ and lags by 41°. Similarly, Eq. (12.15) can be used to show that the output volt-age is larger than the input voltage by the factor 1.325 and also lags by 41°.

For the iterative transfer constant of the network, we refer to Fig. 12.4c. Then, using elementary circuit theory for the current ratio, we have

$$\epsilon^P = \frac{I_1}{I_2} = \frac{433 + j150 - j700}{-j700} = 1.00\underline{/38.2^\circ}$$

With the iterative termination, the output current has the same magnitude as the input current but is shifted in the lagging direction by 38.2°. The iterative atten-uation constant is zero, and the iterative phase shift is 38.2°.

12.4. Transfer Constants and Design Formulas for the L, T, and π Networks.

Because of the practical importance of the L section and of the summetrical T and π, we shall derive explicit formulas for their transfer constants. Like the formulas for the image and iterative imped-ances derived in Sec. 12.2, these will be expressed in terms of the impedances of the individual arms. Often, for design purposes, it is convenient to express the results the other way around, by solving for the impedance of each arm in terms of the transfer constant and the image or iterative impedances. These "design formulas" will also be developed in this section.

L Network—Image Termination.   For the L network, we shall use the notation of Fig. 12.7, where the series arm was denoted by Z_a and the shunt arm by Z_b.   From end 1, the open- and short-circuit impedances are, respectively, $Z_{1o} = Z_a + Z_b$ and $Z_{1s} = Z_a$.   Then, using Eq. (12.20), we obtain immediately for the image transfer constant:

$$\epsilon^{2\theta} = \frac{\sqrt{1 + Z_b/Z_a} + 1}{\sqrt{1 + Z_b/Z_a} - 1} \tag{12.23}$$

An alternative form can be obtained from Eq. (12.21):

$$\tanh \theta = \sqrt{\frac{Z_a}{Z_a + Z_b}} = \sqrt{\frac{Z_a/Z_b}{1 + Z_a/Z_b}} \tag{12.24}$$

In the design of an L network, the desired values of the image impedances are frequently known, and it is required to find the values of the elements Z_a and Z_b.   Referring to the formulas for the image impedances of the L network, Eqs. (12.9) and (12.10), it will be seen that

$$Z_a = Z_{I1} \tanh \theta$$

and

$$Z_b = \frac{Z_{I2}}{\tanh \theta}$$

also that

$$\tanh^2 \theta = 1 - \frac{Z_{I2}}{Z_{I1}}$$

$$\left.\right\} \tag{12.25}$$

Two of the three quantities Z_{I1}, Z_{I2}, and θ can be specified; these two determine the third quantity and the values of the elements.

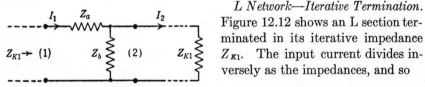

FIG. 12.12.   An L network terminated in its iterative impedance.

L Network—Iterative Termination. Figure 12.12 shows an L section terminated in its iterative impedance Z_{K1}.   The input current divides inversely as the impedances, and so

$$\epsilon^P = \frac{I_1}{I_2} = \frac{Z_b + Z_{K1}}{Z_b} \tag{12.26}$$

where P is the iterative transfer constant of the section.   Solving for Z_b, we obtain

$$Z_b = \frac{Z_{K1}}{\epsilon^P - 1} \tag{12.27}$$

If this is substituted into the expression for Z_{K1} given by Eq. (12.11a) and the result solved for Z_a, we obtain

$$Z_a = Z_{K1}(1 - \epsilon^{-P}) \tag{12.28}$$

The foregoing two equations can be used to determine the values of the elements when the iterative impedance and the iterative transfer constant are given.

If the L is reversed and the power flows from end 2 to end 1, the following formulas can be derived for the elements:

and

$$Z_a = Z_{K2}(\epsilon^P - 1) \left.\begin{array}{c}\\ \\ \\ \end{array}\right\}$$

$$Z_b = \frac{Z_{K2}}{1 - \epsilon^{-P}}$$

(12.29)

where Z_{K2} is the load impedance terminating end 1.

The Symmetrical T and π Networks. To determine an expression for the image transfer constant of the symmetrical T or π section, we have only to note from Fig. 12.8 that each is made up of two L sections which have, in the new notation, $Z_a = Z_1/2$ and $Z_b = 2Z_2$. Because of the additive property of the transfer constant, the T and π have image transfer constants just double that of the L. Hence, changing symbols in Eqs. (12.23) and (12.24), we obtain the following alternative expressions for the T and π:

$$\epsilon^\theta = \frac{\sqrt{1 + 4Z_2/Z_1} + 1}{\sqrt{1 + 4Z_2/Z_1} - 1}$$

(12.30)

and

$$\tanh \frac{\theta}{2} = \sqrt{\frac{Z_1/4Z_2}{1 + Z_1/4Z_2}}$$

(12.31)

Another useful method of expressing this transfer constant is by use of the hyperbolic cosine:

$$\cosh \theta = \frac{1}{2}\left(\epsilon^\theta + \frac{1}{\epsilon^\theta}\right)$$

Substituting from Eq. (12.30), the result simplifies to

$$\cosh \theta = 1 + \frac{Z_1}{2Z_2}$$

(12.32)

Many other equivalent forms can be obtained by the use of the various hyperbolic identities.

Because of the symmetry of these networks, their iterative and image transfer constants are equal, and so we can interpret ϵ^θ as a voltage or current ratio provided that the network is terminated in its characteristic impedance:

$$\epsilon^\theta = \frac{E_1}{E_2} = \frac{I_1}{I_2}$$

(12.33)

Since $\epsilon^\theta = \epsilon^\alpha \epsilon^{j\beta} = \epsilon^\alpha /\underline{\beta}$, the absolute value of (12.30) is equal to ϵ^α, and its angle is equal to β.

Formulas for the elements Z_1 and Z_2 can be obtained in terms of the characteristic impedance and transfer function by combining Eq. (12.31) with the formulas for Z_T and Z_π given by Eqs. (12.12) and (12.13). For the T section, we multiply Z_T by $\tanh \theta/2$ and obtain

$$Z_1 = 2Z_T \tanh \frac{\theta}{2} = 2Z_T \left(\frac{\epsilon^\theta - 1}{\epsilon^\theta + 1} \right) \tag{12.34}$$

Substitution of the foregoing result into the expression for Z_T gives

$$Z_2 = \frac{2Z_T \epsilon^\theta}{\epsilon^{2\theta} - 1} \tag{12.35}$$

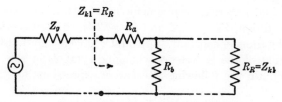

$$Z_{k1} = R_R$$

FIG. 12.13. An L-type attenuator.

A corresponding derivation for the π network yields

$$Z_1 = \frac{Z_\pi (\epsilon^{2\theta} - 1)}{2\epsilon^\theta} \Bigg\}$$

and

$$Z_2 = \frac{Z_\pi}{2} \left(\frac{\epsilon^\theta + 1}{\epsilon^\theta - 1} \right) \Bigg\} \tag{12.36}$$

12.5. Attenuators. An attenuator is a resistive four-terminal network designed to provide a known amount of attenuation between its input and output terminals, while maintaining the impedance level at a given value. Some attenuators are fixed, others are adjustable. The most commonly used types are the L, the symmetrical T, and the ladder.[1]

The L-type attenuator is operated in iterative fashion. A common type of connection is shown in Fig. 12.13. The receiving-end resistance, R_R, is made equal to the iterative impedance Z_{K1}, and consequently the generator works into an impedance of this value regardless of the attenua-

[1] For design information on these and other types of attenuators, see P. K. Mc-Elroy, "Designing Resistive Attenuating Networks," *Proc. IRE*, Vol. 23, pp. 213–233, March, 1935. This paper also discusses the bridged-T attenuator and the unsymmetrical T and π for operation between unequal impedances.

tion. But, with this type of attenuator, the impedance seen by the load varies with the attenuation. The design formulas for iterative operation are obtained from Eqs. (12.27) and (12.28) and are

and

$$R_a = R_R(1 - \epsilon^{-P})$$

$$R_b = \frac{R_R}{\epsilon^P - 1}$$

(12.37)

where P is the attenuation in nepers and ϵ^P is the input-output current (or voltage) ratio of the attenuator. The attenuation can be varied by changing R_a and R_b simultaneously. Another method, which provides changes in finite steps, is to arrange a number of sections so that any desired number of them can be connected in cascade between the generator and the load.

Other variations of the L attenuator are possible. For example, the L can be operated in the reverse direction, with the load made equal to the other iterative impedance, Z_{K2}. Also, if the impedance seen by the load is to be kept constant, the network of Fig. 12.13 can be designed so that Z_{K2} matches the generator impedance. Then the load views the impedance Z_{K2}. But now the impedance seen by the generator will vary with the attenuation and, furthermore, the quantity ϵ^P is no longer the input-output current ratio.

Example. Design an L-type attenuator as shown in Fig. 12.13 to operate into a resistance of 500 ohms and to provide an attenuation of 15 db. The iterative attenuation constant in nepers is therefore $P = 15/8.686 = 1.727$, and so the required input-output current ratio is $\epsilon^P = 5.624$. Then, using Eqs. (12.37), we compute

$$R_a = 500 \left(1 - \frac{1}{5.624}\right) = 411 \text{ ohms}$$

and

$$R_b = \frac{500}{5.624 - 1} = 108 \text{ ohms}$$

The input impedance of the attenuator will be 500 ohms because of the iterative termination on the right. The impedance viewed by the load depends on the generator impedance.

The L network has only two elements and therefore suffers from the disadvantage that only two characteristics can be chosen independently. Generally, these are the input impedance and the attenuation, leaving the impedance viewed by the load to be what it will. The symmetrical T, on the other hand, can be used to provide equal input and output impedances. The symmetry of this network makes image operation and iterative operation the same. The symmetrical T attenuator is shown in Fig. 12.14. For its design, we use Eqs. (12.34) and (12.35) with $Z_T = R$. Then

and

$$\left.\begin{array}{l} \dfrac{R_1}{2} = R \left(\dfrac{\epsilon^\theta - 1}{\epsilon^\theta + 1} \right) \\[4mm] R_2 = \dfrac{2R\epsilon^\theta}{\epsilon^{2\theta} - 1} \end{array}\right\} \qquad (12.38)$$

where θ is the image (and iterative) attenuation constant in nepers, and ϵ^θ is the input-output current ratio. As with the L-type attenuator, the

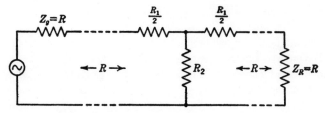

Fig. 12.14. The symmetrical T attenuator.

symmetrical T can be arranged with adjustable resistances to provide variable attenuation, or several fixed sections can be arranged for cascade connection to provide steps of attenuation. The π section can also be made the basis of an attenuator, but the T is generally preferred because, when the elements are varied in steps by switches, all the switch arms can be connected to a common electrical point.

A ladder-type attenuator is made up of a number of symmetrical sections connected in cascade. The load terminates one end in its characteristic impedance; the other end is terminated in a balancing resistance also equal to the characteristic impedance. An example using three π sections is shown in Fig. 12.15a. The source is connected to one of the numbered points, depending on the attenuation required. At any of these points, e.g., the one numbered 3, the impedance looking in either direction along the chain is equal to Z_π; these two paths are in parallel as viewed from the source, and hence the input impedance is equal to $Z_\pi/2$ at any position. In the actual construction, the elements which are in parallel are combined into one element, as shown in Fig. 12.15b. The ladder attenuator can also be based on the T section, but one more element is required after combination. The appropriate design formulas for each section are obtained from Eqs. (12.36) by using $Z_\tau = R_R$.   Then

$$R_1 = R_R \left(\frac{\epsilon^{2\theta} - 1}{2\epsilon^\theta}\right)$$

and

$$2R_2 = R_R \left(\frac{\epsilon^\theta + 1}{\epsilon^\theta - 1}\right)$$

(12.39)

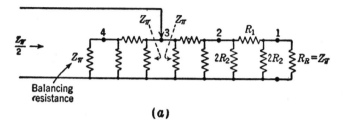

Balancing resistance

(a)

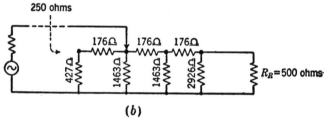

250 ohms

R_R=500 ohms

(b)

FIG. 12.15.   A ladder attenuator.

where θ is the attenuation constant per section of the network.   Because the input current splits into two equal parts, one going to the left and the other to the right, the input-output current ratio is

$$\frac{\text{Input current}}{\text{Output current}} = 2\epsilon^{N\theta}$$

(12.40)

where N is the number of sections included between the load and the source.   The ratio of output currents at two successive positions is simply ϵ^θ.

Figure 12.15b shows a numerical example of a ladder attenuator designed for a load resistance of 500 ohms and an attenuation of 3 db per step. The design formulas give $R_1 = 176$ ohms and $2R_2 = 2,926$ ohms.   Adjacent shunt elements are combined, the resistance of 427 ohms on the left being the parallel combination of the 500-ohm balancing resistance and the adjacent shunt arm of 2,926 ohms.

12.6. Impedance-matching Networks.   The iron-cored transformer is a familiar means of matching impedances at the lower frequencies.   At higher frequencies, the air-cored transformer can be used. It is generally tuned to resonance and so is suitable only over a narrow band of frequencies, but this is satisfactory for most applications at radio frequencies. Another method of matching in this frequency range is to use a four-terminal network which is composed of reactive elements to avoid the dissipation of power. The network can be designed on an image basis to match the resistance of a source to the resistance of the load. Since the reactive impedances in the network will be correct at only one frequency, this type of network is suitable only for operation over a reasonably narrow band.

The simplest impedance-matching network consists of a reactive L structure. With its two elements, it can match any one resistance to any other, but the phase shift must be allowed to become what it will. An unsymmetrical T or unsymmetrical π, with three elements to be chosen, can provide any desired phase shift between input and output in addition to matching the two resistances. The unsymmetrical T and π will not be discussed here but can be found in the literature.[1]

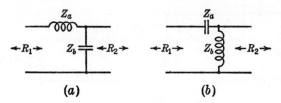

Fig. 12.16.   Alternative forms of the L-type impedance-matching network.

An L-type impedance-matching network is shown in Fig. 12.16. The resistance connected to end 2 must be greater than that connected to end 1, but this, of course, is no real restriction, for either end can be connected to the generator. If the generator impedance is higher than that of the load, the load is connected to end 1 and the generator is connected to end 2. The appropriate design formulas are obtained from Eqs. (12.25) for the image operation of the L network. For an attenuation constant of zero, we have

$$\tanh \theta = \tanh j\beta = j \tan \beta$$

Then, from the last of Eqs. (12.25),

$$j \tan \beta = \pm\sqrt{1 - \frac{R_2}{R_1}}$$

[1] Formulas and design charts for the unsymmetrical T and π will be found in F. E. Terman, "Radio Engineers' Handbook," pp. 210–215, McGraw-Hill Book Company, Inc., New York, 1943.

or

$$\tan \beta = \pm \sqrt{\frac{R_2}{R_1} - 1} \qquad (12.41)$$

where β is the phase shift between input and output (positive if the output lags the input). Then, Eqs. (12.25) become

and
$$\left.\begin{array}{l} Z_a = \pm jR_1 \sqrt{\dfrac{R_2}{R_1} - 1} \\[2.5ex] Z_b = \dfrac{\mp jR_2}{\sqrt{R_2/R_1 - 1}} \end{array}\right\} \qquad (12.42)$$

The signs must be chosen oppositely; i.e., Z_a and Z_b must be reactances of opposite types. The circuit of Fig. 12.16a is often preferred because the series inductance and shunt capacitance help to block the passage of unwanted harmonics.

If the impedances to be matched are not pure resistances, their reactive components can first be canceled by connecting appropriate reactive elements. The resistive impedances then remaining can be matched by the L network.

Example. Design an L-type network to match between a 500- and a 100-ohm resistance. (Either one may be the load, the other the generator. Whichever function it is associated with, the higher resistance must be connected to end 2 in Fig. 12.16.) We use Eqs. (12.42) and find

$$Z_a = \pm j100 \sqrt{5 - 1} = \pm j200 \text{ ohms}$$

and

$$Z_b = \frac{\mp j500}{\sqrt{5 - 1}} = \mp j250 \text{ ohms}$$

From Eq. (12.41),

$$\tan \beta = \pm \sqrt{5 - 1} = \pm 2$$

and

$$\beta = \pm 63.4°$$

Thus, we can use $Z_a = j200$ ohms and $Z_b = -j250$ ohms, in which case the output will lag the input by 63.4°. Or, we can use $Z_a = -j200$ ohms and $Z_b = j250$ ohms, and the output will *lead* the input by 63.4°. The image attenuation constant is zero, and so $\theta = j\beta$ and $\epsilon^\theta = 1.00\underline{/\pm 63.4°}$. The magnitude of the volt-ampere ratio of the network is unity for image operation, and the power output is equal to the power input.

PROBLEMS

1. A certain four-terminal network has the following terminal impedances: With end 2 open-circuited, the impedance at end 1 is $600 + j0$ ohms. With end 2 short-circuited, the impedance at end 1 is $466.7 + j0$ ohms. With end 1 open-circuited, the impedance at end 2 is $300 + j0$ ohms. With end 1 short-circuited, the impedance at end 2 is $233.3 + j0$ ohms.

  a. Determine the image impedances of the network.

  b. Determine the iterative impedances in both directions.

2. An L network (see Fig. 12.7) has $Z_a = -j500$ ohms and $Z_b = j1,000$ ohms. Determine the image and iterative impedances of the network.

3. A T network (see Fig. 12.9) has $Z_1 = j\omega L$ and $Z_2 = -j/\omega C$.

  a. Write the equation for Z_T as a function of frequency.

  b. For $L = 0.100$ henry and $C = 2.0 \times 10^{-6}$ farad, sketch the resistive and reactive components of Z_T over the frequency range $0 < f < 1,500$ cps.

4. Use the data of Prob. 3 for a π network. Determine and sketch Z_π.

5. An L network (see Fig. 12.7) has $Z_a = Z_b = 707 + j0$ ohms.

  a. Compute the image impedances of the network.

  b. Determine the image transfer constant.

  c. The network is terminated at end 2 in $Z_R = Z_{I2}$. At end 1, it is driven by a generator which has $E_g = 10$ volts rms and $Z_g = Z_{I1}$. Compute the resulting values of E_1, I_1, E_2, and I_2. Compute the input-output volt-ampere ratio and check with the quantity $\epsilon^{2\theta}$.

6. *a.* Compute the iterative impedance Z_{K1} for the network of Prob. 5.

  b. If $Z_R = Z_g = Z_{K1}$ and $E_g = 10$ volts, compute E_1, I_1, E_2, and I_2. Compute the iterative transfer constant, P.

7. A certain four-terminal network has the following open- and short-circuit impedances: $Z_{1o} = -j400$ ohms, $Z_{1s} = j267$ ohms, $Z_{2o} = -j600$ ohms, and $Z_{2s} = j400$ ohms.

  a. Check whether the data are consistent.

  b. Compute the two image impedances and the image transfer constant.

  c. The network is terminated at end 2 in $Z_R = Z_{I2}$. It is driven at end 1 with a generator that has $E_g = 10$ volts rms and $Z_g = Z_{I1}$. Compute E_1 and I_1. Compute E_2 and I_2 from ϵ^θ.

8. An L network (see Fig. 12.7) has $Z_a = j500$ ohms and $Z_b = -j1,000$ ohms.

  a. Compute the image impedances and the image transfer constant.

  b. The network is terminated at end 2 in $Z_R = Z_{I2}$. It is driven at end 1 with a generator that has $E_g = 20$ volts rms and $Z_g = Z_{I1}$. Compute E_1 and I_1. Compute E_2 and I_2 from ϵ^θ.

9. Show that the iterative transfer constant, P, is the same for either direction of transmission through a network.

10. Show that the following relations hold for a four-terminal network: $Z_{1o} = Z_{I1}/\tanh \theta$, $Z_{2o} = Z_{I2}/\tanh \theta$, and $Z_{2s} = Z_{I2} \tanh \theta$. Then show that, if the network is terminated in a receiving-end impedance Z_R, the sending-end impedance is given by

$$Z_s = Z_{I1} \frac{Z_R + Z_{I2} \tanh \theta}{Z_{I2} + Z_R \tanh \theta}$$

Compare this expression with Eq. (4.42), which gave the impedance of a uniform transmission line. Also, compare the above expressions for the open- and short-circuit impedances with the corresponding ones for a transmission line.

11. Use the result of Prob. 10 to show that, if $\alpha \gg 1$, the sending-end impedance of the network is approximately equal to Z_{I1} regardless of the value of Z_R.

12. An L network (see Fig. 12.7) is to have $Z_{I1} = 500$ ohms and $Z_{I2} = 100$ ohms (both resistive). Determine the values of the elements Z_a and Z_b; also, compute the image transfer constant.

13. Repeat Prob. 12, using $Z_{I1} = 100$ ohms and $Z_{I2} = 500$ ohms (both resistive).

14. A symmetrical network (T or π; see Fig. 12.9) has $Z_1 = 400 + j0$ ohms and $Z_2 = 500 + j0$ ohms. Compute Z_T, Z_π, and the image transfer constant.

15. An L-type attenuator (see Fig. 12.13) is to be designed to operate on an iterative basis with a load resistance of 500 ohms. The input-output current ratio is to be 10:1. Determine the values of the elements for the attenuator.

16. Repeat Prob. 15, but with the L reversed from the position shown in Fig. 12.13.

17. An L-type attenuator, as shown in Fig. 12.13, is to operate with a load resistance of 70 ohms. The input impedance of the attenuator is also to be 70 ohms. Determine the values of the attenuator elements for attenuations of (a) 3 db, (b) 6 db, and (c) 9 db. If the impedance of the generator is 70 ohms, determine the impedance viewed by the load when the attenuation is 3 db, also when the attenuation is 9 db.

18. A symmetrical T attenuator is to operate with a load resistance of 650 ohms. Determine the proper values for the attenuator elements for attenuations of (a) 3 db, (b) 6 db, and (c) 9 db. If the generator impedance is 650 ohms, what is the impedance viewed by the load?

19. Design a ladder-type attenuator with four steps of 3 db each. The load resistance is 70 ohms. Sketch the attenuator and show the values of the elements.

20. An L-type impedance-matching network is to operate into a 700-ohm resistive load. The internal resistance of the generator is 5,000 ohms, and the frequency is 1 Mc. Sketch the circuit and give the values of the elements in henrys and microfarads: (a) for a circuit configuration that will help eliminate harmonics of the 1-Mc signal, and (b) for a circuit configuration that will help eliminate a 120-cps interfering signal. Compute the phase shift for each of these networks.

21. An L-type impedance-matching network is to match between a 700-ohm resistive load and a 70-ohm source. The frequency is 560 kc. Sketch the circuit and give the values of the elements in henrys and microfarads: (a) for a circuit configuration that will help eliminate harmonics of the 560-kc signal, and (b) for a circuit configuration that will help eliminate a 120-cps interfering signal. Compute the phase shift for each of these networks.

22. *a.* Show that the following relations hold for a four-terminal network (see also Prob. 10). $Z_{1o} = Z_{I1}/\tanh\theta$, $Z_{2o} = Z_{I2}/\tanh\theta$, $Z_{1s} = Z_{I1}\tanh\theta$, and $Z_{2s} = Z_{I2}\tanh\theta$.

b. Refer to Fig. 11.5 and Eqs. (11.10) to (11.12). Show that a T network will have the image impedances Z_{I1} and Z_{I2} and will have an image transfer constant θ if the impedances of its arms are

$$Z_c = \frac{\sqrt{Z_{I1}Z_{I2}}}{\sinh\theta}$$

$$Z_a = \frac{Z_{I1}}{\tanh\theta} - Z_c$$

and

$$Z_b = \frac{Z_{I2}}{\tanh\theta} - Z_c$$

This T network may merely be the equivalent of another four-terminal network at a given frequency, or it may be designed specifically to provide certain values of Z_{I1}, Z_{I2}, and θ.

c. Write the foregoing equations for the special case of a symmetrical network which has $Z_{I1} = Z_{I2} = Z_0$, and compare the results with those obtained for a uniform transmission line (Fig. 4.15).

CHAPTER 13

INSERTION LOSS AND REFLECTION FACTORS

13.1. Insertion Loss. In Sec. 4.9, the effect of inserting a four-terminal network between a source and a load was described in terms of the insertion ratio and the insertion loss. These quantities compare the receiving-end voltage (or current) under two conditions: (1) with a direct connection between source and load and (2) with the network inserted. The definition of insertion ratio is

$$\text{Insertion ratio} = \frac{E_2'}{E_2} = \frac{I_2'}{I_2} \qquad (13.1)$$

where E_2 and I_2 are the voltage and current on the load side of the network, and E_2' and I_2' are the corresponding quantities that would be obtained if the network were replaced by a direct connection. The name *insertion loss* denotes the magnitude of the above effect as expressed in nepers or in decibels, *i.e.*,

$$\text{Insertion loss} = \log_\epsilon \left| \frac{I_2'}{I_2} \right| \qquad \text{nepers}$$

or $\qquad\qquad\qquad\qquad\qquad\qquad\qquad\qquad\qquad\qquad\qquad$ (13.2)

$$\text{Insertion loss} = 20 \log_{10} \left| \frac{I_2'}{I_2} \right| \qquad \text{db}$$

The insertion loss may be negative, in which case it is an insertion gain. The phase angle of the insertion ratio is called the insertion phase shift, for it is the amount by which the receiving-end current is shifted in the lagging direction by the introduction of the network.

The performance of a network depends upon the impedances between which it operates; hence, the insertion loss is not a property of the network alone but of the combination of network, load, and generator impedance taken together.

13.2. Network Terminated in Its Image Impedances. For a network operating between its image impedances, the insertion loss can be expressed rather simply in terms of the image impedances and transfer constant. If the network were removed and a direct connection substituted for it, as in

Fig. 13.1*a*, the receiving-end current would be

$$I_2' = \frac{E_g}{Z_g + Z_R} \tag{13.3}$$

Now we insert the network, which is assumed to have $Z_{I1} = Z_g$ and $Z_{I2} = Z_R$, as in Fig. 13.1*b*. The generator voltage operates into an im-

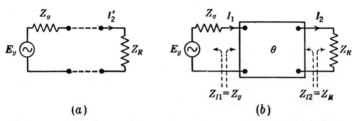

(a)   (b)

F$_{IG}$. 13.1. For the computation of the insertion loss of a network operating between its image impedances.

pedance $2Z_g$, and so $I_1 = E_g/2Z_g$. According to Eq. (12.15), the current ratio for the image termination is

$$\frac{I_2}{I_1} = \epsilon^{-\theta} \sqrt{\frac{Z_{I1}}{Z_{I2}}} \tag{13.4}$$

Hence, the output current is

$$I_2 = \frac{E_g \epsilon^{-\theta}}{2\sqrt{Z_g Z_R}} \tag{13.5}$$

For the image-connected network we now have, by use of Eqs. (13.3) and (13.5),

$$\text{Insertion ratio} = \frac{I_2'}{I_2} = \frac{2\sqrt{Z_g Z_R}}{Z_g + Z_R} \epsilon^{\theta} \tag{13.6}$$

The phase angle of this is the insertion phase shift; the logarithm of its magnitude, as indicated by Eq. (13.2), is the insertion loss.

For reasons that will be explained below, the factor that multiples ϵ^{θ} in the foregoing equation is called a *mismatch factor* or a *reflection factor*:

$$\text{Mismatch or reflection factor} = F_r = \frac{2\sqrt{Z_g Z_R}}{Z_g + Z_R} = \frac{2\sqrt{Z_g/Z_R}}{1 + Z_g/Z_R} \tag{13.7}$$

Suppose that the impedances Z_g and Z_R are directly connected and that they are equal in both magnitude and angle. They are already matched in the same sense that a transmission line is matched and has no reflection when its load is equal to Z_0. (This, of course, is not the same as a "conjugate match" unless the impedances are pure resistances.) The mismatch factor (13.7) is now unity and the insertion ratio (13.6) reduces to ϵ', for the network cannot improve the image match previously existing and can simply insert its transfer constant. Then the insertion loss is simply α nepers. But if the impedances are unequal, the network must insert an impedance transformation to produce an image match and the mismatch factor (13.7) will be different from unity. This will modify the insertion loss.

The name "reflection factor" comes from an analogy with transmission lines. To show the analogy most clearly, we compute the load voltage for the direct connection of Fig. 13.1a and arrange the result as follows:

$$E_2' = \frac{E_g Z_R}{Z_g + Z_R} = \frac{E_g}{2}\left(1 + \frac{Z_R - Z_g}{Z_R + Z_g}\right) = E_0(1 + k) \qquad (13.8)$$

where $E_0 = E_g/2$ is the load voltage that would be obtained if the load impedance were equal to Z_g, and k is a reflection coefficient similar to the one defined for reflections at the end of a transmission line. Although the concept of a reflection was based on the idea of a traveling wave and may seem artificial when applied to lumped impedances, the mathematical relations are similar in both problems. The reflection factor of Eq. (13.7) is equal to $\sqrt{1 - k^2}$.

The reflection factor is the geometric mean of the two impedances divided by their arithmetic mean. If the impedances have different phase angles, the factor may be either larger or smaller than unity. But if they have the same phase angle but different magnitudes, the reflection factor is always less than unity and contributes a negative insertion loss, $i.e.$, an insertion gain. (Of course, this may be outweighed by a large value of ϵ^α, resulting in a net insertion loss for the network.) An interesting example is an ideal transformer used to match two resistances. The volt-ampere ratio of the transformer is unity, and so, by the definition (12.14), $\epsilon^\theta = 1$. When the turns ratio is selected to provide an image match, the results are a reflection factor smaller than unity and an insertion gain.

Example. In the example of Sec 12.6, an impedance-matching network was designed to operate between resistances of 100 and 500 ohms. This network, operating between its image impedances, is shown in Fig. 13.2. In Sec. 12.6, it was found that its image transfer constant was given by $\epsilon^\theta = 1\underline{/63.4°}$. We shall now compute the insertion ratio and the insertion loss of the network.

Using Eq. (13.7), we first compute the reflection factor:

$$F_r = \frac{2\sqrt{100 \times 500}}{100 + 500} = 0.741$$

The insertion ratio (13.6) is, therefore,

$$\text{Insertion ratio} = \frac{I_2'}{I_2} = F_r \epsilon^\theta = 0.741\underline{/63.4°}$$

Insertion of the network causes the output current to lag 63.4° behind the phase position that it would have with a direct connection. The insertion loss is

$$\text{Insertion loss} = 20 \log_{10} 0.741 = -2.6 \text{ db}$$

The insertion of the network improves the output power by 2.6 db, which represents a power ratio of $(1/0.741)^2 = 1.82$.

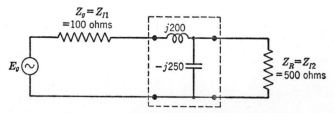

FIG. 13.2.   For the example of Sec. 13.2.

13.3. Network with Mismatched Terminations. Perhaps the most illuminating method of handling the case of a network with unmatched terminations is that followed by Shea.[1] The network, which operates out of an impedance Z_g and into an impedance Z_R, is shown in Fig. 13.3a. For

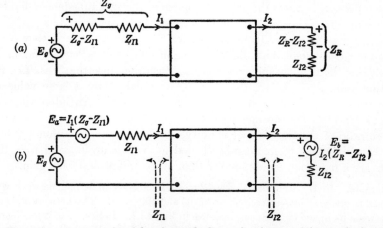

FIG. 13.3.   A network with mismatched terminations and its equivalent.

purposes of analysis, the impedance of the source is broken into two parts, Z_{I1} and $Z_g - Z_{I1}$. Similarly, the receiving-end impedance is expressed as the sum of Z_{I2} and $Z_R - Z_{I2}$. Next, the voltage drop across the impedance $Z_g - Z_{I1}$ is replaced by a fictitious generator that has the correct

[1] See T. E. Shea, "Transmission Networks and Wave Filters," pp. 114–120, D. Van Nostrand Company, Inc., New York, 1929.

voltage $E_a = I_1(Z_g - Z_{I1})$, as shown in Fig. 13.3b. A generator with a voltage $E_b = I_2(Z_R - Z_{I2})$ similarly replaces the extra impedance in the output loop, as shown in the illustration.

Next, the output current caused by each of the three generators is found separately. By the principle of superposition, the sum of these three currents is the correct output current. This method is simpler than a direct computation because, for each of the three generators, the network is image-matched. Furthermore, the equivalent circuit of Fig. 13.3b brings into evidence the two voltages caused by the mismatched impedances which, by analogy with transmission lines, can be ascribed to reflections that take place at the terminals. In fact, the expression for the insertion ratio that we shall derive applies equally well to the unmatched transmission line, and, by placing $Z_{I1} = Z_{I2} = Z_0$, it can be reduced to the expression derived for the uniform line in Chap. 4.

The voltages E_g and E_a in Fig. 13.3b operate into an impedance equal to $2Z_{I1}$, and so we have

$$\text{First component of } I_1 = \frac{E_g - E_a}{2Z_{I1}}$$

Because the equivalent network is image-matched, we can use the current ratio (13.4). Then

$$\text{First component of } I_2 = \frac{(E_g - E_a)\epsilon^{-\theta}}{2\sqrt{Z_{I1}Z_{I2}}}$$

The voltage E_b operates into the impedance $2Z_{I2}$, and so, taking its direction into account, we have

$$\text{Second component of } I_2 = -\frac{E_b}{2Z_{I2}}$$

The component of I_1 produced by E_b can be obtained by using the current ratio (13.4) with the subscripts 1 and 2 interchanged. Taking its direction into account, we obtain

$$\text{Second component of } I_1 = -\frac{E_b\epsilon^{-\theta}}{2\sqrt{Z_{I2}Z_{I1}}}$$

Adding the components together and substituting the values of E_a and E_b from Fig. 13.3, we have

$$I_1 = \frac{E - I_1(Z_g - Z_{I1})}{2Z_{I1}} - \frac{I_2(Z_R - Z_{I2})\epsilon^{-\theta}}{2\sqrt{Z_{I1}Z_{I2}}} \tag{13.9}$$

and

$$I_2 = \frac{E - I_1(Z_g - Z_{I1})\epsilon^{-\theta}}{2\sqrt{Z_{I1}Z_{I2}}} - \frac{I_2(Z_R - Z_{I2})}{2Z_{I2}} \tag{13.10}$$

We now have two equations with two unknowns, I_1 and I_2. These can be solved for the output current, I_2. Then, to obtain the insertion ratio, the current I_2 is divided into the current that would be obtained with a direct connection, I_2', as given by Eq. (13.3). The conventional way of expressing the result is as a product of several factors, each of which can be given a physical meaning:

$$\text{Insertion ratio} = \frac{I_2'}{I_2} = \frac{F\epsilon^\theta}{F_1 F_2 \sigma} \tag{13.11}$$

where $F = 2\sqrt{Z_g Z_R}/(Z_g + Z_R)$, the reflection factor of Z_g and Z_R
$F_1 = 2\sqrt{Z_g Z_{I1}}/(Z_g + Z_{I1})$, the reflection factor of Z_g and Z_{I1}
$F_2 = 2\sqrt{Z_R Z_{I2}}/(Z_R + Z_{I2})$, the reflection factor of Z_R and Z_{I2}
σ is an "interaction factor" defined by

$$\frac{1}{\sigma} = 1 - \left(\frac{Z_g - Z_{I1}}{Z_g + Z_{I1}}\right)\left(\frac{Z_R - Z_{I2}}{Z_R + Z_{I2}}\right)\epsilon^{-2\theta} \tag{13.12}$$

The interaction factor is unity if any one of the following conditions is obtained: (a) if the attenuation of the network is very large, (b) if Z_R matches Z_{I2}, or (c) if Z_g matches Z_{I1}. We can regard it as caused by a wave that is reflected at the load, transmitted back through the network, rereflected at the generator, and transmitted again to the load.

As compared with the insertion ratio (13.6) for the image-matched network, the effects of mismatching at the terminals of the network are given by the product of an input-end reflection factor, an output-end reflection factor, and an interaction factor which is generally near unity in value.

Example. We shall change the example of the preceding section (see Fig. 13.2) by assuming that the receiving-end resistance is reduced to $Z_R = 200$ ohms, and shall compute the insertion loss under this condition. The reflection factor of Z_g and Z_R is now

$$F = \frac{2\sqrt{100 \times 200}}{100 + 200} = 0.941$$

The reflection factor of Z_g and Z_{I1} is unity, because these two quantities are still equal. (Of course, the input impedance of the network is no longer equal to Z_{I1}.) The reflection factor of Z_{I2} and Z_R is

$$F_2 = \frac{2\sqrt{200 \times 500}}{200 + 500} = 0.903$$

The interaction factor is unity because Z_g matches Z_{I1}. As before, $\epsilon^\theta = 1/\underline{63.4°}$. Now, by Eq. (13.11), we obtain

$$\text{Insertion ratio} = \frac{0.941 \times 1/\underline{63.4°}}{1 \times 0.903 \times 1} = 1.042/\underline{63.4°}$$

The insertion phase shift is still 63.4°, but now we have

$$\text{Insertion loss} = 20 \log_{10} 1.042 = 0.36 \text{ db}$$

The receiving-end impedance is so much below the value for which the network was designed that the result is now a power loss instead of a power gain.

PROBLEMS

1. An ideal transformer is used to match on an image basis between a 5-ohm resistive load and a generator with an internal resistance of 5,000 ohms. Find the insertion ratio and the insertion gain provided by the transformer.

2. An impedance-matching network is designed to match on an image basis between a load resistance of 70 ohms and a generator that has an internal resistance of 5,000 ohms. The image transfer constant of this network is $\theta = j1.45$ rad. Compute the insertion ratio, the insertion phase shift, and the insertion gain provided by this network. Compute the ratio of the load power with the network inserted to the load power with a direct connection.

3. The impedance-matching network of Prob. 2 is operated with a load resistance of 70 ohms, but the internal resistance of the generator is now 1,500 ohms. Determine the insertion loss and the insertion phase shift for this condition of operation.

4. An unsymmetrical attenuating network is designed with the image impedances $Z_{I1} = 5,000$ ohms and $Z_{I2} = 977$ ohms. The image transfer constant is $\theta = 2.19 + j0$ nepers. The network operates into a 977-ohm load resistance and is driven by a generator that has an internal resistance of 5,000 ohms. Determine the insertion ratio and the insertion loss in both nepers and decibels. Observe the difference in meaning between the insertion loss and the image attenuation constant of 2.19 nepers.

5. The attenuating network of Prob. 4 is now operated with a load resistance of 200 ohms. The internal resistance of the generator is still 5,000 ohms. Determine the insertion loss for this condition of operation.

6. An impedance-matching network with the image impedances Z_{I1} and Z_{I2} is operating out of a generator which has the correct value of internal resistance: $Z_g = Z_{I1}$. However, the load resistance has an incorrect value: $Z_R \neq Z_{I2}$. Show that a load resistance $Z_R = \sqrt{Z_{I1}Z_{I2}}$ will cause the magnitude of the insertion ratio to be precisely unity.

7. Show that, for a uniform transmission line, the insertion ratio given by Eq (13.11) reduces to the expression given by Eq. (4.76).

CHAPTER 14

FILTERS

14.1. Types of Filters. A filter is a four-terminal network which transmits energy freely within certain bands of frequency and which attenuates outside these bands. Filters are generally used to separate signals on the basis of frequency, but they also find use in other ways, *e.g.*, as delay networks and as coupling networks in vacuum-tube amplifiers.

The name "filter" has slightly different meanings in connection with different problems. The simplest circuit to which the name is applied is the *R-C* filter, which consists of series resistance and shunt capacitance, sometimes built in several sections arranged into a ladder network. This network attenuates the higher frequencies more than the low ones, but the pass band is not sharply defined and this filter is suitable only where the requirements are not severe. The simple tuned circuit is often used for its frequency-selective properties, and the tuned coupled circuit is useful because it can be designed to have a flat-topped band-pass characteristic, but these circuits are not generally called filters. The name is sometimes applied to the bridged-T and parallel-T circuits shown in Fig. 14.1, which can be designed to have zero transmission at a single frequency. The parallel T is particularly useful as a band-elimination circuit at very low frequencies, for it avoids the difficulty of obtaining a suitable low-frequency inductive reactance. When used as the feedback circuit of an amplifier, an over-all band-pass action can be obtained.[1] The filters that we shall describe in this chapter and the next are less specialized and more flexible than any of the foregoing ones, and they offer much greater freedom in the characteristics that can be obtained. The number and widths of the pass bands, the sharpness of cutoff, the attenuation outside the pass bands, and the impedance level, all can be selected within wide limits. In contrast with the simple tuned circuit, the Q's of the elements are kept high; low Q's are not necessary in order to obtain broad pass bands.

Although one of the low-pass filters which we shall describe shortly is

---

[1] See Leonard Stanton, Theory and Application of Parallel-T Resistance-capacitance Frequency-selective Networks, *Proc. IRE*, Vol. 34, pp. 447–456, July, 1946; also, W. N. Tuttle, "Bridged-T and Parallel-T Null Circuits for Measurement at Radio Frequencies," *Proc. IRE*, Vol. 28, pp. 23–29, January, 1940.

similar in structure to the filters that are employed to smooth the output voltage of rectifier power supplies, the design of the latter follows a considerably different point of view. We shall not discuss here the special problems of filters for rectifier power supplies but shall refer the reader to the literature.[1]

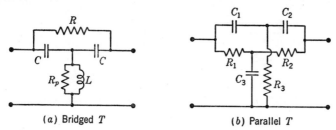

(a) Bridged T (b) Parallel T

FIG. 14.1. Bridged-T and parallel-T circuits which provide zero transmission at a single frequency.

A filter may in principle have any number of pass bands separated by attenuation bands; however, there are only four common types: low pass, high pass, band pass, and band elimination.

The filters most generally used are made up of T or π sections and L "half sections" connected on an image basis to form a ladder network, as

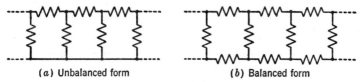

(a) Unbalanced form   (b) Balanced form

FIG. 14.2. Ladder networks.

shown in Fig. 14.2. A more general structure, called the *lattice*, is shown in Fig. 14.3. Not only can the performance of any T or π be duplicated at all frequencies by a lattice, but a lattice can be designed to provide characteristics unobtainable with the T or π. An unbalanced equivalent of the lattice, which uses a unity-turns-ratio transformer, is shown in Fig. 14.3b. The difficulties associated with the use of a transformer can be avoided if the lattice can be reduced to a T or π equivalent, or, if this is not possible, a bridged-T equivalent without a transformer can sometimes be found. While the lattice is more general and flexible than the ladder-type structure, the ladder filter is adequate for most purposes. Some of the characteristics of the lattice will be described briefly in this chapter, but a

[1] See, for example, F. E. Terman, "Radio Engineering," 3d ed., Chap. 11, McGraw-Hill Book Company, Inc., New York 1947.

reasonably full discussion of its theory is beyond the scope of this book and the reader is referred to the literature.[1] We shall concentrate here mainly on the ladder type.

The impedance links indicated by zigzag lines in the diagrams are made

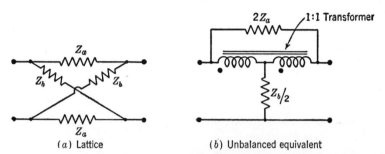

(a) Lattice   (b) Unbalanced equivalent

FIG. 14.3.   A lattice section and a bridged-T equivalent using a transformer.

as nearly pure reactance as possible, so that the filter will absorb very little power in the pass bands. The filter operates not by the absorption and dissipation of energy at the unwanted frequencies, but by presenting a reactive input impedance at these frequencies and thus rejecting the power at its input terminals.

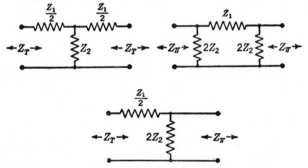

FIG. 14.4.   The T and π sections, and the L half section, of a ladder-type filter. The L can be faced either way.

14.2. Transmission and Attenuation Bands of the Ladder Filter. The characteristics of the ladder-type filter will be investigated with the aid of the formulas for the L, T, and π sections as developed in Chap. 12. The notation is shown in Fig. 14.4. To determine a criterion for the location

[1] An excellent short discussion is given by F. E. Terman, "Radio Engineers' Handbook," pp. 238–244, McGraw-Hill Book Company, Inc., New York, 1943. A more complete treatment, based on the work of Cauer and Bode, is given by E. A. Guillemin, "Communication Networks," Vol. II, Chap. X, John Wiley & Sons, Inc., New York, 1935.

of the pass and attenuation bands, we examine the expression for the image transfer constant. This was given in Eq. (12.32) as

$$\cosh \theta = 1 + \frac{Z_1}{2Z_2} \qquad (14.1)$$

where θ is the image transfer constant of a symmetrical T or π section, and is double the image transfer constant of an L half section. For a symmetrical T or π, the quantity ϵ^θ represents the complex ratio of input to output current when the section is terminated in its image impedance. For an L half section, the volt-ampere definition must be used [see Eqs. (12.14) and (12.15)].

In analyzing the performance of a filter, we shall neglect the losses in the elements entirely. Even if the Q's of the coils are only moderately high, the losses affect the performance in a minor way. Further, we shall assume, at least temporarily, that the filter is always terminated in its image impedance. The effect of this assumption will be discussed in the next section.

To investigate the image-terminated behavior, we express θ as $\alpha + j\beta$ and expand Eq. (14.1) by the hyperbolic identity:

$$\cosh \alpha \cos \beta + j \sinh \alpha \sin \beta = 1 + \frac{Z_1}{2Z_2} \qquad (14.2)$$

Since Z_1 and Z_2 are pure reactances, their quotient is real: positive if both reactances have the same sign, and negative if they have opposite signs. Therefore, the imaginary part of Eq. (14.2) must always be zero, and so we have

$$\sinh \alpha \sin \beta = 0 \qquad (14.3)$$

and

$$\cosh \alpha \cos \beta = 1 + \frac{Z_1}{2Z_2} \qquad (14.4)$$

The first of these relations shows that either α is zero or else β is a multiple of π rad. We shall consider Eq. (14.4) in the light of these two possibilities. The same conclusions will apply to all three types of section shown in Fig. 14.4, for their image transfer constants are all expressed in the same way.

1. *Pass Band.* The condition $\alpha = 0$ corresponds to a pass band. Then, $\cosh \alpha = 1$, and Eq. (14.4) gives for the image phase shift of the section:

$$\cos \beta = 1 + \frac{Z_1}{2Z_2} \qquad (14.5)$$

Since the cosine of a real angle is limited to values between -1 and $+1$, the pass band falls in the region where

$$-1 < 1 + \frac{Z_1}{2Z_2} < 1$$

Subtracting unity from each member and dividing through by 2, we obtain the condition for a pass band:

$$-1 < \frac{Z_1}{4Z_2} < 0 \qquad (14.6)$$

2. *Attenuation Bands.* If Eq. (14.3) is made equal to zero by having $\beta = -\pi$, 0, or π rad, the attenuation is not in general zero and the result is an attenuation band. We must distinguish between two cases as follows:

   a. Suppose that β is zero. Then $\cos \beta = 1$ and Eq. (14.4) gives for the image attenuation constant of the section:

$$\cosh \alpha = 1 + \frac{Z_1}{2Z_2} \qquad (14.7)$$

Since the hyperbolic cosine of a real argument is always greater than unity, the quantity $Z_1/4Z_2$ must be positive here.

   b. Suppose that Eq. (14.3) is made zero by having $\beta = \pm \pi$. Then $\cos \beta = -1$, and Eq. (14.4) yields

$$\cosh \alpha = -1 - \frac{Z_1}{2Z_2} \qquad (14.8)$$

The necessity of having a hyperbolic cosine greater than unity restricts $Z_1/4Z_2$ to negative values less than -1.

The foregoing three conditions are summarized graphically in Fig. 14.5, which shows the attenuation and phase constants per section for the condition of image termination, plotted against the ratio $Z_1/4Z_2$ by use of Eqs. (14.5), (14.7), and (14.8). The algebraic sign of β cannot be determined from Eq. (14.5). However, it is shown in the next section that β is positive if Z_1 is an inductive reactance; negative, if Z_1 is a capacitive reactance.

As the frequency varies, the reactances which compose the series and shunt arms change in value. Therefore, the ratio $Z_1/4Z_2$ changes with frequency. Some ranges of frequency will lie in the region of free transmission, while others will fall in a band of attenuation. The arrangement of the pass and attenuation bands along the frequency scale will depend on the particular configuration of reactive elements used in the two sets of arms. This will be illustrated shortly by examples.

The critical frequencies where the filter passes from a pass band to an attenuation band are called the *cutoff frequencies*. This name is sometimes qualified by the word "nominal," because the filter does not cut off abruptly at this point. Instead, as shown by Fig. 14.5, the attenuation rises smoothly beyond nominal cutoff and reaches a respectable value only when $Z_1/4Z_2$ is well within the attenuation band.

The criterion for a pass band was given by Eq. (14.6): the ratio $Z_1/4Z_2$ must lie between -1 and 0. A convenient graphical method for applying this criterion is to sketch Z_1 and $-4Z_2$ as functions of frequency (strictly speaking, one sketches the reactances X_1 and $-4X_2$, but we shall tacitly understand this to be the case and shall not introduce further symbols)

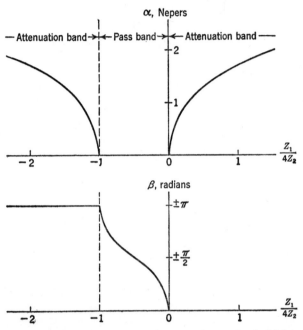

FIG. 14.5. The image attenuation and phase constants for ladder-type sections, plotted as functions of the impedance ratio $Z_1/4Z_2$.

Then, for a pass band, Z_1 and $-4Z_2$ must have the same sign; also, the magnitude of Z_1 must be smaller than the magnitude of $-4Z_2$.

Example 1. Consider a ladder section made up of an inductance L in the series arm and a capacitance C in the shunt arm, so that $Z_1 = j\omega L$ and $Z_2 = -j/\omega C$. The section may be arranged in either the T or π form, as indicated in Fig. 14.6, or it may be an L half section. Since the details of this arrangement do not affect the location of the pass and attenuation bands, the configuration is often shown only vaguely by drawing the full series and shunt arms separately, as in the first sketch. Figure 14.7 is a plot of the variation of Z_1 and $-4Z_2$ with frequency. The first is a straight line through the origin; the second is a rectangular hyperbola. Now we apply the criterion given above. The impedance Z_1 always has the same sign as

$-4Z_2$, but it has a smaller magnitude only to the left of the intersection of the two curves. Hence, the pass band lies between zero frequency and this intersection, and the network is a low-pass filter. The cutoff frequency, ω_c, is determined by

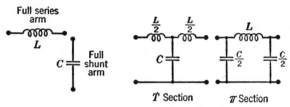

FIG. 14.6.   Filter section for Example 1.

the intersection, where we have

$$Z_1 = -4Z_2$$

or

$$\omega_c L = -4\left(\frac{-j}{\omega_c C}\right)$$

The nominal cutoff frequency is, therefore, given by

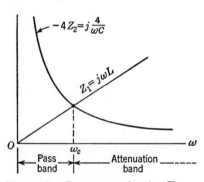

FIG. 14.7.   Reactance plot for Example 1.

$$\omega_c = \frac{2}{\sqrt{LC}} \qquad \text{rad/sec} \qquad (14.9)$$

or

$$f_c = \frac{1}{\pi\sqrt{LC}} \qquad \text{cps} \qquad (14.10)$$

The impedance ratio for use in Fig. 14.5 is

$$\frac{Z_1}{4Z_2} = -\frac{\omega^2 LC}{4} = -\left(\frac{\omega}{\omega_c}\right)^2$$

$$= -\left(\frac{f}{f_c}\right)^2 \qquad (14.11)$$

Referring to Fig. 14.5, we can imagine a point which starts at the origin at zero frequency and which moves leftward as the frequency increases. It reaches the edge of the pass band at a frequency $f = f_c$ and, above this frequency, moves into the attenuation band. The ratio $Z_1/4Z_2$ is always negative for this particular filter; consequently, it never reaches the attentuation band located to the right of the origin in Fig. 14.5.

Example 2.   A second example is shown in Fig. 14.8. The illustration shows the full shunt and series arms and also the T and π configurations. The reactance plot for this filter section is given in Fig. 14.9. At low frequencies, the reactance of the series arm nearly follows the hyperbolic variation of the capacitance alone. Resonance is obtained at ω_1. At high frequencies, the reactance approaches the linear variation which would be caused by the inductance alone. The impedance Z_1 has the same sign as $-4Z_2$ only above ω_1; its magnitude is smaller than that of $-4Z_2$ only below the intersection at ω_2. Consequently, this is a band-pass filter.

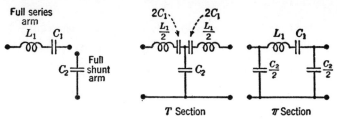

FIG. 14.8. Filter section for Example 2.

The lower cutoff frequency is determined by $Z_1 = 0$, or

$$\omega_1 L_1 - \frac{1}{\omega_1 C_1} = 0$$

from which

$$\omega_1 = \frac{1}{\sqrt{L_1 C_1}} \qquad \text{rad/sec} \qquad (14.12)$$

The upper cutoff frequency is determined by the intersection of the two graphs:

$$j\left(\omega_2 L_1 - \frac{1}{\omega_2 C_1}\right) = j\frac{4}{\omega_2 C_2}$$

which gives

$$\omega_2 = \sqrt{\frac{4C_1 + C_2}{L_1 C_1 C_2}} \qquad \text{rad/sec} \qquad (14.13)$$

The impedance ratio $Z_1/4Z_2$ is given by

$$\frac{Z_1}{4Z_2} = \frac{1}{4}\left(\frac{C_2}{C_1} - \omega^2 L_1 C_2\right) \qquad (14.14)$$

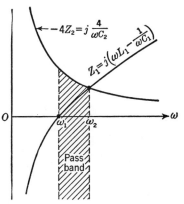

FIG. 14.9. Reactance plot for Example 2.

Referring to the graphs of Fig. 14.5, the impedance ratio starts in the right-hand attenuation band at the value $Z_1/4Z_2 = C_2/4C_1$ when the frequency is zero. As the frequency increases, the point representing the impedance ratio moves to the left through the pass band and finally enters the left-hand attenuation band.

14.3. The Image Impedances of the Ladder Filter. We shall now examine the image impedances of the ladder-type sections (see Fig. 14.4). The "mid-series" image impedance was given by Eq. (12.12) and is

$$Z_T = \sqrt{Z_1 Z_2}\sqrt{1 + \frac{Z_1}{4Z_2}} \qquad (14.15)$$

The "mid-shunt" image impedance is given by

$$Z_\pi = \frac{\sqrt{Z_1 Z_2}}{\sqrt{1 + Z_1/4Z_2}} \qquad (14.16)$$

Now, for a pass band, the ratio $Z_1/4Z_2$ must lie between -1 and 0. This means that $1 + Z_2/4Z_2$ is positive, and its square root is real. Also, the negative ratio $Z_1/4Z_2$ means that Z_1 and Z_2 are reactances of opposite type; hence, their product is positive, and $\sqrt{Z_1 Z_2}$ is real. The product of these two quantities (Z_T) and their quotient (Z_π) are therefore real impedances, i.e., resistances, in a pass band.

In the right-hand attenuation band of Fig. 14.5, the quantity $1 + Z_1/4Z_2$ is still positive, but now Z_1 and Z_2 are reactances of the same type and their product is negative. Hence, both Z_T and Z_π are reactances in this band.

In the left-hand attenuation band, the product $Z_1 Z_2$ is positive, but $1 + Z_1/4Z_2$ is negative. Again, in this band, both Z_T and Z_π are reactances.

From the foregoing we draw the conclusion: the image impedances of the filter are resistive in a pass band are are reactive in an attenuation band.

Some filters are constructed with only a single section if the requirements are not severe, but those built for more exacting purposes have a number of sections connected in cascade. All the sections are designed with the same cutoff frequencies and are joined so that their image impedances match exactly at all frequencies. However, the different sections are usually designed to have different attenuation characteristics so that one will be highly effective where the others are deficient.

The terminating load of the filter generally approximates a resistance. Since the image impedances vary widely with frequency, it is not possible to achieve an image termination at all frequencies. Consequently, reflections are set up at the terminals; and the attenuation, phase shift, and insertion loss are not the same as would be computed on the image basis. Through a large part of the pass band, it is possible to match the load to the resistive image impedance reasonably well. Also, deep in the attenuation band, the reflections are attenuated greatly as they pass through the filter, and so their effect is not serious. But near the edges of the pass bands, where the image impedance is turning from real to imaginary and the attenuation is small, the reflections can alter the performance to a considerable degree. The effect of the reflections can be computed in detail by the use of Eq. (13.11) of the preceding chapter, which gave the insertion ratio for mismatched terminations of a four-terminal network. For many purposes, it is unnecessary to go to this refinement, and the behavior as computed for the image termination will serve adequately.

In the preceding section, the sign of β could not be determined from the cosine relation given by Eq. (14.5). The sign can now be found by a direct consideration of either a T or a π section, using the fact that the image impedance is resistive in a pass band. Imagine the T section of Fig. 14.4 to be terminated in its image impedance. Since a current in a parallel circuit divides inversely as the impedances, the input-output current ratio is

$$\frac{I_1}{I_2} = \frac{Z_2 + Z_1/2 + Z_T}{Z_2} = 1 + \frac{Z_1}{2Z_2} + \frac{Z_T}{Z_2} \qquad (14.17)$$

For the symmetrical T section, the ratio I_1/I_2 is equal to $\epsilon^\theta = \epsilon^{\alpha+j\beta} = \epsilon^\alpha \underline{/\beta}$, and so the angle of the current ratio is the image phase shift β. Now we use the expression (14.17) to determine the quadrant in which β will fall in the pass band. In a pass band, $Z_1/4Z_2$ is real and must lie between 0 and -1. Hence, the sum of the first two terms in Eq. (14.17), $1 + Z_1/2Z_2$, must be real and must lie between $+1$ and -1. The last term, Z_T/Z_2, consists of a resistive impedance divided by a reactive impedance; therefore, this term is imaginary and has a sign opposite to that of Z_2. The current ratio (14.17) thus consists of two parts: a real component which lies between $+1$ and -1, and an imaginary component that has a sign opposite to Z_2. If Z_2 is a negative reactance, the current ratio I_1/I_2 will lie in the first or second quadrant; if Z_2 is a positive reactance, the ratio will lie in the third or fourth quadrant. Thus, considered as an angle smaller than $180°$, β is positive if Z_2 is a negative reactance, and is negative if Z_2 is a positive reactance. But the reactances Z_1 and Z_2 have opposite signs in the pass band, and so we conclude that the angle β will have the same sign as the reactance Z_1.

It was mentioned above that the image impedances of the ladder filter are reactive in an attenuation band, but the sign of the reactance, i.e., whether inductive or capacitive, was not indicated. The matter of sign is not generally of practical interest here, as no attempt is generally made to terminate the filter in its reactive image impedance in an attenuation band. However, it may occasionally be of interest, and so we shall consider the matter briefly. One can use Eq. (14.17) to show that, with the reactive image-impedance termination, the current ratio I_1/I_2 will have a magnitude greater than unity if Z_T has the same sign as Z_1 in the attenuation band. On the other hand, if Z_T has the opposite sign, the magnitude of the current ratio I_1/I_2 will be smaller than unity and the attenuation is negative, i.e., the device is not actually a filter.[1] The formula (14.7) or (14.8) for the attenuation still holds, but α is the negative of the value that is obtained when Z_T has the same sign as Z_1 [note that $\cosh \alpha = \cosh(-\alpha)$]. In a similar way it can be shown that a positive attenuation

---

[1] This difficulty was pointed out to the author by Preston R. Clement.

is obtained with the reactive image termination when Z_π has the same sign as Z_2; otherwise the negative of this attenuation is obtained. A negative attenuation here does not violate the conservation of energy, for the load is reactive and a negative attenuation merely means that the output volt-amperes are greater than the input volt-amperes, the power being zero in both places.

When a filter of unknown design is available for test, its characteristics can be found by measuring the open- and short-circuit impedances over a range of frequency at either set of terminals. The image impedance at those terminals is given by Eq. (12.4):

$$Z_I = \sqrt{Z_{\text{open}} Z_{\text{short}}} \tag{14.18}$$

A pass band is obtained when the two measured reactances have opposite signs, for this makes Z_I real. The same sign for both reactances indicates an attenuation band.

The preceding section developed a criterion for locating the pass and attenuation bands of a ladder-type filter and derived formulas for the attenuation and phase constants for image termination. The present section has analyzed the image impedances and has indicated the use of the material of Chap. 13 when the departure from image termination must be taken into account. We are, therefore, in a position to analyze the performance of any given type of ladder filter structure. We have not yet covered the converse matter of selecting a filter structure to perform a given function and of determining the values of its elements so that the cutoff frequencies and image impedances have the correct values. This will be discussed in the next chapter. The remainder of the present chapter will be devoted to a consideration of the properties of the two-terminal reactive elements which form the arms of the filter, and also to a brief treatment of some of the properties of lattice sections.

14.4. The Properties of Two-terminal Reactive Networks. The design of filters is based on the properties of the two-terminal reactive arms of which the filter network is composed; particularly, on the variation of driving-point impedance with frequency. The most commonly used arms are the simple ones shown in Fig. 14.10. This illustration also shows the variation of reactance with frequency for each of the circuits. In Figs. 14.10c and d, the dashed curves show the reactances of the inductance and capacitance alone. The series LC circuit approaches the capacitive variation at low frequencies; far above resonance, the impedance is governed by the inductance alone. The parallel LC circuit follows the inductive reactance at low frequencies; above antiresonance it approaches the hyperbolic variation of the capacitance.

More complicated reactive networks will have additional points of

resonance and antiresonance. These are designated, respectively, as the
zeros and the *poles* of the impedance function. The impedance of a purely
reactive network is always either zero or infinite at both zero frequency and
at infinite frequency, and these are known as the "external" zeros or poles
of the impedance. The zeros and poles between these two extreme frequencies are said to be "internal." The *LC* series circuit has one internal
zero and no internal poles. The *LC* parallel circuit has one internal pole
and no internal zeros.

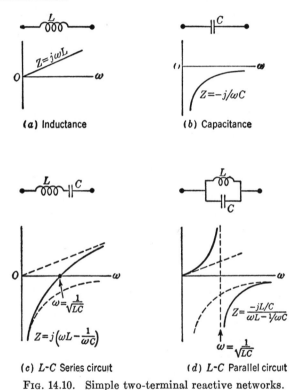

(a) Inductance   (b) Capacitance

(c) *L-C* Series circuit   (d) *L-C* Parallel circuit

FIG. 14.10.   Simple two-terminal reactive networks.

An important general rule for passive, reactive, two-terminal networks
is illustrated by the graphs of Fig. 14.10: the slope of the reactance *vs.*
frequency curve is always positive. With increasing frequency, the reactance always increases in the inductive direction. When it reaches a
pole, it descends discontinuously to infinity in the capacitive region and
then proceeds smoothly upward again. A general proof of this property
is beyond the scope of this book, and we shall content ourselves here with
showing that it is true for networks which can be broken down into series

and parallel combinations of reactive elements.[1] The graphs of Fig. 14.10
show that the slope of the reactance function is always positive for these
simple networks.   Now consider a series connection of two networks, each
of which has a reactance function with a positive slope everywhere.   Denoting these reactances by X_1 and X_2, we have for the total reactance

$$X = X_1 + X_2$$

or, taking the derivative with respect to frequency,

$$\frac{dX}{d\omega} = \frac{dX_1}{d\omega} + \frac{dX_2}{d\omega}$$

Each of the terms on the right is positive; hence, $dX/d\omega$ must be positive
also.   Similarly, consider a circuit made up of two networks in parallel,
each of which has an upward-sloping reactance function.   The net reactance of the combination is

$$X = \frac{X_1 X_2}{X_1 + X_2}$$

The derivative with respect to frequency is

$$\frac{dX}{d\omega} = \frac{X_2^2\, dX_1/d\omega + X_1^2\, dX_2/d\omega}{(X_1 + X_2)^2}$$

The denominator and both terms of the numerator are always positive,
and again the slope of the resulting reactance function is positive everywhere.   Thus, if we start with the simple elements of Fig. 14.10 and build
up a network of any complexity by series and parallel combinations taken
in any order, the driving-point reactance function of the resulting structure
will have positive slope everywhere.   Although the foregoing discussion
covers most of the commonly used networks, it is not a general proof, for
not all networks can be resolved into series and parallel combinations of
elements (a bridge circuit is a simple example).   However, the general
proof mentioned in the footnote shows that the result is correct for all reactive networks.

An important conclusion to be drawn from the positive slope of the reactance function is that the zeros and poles of the function must alternate.
For example, there cannot be two zeros in succession without a pole between them.   This is called the "separation property" of the poles and
zeros.

Figure 14.11 shows two different reactive networks, each of which has

[1] A general proof can be based on energy considerations. See Guillemin, *op.
cit.*, pp. 226–229; and H. W. Bode, "Network Analysis and Feedback Amplifier
Design," Secs. 9.2–9.4, D. Van Nostrand Company, Inc., New York, 1945.

the type of reactance function sketched in the illustration. Each has zero reactance at zero frequency because there is a path from one terminal to the other which avoids all condensers. Also, each has a pole at infinite frequency because there is no path which can avoid an inductance. The first network has two internal poles, which occur at the respective antiresonant frequencies of the parallel LC combinations. Now, as shown in Fig. 14.11, we draw a graph of a reactance function that has (1) a zero at zero frequency, (2) two internal poles, and (3) a pole at infinite frequency. To do this with a positive slope everywhere, there must be two internal zeros. Next consider the second network. Here there are two internal zeros, which occur at the resonant frequencies of the series LC branches. It will be found that the graph of the reactance can be drawn only by having two internal poles. Thus, both circuits have the same type of reactance variation. The types

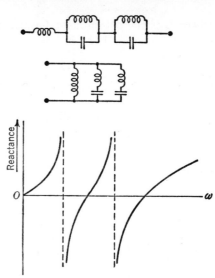

Fig. 14.11. Two reactive networks and their reactance function.

of reactance functions that can be obtained are obviously limited. This is discussed in a more quantitative fashion in the next section.

14.5. Foster's Reactance Theorem.[1] In a paper that is now a classic, Foster showed that the driving-point impedance of a two-terminal reactive network is completely determined as a function of frequency by the location of the internal poles and zeros, except for a constant multiplying factor which can be determined by specifying the impedance at some one frequency.[2] Thus, if two reactive networks have the same poles and zeros and, in addition, have the same impedance at one frequency, they will have equal impedances at all frequencies. Foster also gave a method for determining the elements of a network which would have the least number of elements and which would have poles and zeros at specified frequencies. He gave alternate forms of the network, as shown in Figs. 14.12 and 14.13. These forms are said to be "canonic," for they constitute a basic guide in the analysis and synthesis of two-terminal reactive networks.

---

[1] This section can be omitted without destroying the continuity of the text.

[2] Ronald M. Foster, A Reactance Theorem, *Bell System Tech. J.*, Vol. 3, pp. 259–267, April, 1924. Also see Guillemin, *op. cit.*. Chap. V, and Bode, *op. cit.*, Sec. 9.4.

As was shown in Eq. (10.12), the driving-point impedance of a network can be expressed as

$$Z = \frac{D}{A_{11}} \qquad (14.19)$$

where D is the determinant formed from the coefficients of the equations of the system, and A_{11} is the cofactor of the first row and first column. Each of the branches of the network will have an impedance of the general form $j(\omega L - 1/\omega C) = \dfrac{j}{\omega}\left(\omega^2 L - \dfrac{1}{C}\right)$. Each term in the expansion of an nth-order determinant consists of the product of n of its elements, each element being taken from a different row and column. The cofactor A_{11} has, of course, one less row and one less column than the determinant D. From this, we can reason that the expansion of Eq. (14.19) will be of the form

$$Z = \frac{j}{\omega}\left(\frac{a_0 + a_2\omega^2 + a_4\omega^4 + \cdots + a_{2n}\omega^{2n}}{b_0 + b_2\omega^2 + b_4\omega^4 + \cdots + b_{2n-2}\omega^{2n-2}}\right) \qquad (14.20)$$

Each of the polynomials in the numerator and denominator of this equation contains only even powers of ω, and so they can be factored into

$$Z = j\omega H \frac{(\omega^2 - \omega_1^2)(\omega^2 - \omega_3^2) \cdots (\omega^2 - \omega_{m+1}^2)}{\omega^2(\omega^2 - \omega_2^2)(\omega^2 - \omega_4^2) \cdots (\omega^2 - \omega_m^2)} \qquad (14.21)$$

where $H = a_{2n}/b_{2n-2}$
$m = 2n - 2$

The frequencies ω_1, ω_3, $\cdots$ are the locations of the zeros of impedance, and the frequencies ω_2 ω_4, $\cdots$ are the locations of the poles. As given above, the impedance has a pole at zero frequency. If instead there is to be a zero at zero frequency, ω_1 can be set equal to zero. Also, since the numerator is one degree higher in ω than the denominator, the impedance has a pole at infinite frequency. If the network is to have a zero at infinite frequency, the last factor in the numerator, $(\omega^2 - \omega_{m+1}^2)$, will be omitted, thus making the numerator of lower degree than the denominator. In this case, the constant H will turn out to be a negative number.

Example 1. Determine the expression for the driving-point impedance of a reactive network which has poles at $\omega =$ zero, 2,000 rad/sec, and infinity. Zeros are to be located at $\omega =$ 1,000 and 3,000 rad/sec. The impedance is to be $-j750$ ohms at 500 rad/sec.

From Eq. (14.21) we can write

$$Z = j\omega H \frac{(\omega^2 - 10^6)(\omega - 9 \cdot 10^6)}{\omega^2(\omega^2 - 4 \cdot 10^6)}$$

To determine H, we set Z equal to $-j750$ ohms at $\omega = 500$. Then

$$-j750 = j500H \frac{(0.250 - 1)(0.250 - 9)}{0.250(0.250 - 4)}$$

from which

$$H = 0.214$$

Then,

$$Z = j0.214\omega \frac{(\omega^2 - 10^6)(\omega^2 - 9\cdot 10^6)}{\omega^2(\omega^2 - 4\cdot 10^6)} \tag{14.22}$$

Example 2. A reactive network is to have zeros of impedance at $\omega = 0$, $\omega = 2,000$, and at infinite frequency. Poles are to be located at $\omega = 1,000$ and $\omega = 4,000$. The impedance is to be $j700$ at $\omega = 500$. We can write

$$Z = j\omega H \frac{\omega^2 - 4\cdot 10^6}{(\omega^2 - 10^6)(\omega^2 - 16\cdot 10^6)}$$

To evaluate H, we substitute

$$j700 = j500H \frac{0.25 - 4}{(0.25 - 1)(0.25 - 16)\cdot 10^6}$$

From this,

$$H = -4.41 \times 10^6$$

and

$$Z = -j4.41 \times 10^6 \omega \frac{\omega^2 - 4\cdot 10^6}{(\omega^2 - 10^6)(\omega^2 - 16\cdot 10^6)}$$

As a first step in determining a simple reactive network which will provide a given form of driving-point impedance, we attempt to break Eq. (14.21) apart into the sum of simple components. First, we note that, as ω approaches infinity, Z approaches $j\omega H$. Then we write Z as the sum of $j\omega H$ plus partial fractions, each of which approaches zero as ω approaches infinity. This gives us an expression of the form

$$Z = j\omega H - j\omega \left(\frac{A_0}{\omega^2} + \frac{A_2}{\omega^2 - \omega_2^2} + \cdots + \frac{A_m}{\omega^2 - \omega_m^2}\right) \tag{14.23}$$

where the A's are coefficients as yet undetermined. The minus sign in front of the parentheses has been used for later convenience, and the factor H has been absorbed into the A's.

Now we try to identify each of the terms with a simple circuit configuration. The first term, $j\omega H$, can be recognized immediately as an inductive impedance. The second term, $-jA_0/\omega$, corresponds to a series capacitance.

Each of the remaining terms can be identified with a parallel LC circuit, for the impedance of such a circuit can be written as

$$\frac{j\omega L(-j/\omega C)}{j(\omega L - 1/\omega C)} = \frac{-j\omega/C}{\omega^2 - 1/LC} = \frac{-j\omega/C}{\omega^2 - \omega_a^2}$$

where ω_a is the antiresonant frequency. We can now write Eq. (14.23) as the sum of impedances:

$$Z = j\omega L_{m+2} - j\omega \left(\frac{1/C_0}{\omega^2} + \frac{1/C_2}{\omega^2 - \omega_2^2} + \cdots + \frac{1/C_m}{\omega^2 - \omega_m^2} \right) \qquad (14.24)$$

in which we have written $j\omega L_{m+2}$ in place of $j\omega H$. The circuit that corresponds to this impedance is shown in Fig. 14.12.

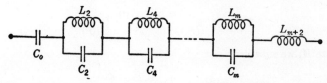

Fig. 14.12. Foster's first canonic form for a reactive driving-point impedance.

Now we use the foregoing equation to find a formula for the capacitances. Consider any term of the form $-j\omega/C_k(\omega^2 - \omega_k^2)$. If we multiply Eq. (14.24) through by $(\omega^2 - \omega_k^2)$ and then set $\omega = \omega_k$, all the terms on the right will go to zero except $-j\omega/C_k$. Thus, we get the formula

$$\frac{1}{C_k} = \left[\frac{(\omega^2 - \omega_k^2)}{\omega} jZ \right]_{\omega = \omega_k} \qquad (14.25)$$

where ω_k is the location of the pole of impedance. The expression for Z in this equation will be used in the form given by Eq. (14.21). For antiresonance at the proper frequency, we must have

$$L_k = \frac{1}{\omega_k^2 C_k} \qquad (14.26)$$

The capacitance C_0 can be regarded as the limit of a parallel LC circuit which is antiresonant at zero frequency and therefore, by Eq. (14.26), has infinite inductance in parallel with C_0. The inductance L_{m+2} can be regarded as the limit of a parallel LC circuit which is antiresonant at infinite frequency, and therefore has zero capacitance in parallel with L_{m+2}. The two formulas just given can be used to find all the elements of the Foster circuit except L_{m+2}. However, for this, we recall that

$$L_{m+2} = H \qquad (14.27)$$

The circuit of Fig. 14.12 produces poles at zero frequency and at infinite frequency. For a zero at zero frequency, we merely replace C_0 by a short circuit. For a zero at infinite frequency, we replace L_{m+2} by a short circuit. The formulas (14.25) and (14.26) still apply for the remaining elements.

Example 3. In Example 1 of this section, Eq. (14.22) gave the expresson for a reactive impedance which had prescribed poles and zeros, and had a given impedance at one frequency. Find the elements of the Foster circuit which will give this impedance.

For the capacitances, we substitute the expression for Z (Eq. 14.22) into the formula (14.25), and obtain

$$\frac{1}{C_k} = \left[-0.214 \frac{(\omega^2 - \omega_k{}^2)(\omega^2 - 10^6)(\omega^2 - 9 \cdot 10^6)}{\omega^2(\omega^2 - 4 \cdot 10^6)} \right]_{\omega = \omega_k}$$

This is to be evaluated at the two poles: $\omega_k{}^2 = 0$ and $\omega_k{}^2 = 4 \cdot 10^6$. For the first of these, we have

$$\frac{1}{C_0} = -0.214 \frac{(-10^6)(-9 \cdot 10^6)}{(-4 \cdot 10^6)} = 0.481 \times 10^6 \text{ farad}^{-1}$$

or

$$C_0 = 2.08 \times 10^{-6} \text{ farad}$$

For $\omega_k{}^2 = 4 \cdot 10^6$, we have

$$\frac{1}{C_2} = -0.214 \frac{(4 \cdot 10^6 - 10^6)(4 \cdot 10^6 - 9 \cdot 10^6)}{(4 \cdot 10^6)} = 0.803 \times 10^6 \text{ farad}^{-1}$$

or

$$C_2 = 1.246 \times 10^{-6} \text{ farad}$$

Using Eq. (14.26),

$$L_2 = \frac{0.803 \times 10^6}{4 \cdot 10^6} = 0.201 \text{ henry}$$

From Eq. (14.27), the series inductance that gives a pole at infinity is

$$L_4 = H = 0.214 \text{ henry}$$

The network consists of the following in series: C_0, the $L_2 C_2$ parallel combination, and L_4.

The second canonic form of Foster's network is shown in Fig. 14.13. Here, L_1 gives a zero at zero frequency, L_3 and C_3 produce a zero at ω_3, etc., and C_{m+1} produces a zero at infinite frequency. To derive this circuit, the admittance is

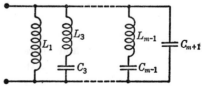

FIG. 14.13. Foster's second canonic form for a reactive driving-point impedance.

written in a form similar to (14.20), and a partial-fraction expansion is used. The terms are then identified with series LC branches. The formulas for the elements are[1]

$$L_k = \left[\frac{-j\omega Z}{\omega^2 - \omega_k{}^2}\right]_{\omega=\omega_k} \Bigg\}$$

$$C_k = \frac{1}{\omega_k{}^2 L_k} \Bigg\} \qquad (14.28)$$

where ω_k is the location of a zero of impedance. The capacitance C_{m+1}, which produces a zero of impedance at infinite frequency, is given by

$$C_{m+1} = \frac{1}{(-H)} \qquad (14.29)$$

where H is the constant (now negative) which multiplies the original expression for impedance. If the impedance has a pole at infinite frequency, C_{m+1} is replaced by an open circuit. If there is a pole at zero frequency, L_1 is infinite and is replaced by an open circuit. The formulas (14.28) still apply for the remaining elements.

14.6. Inverse, or Reciprocal, Impedances. Two impedances are said to be inverse, or reciprocal, if the variation of one with frequency is the inverse of the variation of the other. Then the product of the two is a

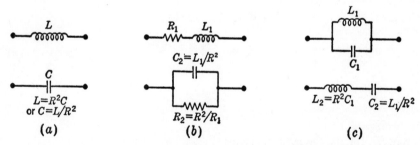

Fig. 14.14. Examples of networks which are reciprocal with respect to R^2.

constant which is independent of frequency. In the most important cases, the product is real and has the dimensions of a resistance squared:

$$Z_1 Z_2 = R^2 \qquad (14.30)$$

Here, Z_1 is said to be the reciprocal of Z_2 with respect to R^2.

Several simple examples of reciprocal networks are shown in Fig. 14.14. For the first network, we have

$$Z_1 Z_2 = (j\omega L)\left(\frac{-j}{\omega C}\right) = \frac{L}{C}$$

[1] See Foster, *op. cit.*, and Guillemin, *op. cit.*

Here, $R^2 = L/C$. If we start with a given value of L, the reciprocal impedance with respect to R^2 is obtained with $C = L/R^2$ farads. On the other hand, if the capacitance C is given, the reciprocal impedance is obtained with $L = R^2C$ henrys.

In the second example, we have

$$Z_1Z_2 = (R_1 + j\omega L_1)\left(\frac{1}{1/R_2 + j\omega C_2}\right)$$

If we make $R_2 = R^2/R_1$ and $C_2 = L_1/R^2$, the product Z_1Z_2 is equal to R^2, and the two impedances are reciprocal. Observe that the series connection of one circuit is replaced by a parallel one in the other, and that each element in one circuit is replaced by its reciprocal with respect to R^2 in the other.

In the third example of Fig. 14.14, the product of the impedances is

$$Z_1Z_2 = \left(\frac{-j}{\omega C_1 - 1/\omega L_1}\right)j\left(\omega L_2 - \frac{1}{\omega C_2}\right)$$

If $L_2 = R^2C_1$ and $C_2 = L_1/R^2$, the product Z_1Z_2 is R^2, and the impedances are reciprocal. The series connection replaces the parallel one, and each element is replaced by its reciprocal with respect to R^2. The pole of Z_1 is replaced by the zero of Z_2.

The principles illustrated in the foregoing examples can be extended to any network which consists of impedances connected in series and parallel, as shown in Fig. 14.15. To "reciprocate" the impedance Z_1 with respect to R^2, each series connection is replaced by a parallel one, and each parallel connection by a series one; in addition, each element is replaced by its reciprocal with respect to R^2.

In general, for two purely reactive networks to be reciprocal, the poles of each one must correspond with the zeros of the other. If the product Z_1Z_2 is then equal to R^2 at any one frequency, it will be equal to R^2 at all frequencies.

14.7. The Lattice Filter. As was stated in Sec. 14.1, we shall give

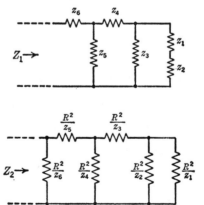

Fig. 14.15. Reciprocal networks.

only a brief introduction to the lattice filter here and shall refer the reader to the books mentioned in the last footnote of that section for a more complete treatment. The lattice, with its associated load impedance, can be regarded as a bridge circuit. This is illustrated in Fig. 14.16. Obviously, the bridge will be balanced and there will be zero transmission, *i.e.*, infinite attenuation, at any frequency where $Z_a = Z_b$.

The lattice is a symmetrical structure and has the same open- and short-circuit impedances at both ends. By inspection of Fig. 14.16a, these are

and

$$\left.\begin{array}{c} Z_{\text{open}} = \dfrac{Z_a + Z_b}{2} \\[4mm] Z_{\text{short}} = \dfrac{2Z_a Z_b}{(Z_a + Z_b)} \end{array}\right\} \qquad (14.31)$$

The image impedance will be obtained from Eq. (12.4). This impedance is the same at either end and is equal to the iterative impedance in either

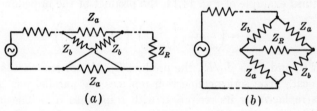

(a)   (b)

FIG. 14.16. The lattice, with its associated load impedance, can be redrawn as a bridge circuit.

direction, and so we shall simply call it the characteristic impedance, Z_0. Then

$$Z_0 = \sqrt{Z_{\text{open}} Z_{\text{short}}} = \sqrt{Z_a Z_b} \qquad (14.32)$$

If Z_a and Z_b are reactances of opposite sign, Z_0 is real—a resistance. Reasoning from the results for the ladder-type sections, we expect this to indicate a pass band. If Z_a and Z_b are reactances of the same sign, Z_0 will be a reactance, corresponding to a region of attenuation.

For this symmetrical network, the image transfer constant is equal to the input-output current or voltage ratio when the network is terminated in Z_0. We use Eq. (12.20) and write

$$\epsilon^{2\theta} = \frac{1 + \sqrt{Z_{\text{short}}/Z_{\text{open}}}}{1 - \sqrt{Z_{\text{short}}/Z_{\text{open}}}}$$

Substituting from Eq. (14.31) and clearing of fractions, we find

$$\epsilon^{2\theta} = \frac{1 + 2\sqrt{Z_a/Z_b} + Z_a/Z_b}{1 - 2\sqrt{Z_a/Z_b} + Z_a/Z_b}$$

$$= \left(\frac{1 + \sqrt{Z_a/Z_b}}{1 - \sqrt{Z_a/Z_b}}\right)^2$$

Then,

$$\epsilon^{\theta} = \epsilon^{\alpha}\epsilon^{j\beta} = \frac{1 + \sqrt{Z_a/Z_b}}{1 - \sqrt{Z_a/Z_b}} \tag{14.33}$$

If Z_a and Z_b are reactances of opposite signs (for which Z_0 is a resistance), Z_a/Z_b is negative and $\sqrt{Z_a/Z_b}$ is imaginary. Then the numerator and denominator of Eq. (14.33) are conjugate complex numbers. The quotient of two conjugate complex numbers has a magnitude of unity and an angle twice that of the numerator, and so we have

$$\left.\begin{aligned} \alpha &= 0 \\[2em] \beta &= 2\tan^{-1}\sqrt{-Z_a/Z_b} \end{aligned}\right\} \tag{14.34}$$

and

Obviously, this is a pass band.

If Z_a and Z_b are reactances of the same sign (which makes Z_0 a reactance), Z_a/Z_b is positive and $\sqrt{Z_a/Z_b}$ is real. Now we must distinguish between two possibilities:

(a) If $Z_a < Z_b$, the right side of Eq. (14.33) is positive. Since ϵ^{α} is always positive, we must have $\epsilon^{j\beta} = +1$. Then $\beta = 0$, and

$$\epsilon^{\alpha} = \frac{1 + \sqrt{Z_a/Z_b}}{1 - \sqrt{Z_a/Z_b}}$$

Solving for $\sqrt{Z_a/Z_b}$, we obtain

$$\sqrt{\frac{Z_a}{Z_b}} = \frac{\epsilon^{\alpha} - 1}{\epsilon^{\alpha} + 1}$$

$$= \frac{\epsilon^{\alpha/2} - \epsilon^{-\alpha/2}}{\epsilon^{\alpha/2} + \epsilon^{-\alpha/2}}$$

or

$$\tanh\frac{\alpha}{2} = \sqrt{\frac{Z_a}{Z_b}} \tag{14.35}$$

(b) On the other hand, if $Z_a > Z_b$, the right side of Eq. (14.33) will be negative. This will require β to be π rad or $-\pi$ rad, since $\epsilon^{j\pi} = \epsilon^{-j\pi} = -1$. Then Eq. (14.33) becomes

$$-\epsilon^{\alpha} = \frac{1 + \sqrt{Z_a/Z_b}}{1 - \sqrt{Z_a/Z_b}}$$

Solving for $\sqrt{Z_a/Z_b}$, we now obtain the reciproal of $\tanh \alpha/2$, and so we have

$$\tanh\frac{\alpha}{2} = \sqrt{\frac{Z_b}{Z_a}} \tag{14.36}$$

We have seen that a pass band is obtained in a lattice filter when Z_a and Z_b are reactances of opposite signs. As an example, consider the simple filter of Fig. 14.17. Here, Z_a and Z_b have opposite signs below the fre-

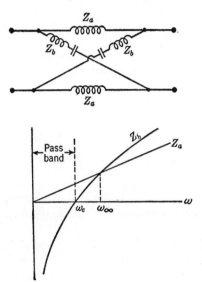

quency ω_c. This is, therefore, a low-pass filter with the cutoff frequency ω_c. At the frequency marked ω_∞ in the sketch, the two curves of reactance cross. Here the lattice, considered as a bridge, is in balance. At this frequency the transmission is zero and the attenuation is infinite.

FIG. 14.17. Illustrating the determination of the pass and attenuation bands of a lattice section.

The impedance Z_a can in general have any number of zeros in a pass band, provided that Z_b has poles at the same frequencies. Also, each pole of Z_a must be matched by a corresponding zero of Z_b. Then the two reactances remain of opposite sign. But in an attenuation band, the two reactances must have like signs, and so their poles and zeros must occur together. This is illustrated schematically in Fig. 14.18 for a low-pass filter. Obviously, there is a great deal of flexibility in the design of a lattice filter, for the number of poles and zeros is limited only by the number of elements that one is willing to put in each arm. Also, for a given number of poles and zeros, the exact placement of each is a matter of design. For these reasons, the attenuation and phase characteristics of the lattice filter and the variation of its characteristic impedance can be adjusted in a much more flexible way than those of the ladder filter.

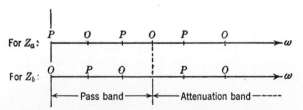

FIG. 14.18. Schematic representation of an arrangement of poles and zeros for a low-pass lattice section. P = pole, O = zero.

Some lattice sections can be reduced to T or π sections, and some others can be reduced to a bridged T without the use of a transformer. But others cannot be so reduced, and these provide characteristics not obtainable with the simpler structures. As an important example of the latter, we shall consider briefly the "all-pass" lattice filter, which is used as a phase-equalizing network to compensate for undesired phase *vs.* frequency characteristics in other parts of a transmission system.[1] The all-pass structure has no attenuation band; instead, it provides a phase shift which is a function of frequency. An all-pass lattice is obtained by making Z_a the inverse of Z_b. Then their product is constant, and Z_0 is a constant resistance. The quotient Z_a/Z_b will be a function of frequency and, as shown by Eq. (14.34), this will provide a phase shift which is a function of frequency. A simple example is shown in Fig. 14.19. At very low frequencies, the two upper terminals are connected directly together. At very high frequencies,

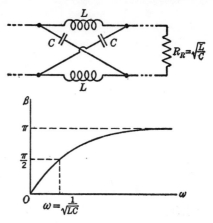

FIG. 14.19. An example of a phase-equalizing network.

the inductances block the flow of current and the capacitances become virtual short circuits, thus reversing the phase at the output terminals. Since $Z_a = j\omega L$ and $Z_b = -j/\omega C$, we have $Z_0 = \sqrt{L/C}$ at all frequencies. The phase shift is, by Eq. (14.34),

$$\beta = 2 \tan^{-1} \omega \sqrt{LC}$$

The variation of this with frequency is given in Fig. 14.19. Other phase *vs.* frequency characteristics can be obtained by more complex arms.

PROBLEMS

The first four problems below describe the full series and shunt arms of ladder-type sections. Sketch T and π sections for each of these. Using a reactance plot, locate the transmission and attenuation bands and determine formulas for the cutoff frequencies.

1. Series arm: capacitance C. Shunt arm: inductance L.

2. Series arm: inductance L_1. Shunt arm: inductance L_2 in series with a capacitance C.

[1] The equalization of attenuation is generally more important than the equalization of phase, although both are sometimes necessary. For a discussion of both attenuation and phase equalizers, see O. J. Zobel, Distortion Correction in Electrical Circuits with Constant-resistance Recurrent Networks, *Bell System Tech. J.*, Vol. VII, pp. 438–534, July, 1928; W. L. Everitt, "Communication Engineering," 2d ed., Chap. IX, McGraw-Hill Book Company, Inc., New York, 1937; F. E. Terman, "Radio Engineers' Handbook," pp. 244–249, McGraw-Hill Book Company, Inc., New York 1943

3. Series arm: inductance L_1 in series with a capacitance C. Shunt arm: inductance L_2.

4. Series arm: inductance L_1. Shunt arm: inductance L_2 in parallel with a capacitance C.

5. A certain ladder-type filter section has a series arm that consists of an inductance in series with a capacitance. The shunt arm consists of an inductance in parallel with a capacitance. The resonant frequency of the series arm is the same as the antiresonant frequency of the shunt arm. Use a reactance plot to determine the number and relative locations of the transmission and attenuation bands. What will happen if the resonant frequency of the series arm is made lower than the antiresonant frequency of the shunt arm?

6. Refer to Example 1 of Sec. 14.2.

a. Write algebraic expressions for Z_T and Z_π of this filter section (as functions of L, C, and ω).

b. If Z_T is to approach 500 ohms at very low frequencies, and if the cutoff frequency is to be 4,000 cps, find the required values of L and C. Sketch T and π sections, showing the values of the elements.

c. Using the results of part *b*, sketch Z_T as a function of frequency in the range $0 < f < 8,000$ cps.

d. Determine α and β at 2,000 cps and at 8,000 cps.

7. Refer to Example 2 of Sec. 14.2. Given $L_1 = 0.0795$ henry, $C_1 = 0.0354 \times 10^{-6}$ farad, and $C_2 = 0.0795 \times 10^{-6}$ farad.

a. Compute the cutoff frequencies in cycles per second.

b. Determine the value of Z_T in the middle of the pass band.

c. With the aid of Fig. 14.5, sketch α and β *vs.* frequency between zero frequency and 1.4 times the upper cutoff frequency.

8. Figure P8 shows a two-terminal reactive network.

a. Sketch the form of the input reactance as a function of frequency.

b. Find formulas for the frequencies at which the zeros occur (express in terms of L and C).

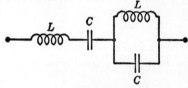

FIG. P8.

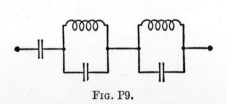

FIG. P9.

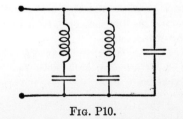

FIG. P10.

9. Figure P9 shows a two-terminal reactive network. Sketch the form of the input reactance as a function of frequency.

10. Sketch the form of the reactance as a function of frequency for the reactive two-terminal network shown in Fig. P10.

11. Sketch the circuit diagram of the second Foster network that will have the same type of driving-point reactance function as the network of Fig. P8.

12. A reactive two-terminal network is to have zero impedance at the following frequencies: zero, $\omega = 5{,}000$ rad/sec, and at infinite frequency. Poles of the impedance are to be located at $\omega = 2{,}000$ rad/sec and at $\omega = 8{,}000$ rad/sec. The impedance is to be $j500$ ohms at $\omega = 1{,}000$ rad/sec.

a. Write an expression for the impedance function. What would happen if we tried to specify a negative reactance between zero frequency and $\omega = 2{,}000$?

b. Design a network of Foster's first form that will have the desired characteristics.

c. Design a network of Foster's second form that will have the desired characteristics.

13. A certain open-circuited "lossless" transmission line has the following zeros and poles of its sending-end impedance: Poles at $f = 0$, $f = 2 \times 10^6$ cps, ... Zeros at $f = 10^6$ cps, $f = 3 \times 10^6$ cps, ... The sending-end impedance at 500,000 cps is $-j400$ ohms.

a. A reactive two-terminal network composed of lumped inductances and capacitances is to reproduce the foregoing data up to and including the zero at 3 Mc. It will have a pole at infinite frequency. Write an expression for the impedance function of this network.

b. Design a network of Foster's first form that will have the desired characteristics.

14. Given an inductance of 0.100 henry. Determine the element that is reciprocal to this with respect to R^2, where $R = 500$ ohms.

15. A resistance R_1 and a capacitance C_1 are connected in series. Find the network that is reciprocal with respect to R^2. Specify the numerical constants of the reciprocal network if $R_1 = 5{,}000$ ohms, $C_1 = 0.5 \times 10^{-6}$ farad, and $R = 700$ ohms.

16. Reciprocate the network of Fig. P16 with respect to R^2. Make a sketch of the reciprocal network and indicate the numerical values of the elements if $L = 0.100$ henry, $C = 2 \times 10^{-6}$ farad, and $R = 500$ ohms. Sketch the form of the impedance function *vs.* frequency for each of the two networks.

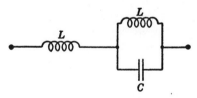

Fig. P16.

17. Refer to the lattice-filter section shown in Fig. 14.17. Let $Z_a = j\omega L$ and $Z_b = j(\omega L - 1/\omega C)$. Determine an expression for the cutoff frequency in terms of L and C. Determine the values of α and β at (*a*) half the cutoff frequency and (*b*) double the cutoff frequency.

18. A lattice filter has Z_a composed of an inductance L_1 in series with a capacitance C_1. The impedance Z_b consists of a capacitance C_2. Determine the location of the transmission and attenuation bands and write the expression for the cutoff frequency.

19. A lattice-filter section has Z_b composed of an inductance L in series with a capacitance C. The impedance Z_a is composed of an inductance $2L$ in series with a capacitance $2C$. Locate the transmission and attenuation bands, and determine formulas for the cutoff frequencies. Determine the values of α and β at the following three frequencies: at the lower cutoff frequency, at the upper cutoff frequency, and at the geometric mean between the two cutoff frequencies.

20. A lattice phase-equalizing section has Z_a composed of an inductance L in parallel with a capacitance C. The impedance Z_b is made up of an inductance L in series with a capacitance C. Determine a formula for the phase shift as a function of frequency. Sketch β vs. frequency for the numerical values $L = 0.050$ henry, $C = 0.25 \times 10^{-6}$ farad. What is the characteristic impedance of this section?

CHAPTER 15

THE DESIGN OF LADDER FILTERS

15.1. Introduction. Ladder-type sections are sometimes used singly but, when the requirements are at all severe, a number of sections are connected in cascade. The individual sections are designed to have the same cutoff frequencies and equal image impedances at their junctions. However, the sections are not all alike but are selected so that the desirable attenuation characteristics of one will compensate for deficiencies in another. In addition, the impedance-transforming properties of a half sec-

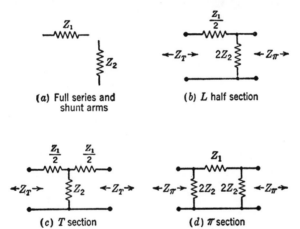

(a) Full series and shunt arms

(b) L half section

(c) T section

(d) π section

FIG. 15.1. Illustrating the notation for the ladder-type section.

tion are utilized at the ends of the filter to provide a more constant image impedance at the terminals. The result is called a *composite filter*.

The simplest ladder section is the so-called *constant-k* type. Derived from this is a more complex type of section, called *m-derived*, which is designed to match the constant-k in image impedance. We shall describe these two types in some detail and shall give brief mention to some other sections of the ladder type.

The notation used for the ladder-type section is shown in Fig. 15.1. Formulas for the image impedances and for the image attenuation and phase constants were derived in Secs. 14.2 and 14.3. These will be repeated here for convenience. The mid-series image impedance was given by Eq. (14.15) and is

$$Z_T = \sqrt{Z_1 Z_2} \sqrt{1 + \frac{Z_1}{4Z_2}} \tag{15.1}$$

The mid-shunt image impedance is, from Eq. (14.16),

$$Z_\pi = \frac{\sqrt{Z_1 Z_2}}{\sqrt{1 + Z_1/4Z_2}} \tag{15.2}$$

A *pass band* is obtained when $-1 < Z_1/4Z_2 < 0$. Then we have

and, from Eq. (14.5),
$$\left. \begin{array}{c} \alpha = 0 \\[2mm] \cos \beta = 1 + \dfrac{Z_1}{2Z_2} \end{array} \right\} \tag{15.3}$$

There are two possibilities for an *attenuation band*:

    a. If $Z_1/4Z_2 > 0$, we have, from Eq. (14.7),

and
$$\left. \begin{array}{c} \cosh \alpha = 1 + \dfrac{Z_1}{2Z_2} \\[2mm] \beta = 0 \end{array} \right\} \tag{15.4}$$

    b. If $Z_1/4Z_2 < -1$, we have, from Eq. (14.8),

and
$$\left. \begin{array}{c} \cosh \alpha = -1 - \dfrac{Z_1}{2Z_2} \\[2mm] \beta = \pm \pi \end{array} \right\} \tag{15.5}$$

The image phase and attenuation constants were plotted as functions of $Z_1/4Z_2$ in Fig. 14.5.

15.2. Constant-k Ladder Sections. A particularly important ladder-type section is obtained by selecting the series and shunt reactance arms so that the product of their impedances is a constant. In the notation originally used,

$$Z_1 Z_2 = k^2$$

where k is a positive real constant which is independent of frequency. From the discussion of Sec. 14.6, we see that the two impedances are reciprocal with respect to k^2, where k is a resistance, and so we shall write instead

$$Z_1 Z_2 = R^2 \tag{15.6}$$

However, we shall continue to use the name "constant-k section." The constant-k section is particularly simple and reasonably effective; in addition, its main faults can be remedied by using it in cascade with the m-derived sections described later.

The inverse reactive impedances Z_1 and Z_2 are always of opposite sign, and so their quotient is negative at all frequencies. Therefore, this type of section always has $Z_1/4Z_2 < 0$, and so case (a) for an attenuation band does not arise. Equations (15.4) do not apply, and the right-hand portion of Fig. 14.5 is not used. The nominal cutoff frequencies always correspond to

$$\frac{Z_1}{4Z_2} = -1 \qquad (15.7)$$

Four common types of constant-k sections will now be described: low-pass, high-pass, band-pass, and band-elimination.

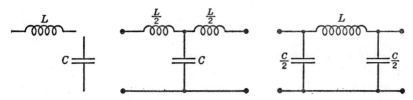

Fig. 15.2. The constant-k low-pass section.

Low-pass. The series arm is an inductance L, and the shunt arm is a capacitance C. The impedances are inverse, and so the product is constant:

$$Z_1 Z_2 = (j\omega L)\left(\frac{-j}{\omega C}\right) = \frac{L}{C} = R^2$$

Hence,

$$R = \sqrt{\frac{L}{C}} \qquad (15.8)$$

The series and shunt arms, and their connection into symmetrical T and π sections, are shown in Fig. 15.2. This was the filter treated in Example 1 of Sec. 14.2, and the reactance plot of Fig. 14.7 indicated that this was a low-pass structure. The nominal cutoff frequency can be found by use of the relation (15.7) and was also given by Eq. (14.10). It is

$$\omega_c = \frac{2}{\sqrt{LC}} \qquad (15.9a)$$

or

$$f_c = \frac{1}{\pi\sqrt{LC}} \qquad (15.9b)$$

The mid-series and mid-shunt image impedances of the section will be obtained from Eqs. (15.1) and (15.2) using $\sqrt{Z_1 Z_2} = R$. We can write

$$\frac{Z_1}{4Z_2} = -\frac{\omega^2 LC}{4}$$

or, using Eq. (15.9a),

$$\frac{Z_1}{4Z_2} = -\left(\frac{\omega}{\omega_c}\right)^2 = -\left(\frac{f}{f_c}\right)^2 \qquad (15.10)$$

Then the equations for the image impedances can be written as

$$Z_T = R \sqrt{1 - \left(\frac{f}{f_c}\right)^2} \qquad (15.11a)$$

and

$$Z_\pi = \frac{R}{\sqrt{1 - (f/f_c)^2}} \qquad (15.11b)$$

Sketches of these impedances as functions of frequency are given in Fig. 15.3. At low frequencies, they approach the value $R = \sqrt{L/C}$. At cutoff, Z_T is zero and Z_π is infinite. Above cutoff, the image impedances are pure reactances.

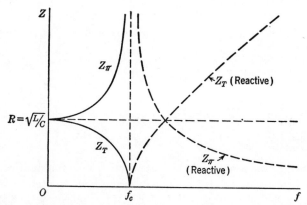

FIG. 15.3. The variation of the mid-series and mid-shunt image impedances of a low-pass constant-k section.

Assuming the section to be terminated on an image basis, we have $\alpha = 0$ in the pass band and, from Eqs. (15.3) and (15.10),

$$\beta = \cos^{-1}\left[1 - 2\left(\frac{f}{f_c}\right)^2\right] \qquad (15.12)$$

Above the cutoff frequency, the phase shift with image termination is $\beta = \pi$ rad and, from Eqs. (15.5) and (15.10),

$$\alpha = \cosh^{-1}\left[2\left(\frac{f}{f_c}\right)^2 - 1\right] \tag{15.13}$$

These quantities are plotted in Fig. 15.4.

The variation of image impedance makes it virtually impossible to terminate the section in its image impedance at all frequencies. The resulting mismatch will give rise to reflections at the terminals; consequently, the actual attenuation and phase shift of the section will be different from the image values given above. A terminating resistance near the value of R can be used, this resistance being selected to provide a reasonably good match over the main range of signals to be passed. In the multisection filter, the use of m-derived half sections at the ends provides a good impedance match over a considerable portion of the pass band. This is described in Sec. 15.4. The most serious reflection effects occur near the edge of the pass band and, for a precise design, the methods of Chap. 13 can be used to correct the results in this region. An exact analysis of a

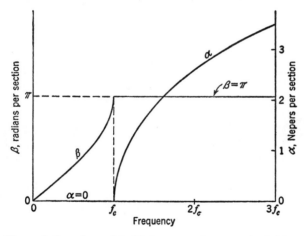

FIG. 15.4.   The variation of α and β with frequency for a constant-k low-pass section.

single low-pass constant-k section is given in Sec. 15.3, assuming that the section works into and out of resistances equal to R. Under this condition, the insertion loss at nominal cutoff is shown to be 3 db, and the insertion phase shift is 135°.

Design formulas for the low-pass constant-k section are obtained by solving Eqs. (15.8) and (15.9b) for L and C, giving

and
$$\left. \begin{aligned} L &= \frac{R}{\pi f_c} \\ C &= \frac{L}{R^2} = \frac{1}{\pi f_c R} \end{aligned} \right\} \qquad (15.14)$$

High-pass. The high-pass constant-k section has a series capacitance and a shunt inductance as shown in Fig. 15.5, and so

$$Z_1 Z_2 = \left(\frac{-j}{\omega C} \right) j\omega L = \frac{L}{C}$$

and so, as for the low-pass section,

$$R = \sqrt{\frac{L}{C}} \qquad (15.15)$$

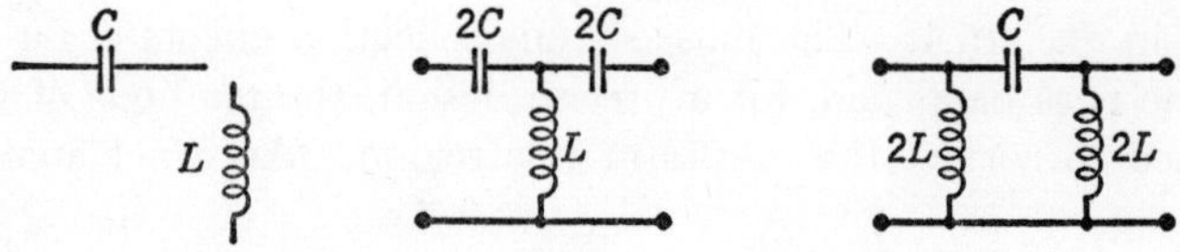

FIG. 15.5. The constant-k high-pass section.

The cutoff frequency is obtained by using Eq. (15.7),

$$\frac{Z_1}{4Z_2} = \frac{-j/\omega_c C}{4j\omega_c L} = -1$$

from which

$$f_c = \frac{\omega_c}{2\pi} = \frac{1}{4\pi\sqrt{LC}} \qquad (15.16)$$

The mid-series and mid-shunt image impedances are obtained from Eqs. (15.1) and (15.2) and can be written as

and
$$\left. \begin{aligned} Z_T &= R \sqrt{1 - \left(\frac{f_c}{f} \right)^2} \\ Z_\pi &= \frac{R}{\sqrt{1 - (f_c/f)^2}} \end{aligned} \right\} \qquad (15.17)$$

The image attenuation and phase constants are obtained from Eqs. (15.3) and (15.5). In the pass band, above f_c, we have

$$\beta = \cos^{-1}\left[1 - 2\left(\frac{f_c}{f} \right)^2 \right] \qquad (15.18)$$

Near the end of Sec. 14.3 it was shown that the sign of β in a pass band is the same as the sign of the series reactance. Hence, β is a negative angle here. In the attenuation band, below f_c, we have $\beta = -\pi$ and

$$\alpha = \cosh^{-1}\left[2\left(\frac{f_c}{f}\right)^2 - 1\right] \tag{15.19}$$

These quantities are sketched in Fig. 15.6.

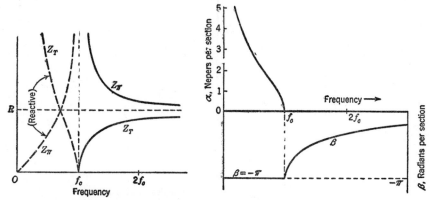

FIG. 15.6.   Characteristics of the high-pass constant-k section.

Design formulas for the high-pass constant-k section are obtained by solving (15.15) and (15.16) for C and L. The results are

$$\left.\begin{array}{l} C = \dfrac{1}{4\pi f_c R} \\[2ex] L = \dfrac{R}{4\pi f_c} \end{array}\right\} \tag{15.20}$$

and

Band-pass. The full series and shunt arms of the constant-k band-pass section are shown in Fig. 15.7. The zero of the series arm coincides with the pole of the shunt arm, and so the two are inverse reactances. The reactance plot of Fig. 15.7b shows that the structure has a single pass band. The product of the impedances is

$$Z_1 Z_2 = j\omega L_1 \left(1 - \frac{1}{\omega^2 L_1 C_1}\right)\left(\frac{-j/\omega C_2}{1 - 1/\omega^2 L_2 C_2}\right)$$

To make the zero of Z_1 coincide with the pole of Z_2, we must have

$$L_1 C_1 = L_2 C_2 \tag{15.21}$$

whereupon the product of the impedances becomes

$$Z_1 Z_2 = \frac{L_1}{C_2} = R^2$$

But, by Eq. (15.21), $L_1/C_2 = L_2/C_1$, and so

$$R = \sqrt{\frac{L_1}{C_2}} = \sqrt{\frac{L_2}{C_1}} \qquad (15.22)$$

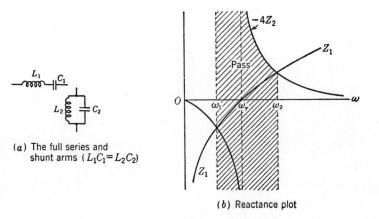

(a) The full series and
    shunt arms ($L_1 C_1 = L_2 C_2$)

(b) Reactance plot

FIG. 15.7. The band-pass constant-k section.

As shown by Eq. (15.7) and also by the reactance plot, the cutoff frequencies are obtained when $Z_1 = -4Z_2$, and so at cutoff we have

$$j\omega_c L_1 \left(1 - \frac{1}{\omega_c{}^2 L_1 C_1} \right) = \frac{4j}{\omega_c C_2 (1 - 1/\omega_c{}^2 L_2 C_2)}$$

If we substitute $L_1 C_1$ in place of $L_2 C_2$, this can be written as

$$\omega_c{}^2 L_1 C_2 \left(1 - \frac{1}{\omega_c{}^2 L_1 C_1} \right)^2 = 4$$

Extracting the square root of both sides, we have

$$\omega_c \sqrt{L_1 C_2} \left(1 - \frac{1}{\omega_c{}^2 L_1 C_1} \right) = \pm 2$$

This can be rearranged into

$$\omega_c{}^2 \pm \frac{2}{\sqrt{L_1 C_2}} \, \omega_c - \frac{1}{L_1 C_1} = 0$$

Solving this for ω_c, we obtain

$$\omega_c = \pm \frac{1}{\sqrt{L_1 C_2}} \pm \sqrt{\frac{1}{L_1 C_2} + \frac{1}{L_1 C_1}} \qquad (15.23)$$

The second negative sign will result in negative values for ω_c; hence, we ignore it and write the upper and lower cutoff frequencies as

$$\omega_1 = 2\pi f_1 = \sqrt{\frac{1}{L_1 C_2} + \frac{1}{L_1 C_1}} - \frac{1}{\sqrt{L_1 C_2}}$$

and

$$\omega_2 = 2\pi f_2 = \sqrt{\frac{1}{L_1 C_2} + \frac{1}{L_1 C_1}} + \frac{1}{\sqrt{L_1 C_2}}$$

(15.24)

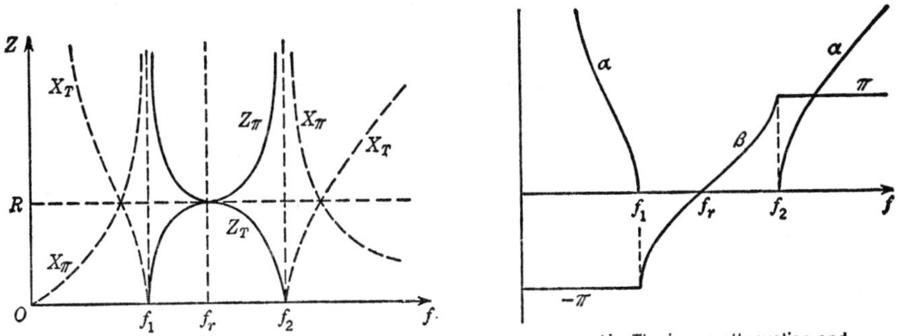

(a) Image inpedances

(b) The image attenuation and phase constants

Fig. 15.8. Characteristics of the band-pass constant-k section.

The product $\omega_1\omega_2$ is merely $1/L_1 C_1$, which is the square of the frequency at which Z_1 is resonant and Z_2 is antiresonant. Denoting this frequency by ω_r, we have the relation

$$\omega_r = \sqrt{\omega_1\omega_2}$$

(15.25)

Therefore, the resonant frequency occurs at the geometric mean of the two cutoff frequencies. If ω_1 and ω_2 are close together, this is nearly the same as the arithmetic mean.

Figure 15.8 shows the variation of Z_T, Z_π, α, and β, as computed from the equations of Sec. 15.1. The design equations are obtained by solving Eqs. (15.22) and (15.24) for the values of the elements. The results are

$$L_1 = \frac{R}{\pi(f_2 - f_1)}$$

$$C_1 = \frac{f_2 - f_1}{4\pi R f_1 f_2}$$

$$L_2 = \frac{R(f_2 - f_1)}{4\pi f_1 f_2}$$

and

$$C_2 = \frac{1}{\pi R(f_2 - f_1)}$$

(15.26)

Band-elimination. The constant-k band-elimination section has the configuration shown in Fig. 15.9, in which, to make the arms inverse, we must have

$$L_1C_1 = L_2C_2 \tag{15.27}$$

The reactance plot of Fig. 15.9*b* shows that there is a single attenuation band which is located between the frequencies ω_1 and ω_2. The attenuation is infinite at the resonant frequency ω_r, for here the series arm is an open circuit and the shunt arm is a short circuit. The product Z_1 and Z_2 is, as for the band-pass filter,

$$Z_1Z_2 = \frac{L_1}{C_2} = \frac{L_2}{C_1} = R^2 \tag{15.28}$$

(*a*) The full series and shunt arms

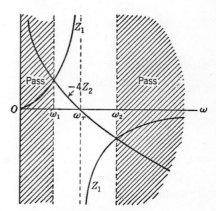

(*b*) Reactance plot

FIG. 15.9. The band-elimination constant-k section.

By following a procedure similar to that used for the band-pass filter, we can obtain the following design equations, where f_1 and f_2 are the cutoff frequencies, and R is the value of the image impedance at zero frequency and at infinite frequency:

$$
\left.
\begin{aligned}
L_1 &= \frac{R(f_2 - f_1)}{\pi f_1 f_2} \\[2mm]
C_1 &= \frac{1}{4\pi R(f_2 - f_1)} \\[2mm]
L_2 &= \frac{R}{4\pi(f_2 - f_1)} \\[2mm]
C_2 &= \frac{f_2 - f_1}{\pi R f_1 f_2}
\end{aligned}
\right\} \tag{15.29}
$$

and

15.3. Insertion Loss of a Single Low-pass Constant-k Section. When a low-pass action is desired and the requirements are not severe, a single constant-k section is sometimes used. The image termination cannot be achieved at all frequencies, and the departure from ideal image operation will be appreciable. This problem is simple enough to be analyzed by elementary circuit theory. As shown in Fig. 15.10, we shall assume a T

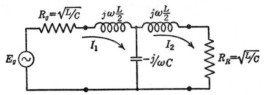

Fig. 15.10. A single low-pass constant-k section operating between resistances equal to $\sqrt{L/C}$.

section operating between resistances each equal to the characteristic impedance of the section at zero frequency, $R = \sqrt{L/C}$.

The mesh equations for the system are

$$\left[\sqrt{\frac{L}{C}} + j\left(\frac{\omega L}{2} - \frac{1}{\omega C}\right)\right] I_1 - \left(\frac{-j}{\omega C}\right) I_2 = E_g$$

and

$$-\left(\frac{-j}{\omega C}\right) I_1 + \left[\sqrt{\frac{L}{C}} + j\left(\frac{\omega L}{2} - \frac{1}{\omega C}\right)\right] I_2 = 0$$

(15.30)

Solving for I_2, the result can be written as

$$I_2 = \frac{E_g}{\sqrt{\frac{L}{C}}\left[2 - \omega^2 LC + j\omega\sqrt{LC}\,(2 - \omega^2 LC/4)\right]}$$

(15.31)

From Eq. (15.9a) we substitute $\sqrt{LC} = 2/\omega_c$, where ω_c is the nominal cutoff frequency of the section. Then we have

$$I_2 = \frac{E_g}{2\sqrt{L/C}}\left\{\frac{1}{1 - 2(\omega/\omega_c)^2 + j(\omega/\omega_c)[2 - (\omega/\omega_c)^2]}\right\}$$

(15.32)

If the network were removed from Fig. 15.10 and a direct connection substituted for it, the load current would be

$$I_2' = \frac{E_g}{R_g + R_R} = \frac{E_g}{2\sqrt{L/C}}$$

(15.33)

Therefore, we have

$$\text{Insertion ratio} = \frac{I_2'}{I_2} = 1 - 2\left(\frac{\omega}{\omega_c}\right)^2 + j\left(\frac{\omega}{\omega_c}\right)\left[2 - \left(\frac{\omega}{\omega_c}\right)^2\right]$$

(15.34)

The magnitude of this is obtained by taking the square root of the sum of the squares, and the result is simply

$$\left| \frac{I_2'}{I_2} \right| = \sqrt{1 + \left(\frac{\omega}{\omega_c} \right)^6} \qquad (15.35)$$

From this we obtain

$$\text{Insertion loss} = 20 \log_{10} \left| \frac{I_2'}{I_2} \right| = 10 \log_{10} \left[1 + \left(\frac{\omega}{\omega_c} \right)^6 \right] \text{ db} \quad (15.36)$$

The insertion phase shift is

$$\text{Insertion phase shift} = \tan^{-1} \left(\frac{\omega}{\omega_c} \right) \left[\frac{2 - (\omega/\omega_c)^2}{1 - 2(\omega/\omega_c)^2} \right] \qquad (15.37)$$

If the network were removed, the load voltage, E_R', would be in phase with E_g. The insertion phase shift is the angle by which the actual load voltage, E_R, lags E_R'; therefore, it is also equal to the angle by which the load voltage in Fig. 15.10 lags E_g.

The results given by Eqs. (15.36) and (15.37) are summarized in the following table. Included for comparison are the results obtained from Eqs. (15.12) and (15.13) for image-impedance operation. Since the two image impedances are equal, the insertion ratio for image operation is ϵ^θ, the insertion loss is α nepers, and the insertion phase shift is β rad (see Sec. 13.2).

COMPARISON BETWEEN RESISTANCE-TERMINATED AND IMAGE-TERMINATED LOW-PASS SECTIONS

$\dfrac{\omega}{\omega_c} = \dfrac{f}{f_c}$	Section operating between $Z_g = Z_R = \sqrt{L/C}$		Section operating between image impedances, Z_T	
	Insertion loss, db	Insertion phase shift, deg	Insertion loss (α), db	Insertion phase shift (β), deg
0	0	0	0	0
0.500	0.06	60.3	0	60
0.707	0.5	90	0	90
1	3.0	135	0	180
2	18.1	210	22.9	180
4	36.1	241	35.9	180
8	54.0	256	48.1	180
16	72.1	263	60.1	180

The tabulated results show that the resistance-terminated single-section filter is reasonably effective. In addition, the results are not so much different from the image-terminated behavior as might be expected. The most important difference is in the phase shift near cutoff and in the attenuation band. The resistance-terminated section has a phase shift that approaches 270° at frequencies far above cutoff. Deep in the attenuation band, the attenuation of the resistance-terminated section rises faster than that of the image-terminated one.

For frequencies somewhat beyond cutoff, the rate at which the insertion loss increases with frequency can be expressed conveniently in *decibels per octave*. When ω is somewhat greater than ω_c, Eq. (15.36) is, approximately,

$$\text{Insertion loss} \approx 10 \log_{10} \left(\frac{\omega}{\omega_c}\right)^6 = 60 \log_{10} \frac{\omega}{\omega_c}$$

If the frequency is now doubled, the argument increases by a factor of 2. Since the logarithm of a product is the sum of the logarithms, this produces an increase in insertion loss of

$$60 \log_{10} 2 \approx 60 \times 0.3 = 18 \text{ db}$$

Therefore, the insertion loss of the network of Fig. 15.10 increases at the rate of 18 db per octave for frequencies well above cutoff.

If the T section of Fig. 15.10 is replaced by the corresponding π section, the results are found to be identical with those derived above. That this is to be expected can be shown by the methods of Chap. 13, the reason being that Z_π is the reciprocal of Z_T with respect to R^2, and the two sections therefore turn out to have equal reflection factors at their terminals.

15.4. The m-derived Ladder Section. The constant-k filter section has two serious defects. First, its image impedance varies over a wide range, giving rise to reflections which are particularly serious near the edges of the pass bands. The reflections may either reinforce the output voltage or subtract from it and are particularly bothersome in a pass band, for they cause the ratio of output to input voltage to vary with frequency. This is in contrast with the idealized behavior for image termination, for which $\alpha = 0$ and $|E_{\text{in}}| = |E_{\text{out}}|$ throughout the pass band. The second defect is the small attenuation just outside the pass band, as illustrated by the graph of Fig. 15.4. In principle, this might be remedied by using a large number of sections so that the sum of their attenuations would be large, but this can be done only at the cost of employing an undesirably large number of circuit elements. (The attenuation is satisfactory, for frequencies well away from a pass band. A low-pass section operating at $f = 4f_c$ has an image attenuation constant of 4.13 nepers, or 35.9 db, per section.)

The so-called m-derived section is devised to fit, on an image-impedance basis, into a chain of constant-k sections and to provide a high attenuation just outside the pass band. It is not generally used alone, as its attenuation characteristics are not satisfactory at frequencies far from the pass band. An m-type section is said to be "series-derived" or "shunt-derived" depending on whether it is designed to match the mid-series or the mid-shunt image impedance of a constant-k section. The constant-k section from which it is derived is sometimes called the *prototype*.

In addition to complementing the attenuation characteristics of the k-type section, the m-derived type also provides an acceptable solution for the first-mentioned problem—that of the variation in image impedance with frequency. This is done by using the impedance-transforming properties of an m-type half section.

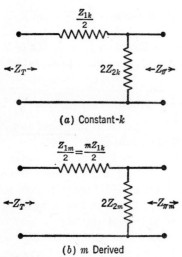

(a) Constant-k

(b) m Derived

FIG. 15.11. A constant-k half section and its corresponding series-derived m-type structure.

15.5. The Series-derived m-type Section and the Composite Filter.

The series-derived m-type structure is to have a mid-series image impedance identical with its constant-k prototype. Figure 15.11 shows a constant-k half section and its corresponding m-type structure. In the latter, the full series and shunt arms are designated as Z_{1m} and Z_{2m}, respectively. The mid-series image impedance can be expressed as

$$Z_T = \sqrt{Z_{\text{short}} Z_{\text{open}}} \qquad (15.38)$$

in which the open- and short-circuit impedances are measured from the left end. Suppose that we alter Z_{short} by a factor m. Then, to maintain the same Z_T, the impedance Z_{open} must be divided by the factor m. Since the short-circuit impedance of the half section is equal to the impedance of the series branch, the proposed alteration requires that

$$Z_{1m} = mZ_{1k} \qquad (15.39)$$

The necessary alteration of the open-circuit impedance is expressed by

$$\frac{Z_{1m}}{2} + 2Z_{2m} = \frac{1}{m}\left(\frac{Z_{1k}}{2} + 2Z_{2k}\right) \qquad (15.40)$$

Using Eq. (15.39) to eliminate Z_{1m} from this expression, we obtain for the new shunt arm,

$$Z_{2m} = \frac{Z_{2k}}{m} + \left(\frac{1 - m^2}{4m}\right) Z_{1k} \qquad (15.41)$$

Thus, the shunt arm has two parts in series, one having the same character as the original shunt arm, and the other like the original series arm. The resulting T sections are shown in Fig. 15.12. The element involving the factor $(1 - m^2)/m$ sets the requirement

$$0 \le m \le 1 \qquad (15.42)$$

in order to avoid negative inductances and capacitances.

Since Z_{1k} and Z_{2k} are inverse with respect to R^2 for the constant-k filter, the two are always opposite reactances. Therefore, the shunt arm of the m section has two reactances of opposite type. At some frequency these

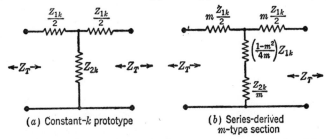

(a) Constant-k prototype   (b) Series-derived m-type section

FIG. 15.12.   Constant-k and series-derived m-type sections. Their mid-series image impedances are equal at all frequencies.

reactances will cancel, short-circuiting the filter and producing theoretically infinite attenuation. When this condition occurs, we have

$$\frac{(1 - m^2)}{4m} Z_{1k} + \frac{Z_{2k}}{m} = 0$$

or

$$\frac{Z_{1k}}{4Z_{2k}} = \frac{-1}{(1 - m^2)} \qquad (15.43)$$

where Z_{1k} and Z_{2k} refer to the arms of the prototype.

This can (theoretically) be made to occur at any value of $Z_{1k}/4Z_{2k}$ from -1 to $-\infty$ by appropriate selection of m. This range of selection covers the entire attenuation band of the prototype section.

We have devised the series-derived m-type section so that its mid-series image impedance is the same as that of its prototype. Its mid-shunt image impedance, however, will be different from that of the prototype. From Eq. (15.2) we can write

$$Z_{\pi m} = \frac{\sqrt{Z_{1m}Z_{2m}}}{\sqrt{1 + Z_{1m}/4Z_{2m}}}$$

Substituting from Eqs. (15.39) and (15.41), we obtain

$$Z_{\pi m} = \frac{\sqrt{Z_{1k}Z_{2k}}}{\sqrt{1 + Z_{1k}/4Z_{2k}}}\left[1 + (1 - m^2)\frac{Z_{1k}}{4Z_{2k}}\right] \qquad (15.44)$$

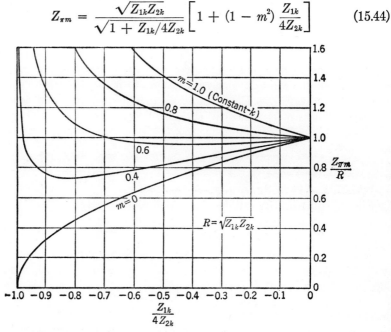

FIG. 15.13.   Variation of the mid-shunt image impedance of a series-derived m-type section.

This is equal to the mid-shunt image impedance of the prototype multiplied by the factor in brackets.   Equation (15.44) is plotted in Fig. 15.13 over the pass range $(-1 < Z_{1k}/4Z_{2k} < 0)$ for various values of m.   When $m = 1$, the m-type section reduces to the constant-k type, and we have the usual variation of Z_{π}.

Figure 15.13 shows that, when $m = 0.6$, the mid-shunt image impedance of the series-derived section is remarkably constant over most of the pass band.   This suggests the use of half sections with $m = 0.6$ at the two ends of a filter, as illustrated by the multisection filter of Fig. 15.14.   The constant-k section provides a high attenuation deep in the attenuation band. In cascade with this is an m-type section which has m selected to provide "infinite" attenuation somewhere just outside the edge of the pass band.

At the two ends are half sections with $m = 0.6$. At all frequencies, all these sections have the same image impedance (Z_T) at their junctions, and they have, of course, the same cutoff frequencies. However, at the input and output terminals of the filter, the image impedance is $Z_{\pi m}$ corresponding to $m = 0.6$. This matches the source and load resistances reasonably well over most of the pass band and minimizes the reflection effects at the

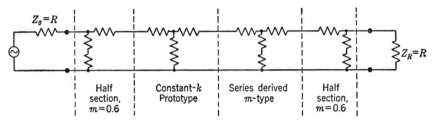

Fig. 15.14.   An example of a composite filter. All sections have the same Z_T.

terminals.   Numerous variations are, of course, possible in selecting internal sections for this type of filter.

15.6. The Series-derived Low-pass Filter.   The properties outlined in the preceding section will be illustrated here for the low-pass filter.   The series-derived m-type section is obtained from its constant-k prototype by

(a) Constant-k prototype   (b) Series-derived m-type section

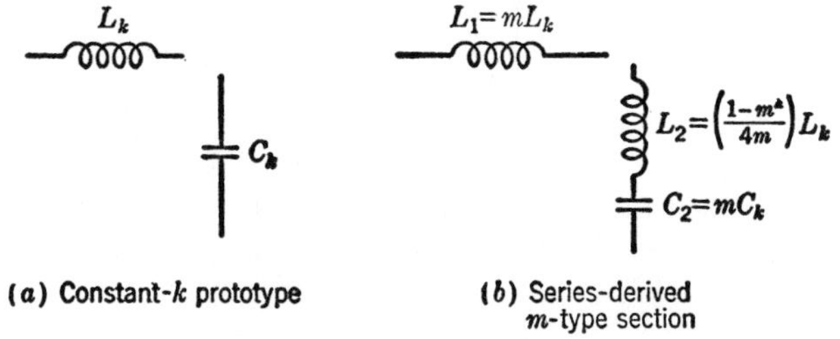

Fig. 15.15.   Low-pass ladder sections.

use of Eqs. (15.39) and (15.41).   Also, see Fig. 15.12.   The resulting configuration for the low-pass filter is shown in Fig. 15.15.

From Eq. (15.43), the shunt arm of the m section will be resonant and will produce infinite attenuation when

$$-\frac{Z_{1k}}{4Z_{2k}} = \frac{\omega^2 L_k C_k}{4} = \frac{1}{1 - m^2}$$

If we denote this frequency by ω_∞ and recall that cutoff occurs at $\omega_c = 2/\sqrt{L_k C_k}$, we can write

$$\frac{\omega_\infty}{\omega_c} = \frac{f_\infty}{f_c} = \frac{1}{\sqrt{1 - m^2}} \tag{15.45}$$

Solving for m, we obtain

$$m = \sqrt{1 - \left(\frac{f_c}{f_\infty}\right)^2} \qquad (15.46)$$

This equation is used to select the value of m which will provide high attenuation in a desired region.

The mid-series image impedance is, of course, the same as for the constant-k section, and is given by Eq. (15.11a) and Fig. 15.3. The mid-shunt image impedance is obtained from Eq. (15.44). Noting that $Z_{1k}/4Z_{2k} = -\omega^2 L_k C_k/4 = -(f/f_c)^2$, the result can be expressed as

$$Z_{\pi m} = \frac{R}{\sqrt{1 - (f/f_c)^2}}\left[1 - (1 - m^2)\left(\frac{f}{f_c}\right)^2\right] \qquad (15.47)$$

By the use of Eq. (15.45), the factor in brackets can be written as $[1 - (f/f_\infty)^2]$.

The value $m = 0.6$ provides a reasonably constant mid-shunt impedance over most of the pass band and, as shown by Eq. (15.45), provides in addition a peak of attenuation at

$$f_\infty = \frac{f_c}{\sqrt{1 - (0.6)^2}} = 1.25 f_c$$

Peaks of attenuation can be obtained at other frequencies by using additional m-type sections with other values of m.

Whereas the constant-k section has inverse arms and always has $Z_1/4Z_2 < 0$, this is not true of the m-derived section. In preparation for computing the image attenuation constant from the basic equations (15.4) and (15.5), we refer to Fig. 15.15 and write the expression for $Z_{1m}/2Z_{2m}$:

$$\frac{Z_{1m}}{2Z_{2m}} = \frac{j\omega m L_k/2}{j\omega\left(\dfrac{1 - m^2}{4m}\right)L_k - \dfrac{j}{\omega m C_k}}$$

Using the relation $\omega_c^2 = 4/L_k C_k$, this reduces to

$$\frac{Z_{1m}}{2Z_{2m}} = \frac{2m^2(f/f_c)^2}{(1 - m^2)(f/f_c)^2 - 1} \qquad (15.48)$$

This ratio is negative for frequencies lying between 0 and f_∞. At f_∞ the denominator becomes zero, and the ratio becomes infinite. Above f_∞ the ratio is positive. Referring back to Fig. 14.5, one can imagine a point that starts at the origin at zero frequency and moves to the left. At cutoff it reaches an abscissa of -1 and moves into the attenuation region. At f_∞

the point moves to extreme values at the left and returns at the extreme right side of the axis of abscissas. As the frequency increases beyond f_∞, the point moves toward the origin from the right side. Equation (15.5) can be used to compute the attenuation for $f_c < f < f_\infty$, and Eq. (15.4) can be used for $f > f_\infty$. The results of the computation are shown in Fig. 15.16 for several values of m. For $m = 1$, the section reduces to the constant-k type, and the peak of attenuation comes at infinite frequency. It will be seen that a small value of m causes a high attenuation just beyond nominal cutoff, but that the price paid for this desirable feature is a low attenuation further along the frequency scale.

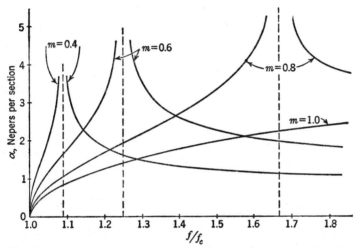

Fig. 15.16.   The image attenuation constants of m-type low-pass sections for several values of m.

Example: Low-pass Composite Filter.   Design a low-pass composite filter which is to work between resistive impedances of 500 ohms and which has a cutoff frequency of 4,000 cps. In addition to the peak of attenuation produced at $f_\infty = 1.25f_c = 5,000$ cps by the terminating half sections, a peak is desired at 4,500 cps.

For the prototype, we use Eqs. (15.14) with $R = 500$ ohms and $f_c = 4,000$ cps. This results in

$$L_k = \frac{R}{\pi f_c} = \frac{500}{4,000\pi} = 39.8 \times 10^{-3} \text{ henry}$$

$$C_k = \frac{1}{\pi R f_c} = \frac{1}{500 \times 4,000\pi} = 0.159 \times 10^{-6} \text{ farad}$$

For the terminating half sections, we refer to Fig. 15.15 and use $m = 0.6$. Then

$$L_1 = mL_k = 0.6 \times 39.8 \times 10^{-3} = 23.9 \times 10^{-3} \text{ henry}$$

$$L_2 = \left(\frac{1 - m^2}{4m}\right) L_k = \left(\frac{0.64}{4 \times 0.6}\right) \times 39.8 \times 10^{-3} = 10.6 \times 10^{-3} \text{ henry}$$

$$C_2 = mC_k = 0.6 \times 0.159 \times 10^{-6} = 0.0954 \times 10^{-6} \text{ farad}$$

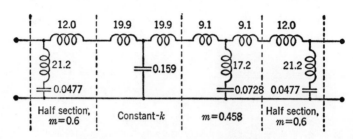

(a) Individual sections in the composite filter

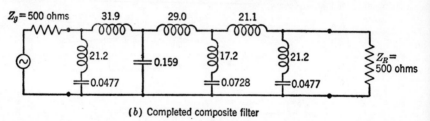

(b) Completed composite filter

FIG. 15.17. A low-pass composite filter. Values are indicated in millihenrys and microfarads.

To design a section which provides a peak of attenuation at 4,500 cps, we use Eq. (15.46):

$$m = \sqrt{1 - \left(\frac{4,000}{4,500}\right)^2} = 0.458$$

Referring again to Fig. 15.15 and using the new value of m, we find for this section,

$$L_1' = 0.458 \times 39.8 \times 10^{-3} = 18.2 \times 10^{-3} \text{ henry}$$

$$L_2' = \left(\frac{1 - (0.458)^2}{4 \times .458}\right) \times 39.8 \times 10^{-3} = 17.2 \times 10^{-3} \text{ henry}$$

$$C_2' = 0.458 \times 0.159 \times 10^{-6} = 0.0728 \times 10^{-6} \text{ farad}$$

The two full sections and the terminating half sections are shown in Fig. 15.17. In assigning values to the elements, it must be remembered that each half of a series arm has the impedance $Z_1/2$. Figure 15.17*b* shows the structure that is obtained when adjacent elements are combined to produce the least number of individual parts. The resulting attenuation of the composite filter, neglecting reflection effects at the terminals and assuming lossless elements, is shown in Fig. 15.18.

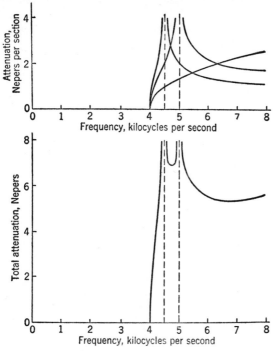

FIG. 15.18. Attenuation of the composite filter of Fig. 15.17, neglecting reflection effects.

15.7. Other Series-derived *m*-type Sections.

The general formulas given in Sec. 15.5 can be used in a manner similar to that just illustrated for the low-pass filter, to design *m*-type sections of the high-pass, band-pass, and band-elimination types. The details of the design relations will not be given here but can be found in a number of books. The tables of design information given in the references below will be found to be particularly useful:

Harold Pender and Knox McIlwain, "Electrical Engineers' Handbook—Communications and Electronics," pp. 7–46 to 7–52, John Wiley & Sons, Inc., New York, 1936.

T. E. Shea, "Transmission Networks and Wave Filters," pp. 291, 306, 315–318, 335; D. Van Nostrand Company, Inc., New York, 1929

F. E. Terman, "Radio Engineers' Handbook," pp. 228–231, McGraw-Hill Book Company, Inc., New York, 1943.

In these tables, it will be found that the configuration of a complicated arm is sometimes altered from that given in the derivations here. This is done to provide more favorable values of inductance and capacitance. In each case, the methods of Secs. 14.4 and 14.5 can be used to show that the new arm has the same reactance *vs.* frequency variation as the old.

15.8. The Shunt-derived m-type Section. The series-derived m-type section was devised to match on an image basis with T sections of the constant-k type. The shunt-derived m-type section is, on the other hand, devised to match with the π section. Figure 15.19 shows the two half sections which are to have the same Z_π. To derive the m section, the open-circuit impedance, as measured from the left, is divided by the factor m. This gives the relation

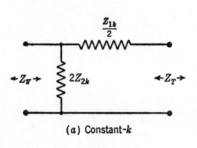

(a) Constant-k

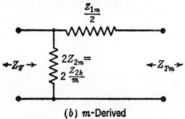

(b) m-Derived

Fig. 15.19. A constant-k half section and its corresponding shunt-derived m-type structure.

$$Z_{2m} = \frac{Z_{2k}}{m} \qquad (15.49)$$

To keep Z_π the same for both sections, the short-circuit impedance must be multiplied by m. This requires that

$$\frac{1}{Z_{1m}} = \frac{1}{mZ_{1k}} + \left(\frac{1 - m^2}{4m}\right)\frac{1}{Z_{2k}} \qquad (15.50)$$

Thus, the series arm consists of two parallel portions, one having the character of Z_{1k} and the other like Z_{2k}. In the attenuation band, the anti-resonance of the two parallel portions produces a (theoretically) infinite attenuation. The configuration is shown in Fig. 15.20.

The mid-shunt image impedance is now the same as that of the con-

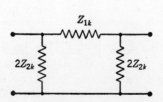

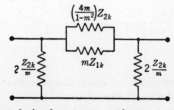

Fig. 15.20. Constant-k and shunt-derived m-type sections.

stant-k prototype.   The mid-series image impedance is found to be

$$Z_{Tm} = \frac{\sqrt{Z_{1k}Z_{2k}}\,\sqrt{1 + Z_{1k}/4Z_{2k}}}{1 + (1 - m^2)Z_{1k}/4Z_{2k}} \qquad (15.51)$$

This is plotted in Fig. 15.21.   The ordinates are reciprocal with respect to R^2 of those shown in Fig. 15.13 for Z_π of the series-derived section.

For a given value of m, the attenuation characteristics are the same as for the corresponding series-derived section.   A composite filter based on the shunt-derived m section is sketched in Fig. 15.22.   The relative con-

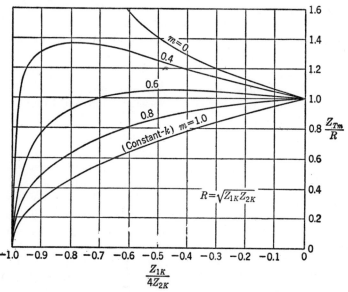

FIG. 15.21.   Variation of the mid-series image impedance of a shunt-derived m-type section.

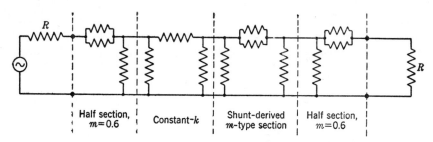

FIG. 15.22.   A composite filter designed around the shunt-derived m-type section. All sections have the same Z_π.

stancy of the mid-series image impedance with $m = 0.6$ is utilized in the terminating half sections. Design information will be found in the references given in the preceding section.

15.9. Other Filter Sections. The preceding sections have discussed the more common types of filters, but they by no means exhaust the possibilities. By starting with the m-type section, one can derive from it a so-called m-m' type, by a process similar to that used in going from the constant-k to the m-derived type. The "double m-derived" half section can be designed to have a more constant image impedance than that of the "single m-derived" section. Also, it can provide improved attenuation characteristics.[1]

Filter sections are sometimes employed which are neither constant-k nor m-derived. The sections are generally designed to fit either mid-series or mid-shunt with sections of the constant-k type. Example 2 of Sec. 14.2 is an illustration of this. Its mid-series image impedance is the same as that of the constant k. The more important possibilities are shown in the design tables of the references given in Sec. 15.7.

15.10. The Lump-loaded Cable as a Low-pass Filter. As was discussed in Sec. 2.9, an audio-frequency cable has an abnormally low inductance and high capacitance because of the proximity of the conductors within the sheath. The attenuation of the line can be reduced considerably, and the line can be made to approach more nearly a distortionless condition, by "loading" it with series inductance. On land lines, this is generally done by connecting series inductance coils at intervals. The effect of this was discussed in Sec. 2.9 under the assumption that the inductance was uniformly distributed. This analysis is reasonably accurate up to frequencies at which the number of coils per wavelength is small; beyond this, the lumpiness of the loading produces an effect like that of a low-pass filter.

In most lump-loaded cables, the inductance of the loading coils is much greater than the inductance of the line itself, and so the main effect of a section of line between two coils is only that of the capacitance. Viewed in this way, the line consists of a number of cascaded sections, each of which contains a lumped series inductance and an approximately lumped shunt capacitance. This makes the line a constant-k low-pass filter. We shall use the following notation:

L_l = the inductance added by the loading coils per mile of line

C = the capacitance of the line per mile

n = the number of loading coils, or "loading sections," per mile

[1] See Otto J. Zobel, Extensions to the Theory and Design of Electric Wave Filters, *Bell System Tech. J.*, Vol. 10, p. 284, April, 1931; and E. A. Guillemin, "Communication Networks," Vol. II, pp. 341–352, John Wiley & Sons. Inc., New York, 1935.

The inductance of one loading section is then L_l/n henrys, and the capacitance per section is C/n farads. Using Eq. (15.9b) for the cutoff frequency, we find

$$f_c = \frac{1}{\pi \sqrt{(L_l/n)(C/n)}} = \frac{n}{\pi \sqrt{L_l C}} \qquad (15.52)$$

At frequencies well below cutoff, the line behaves very much as though the inductance were distributed, and the usual smooth-line formulas can be used. At frequencies nearer cutoff, the performance can be analyzed by reducing the section of line between coils to an equivalent T section. The method of doing this was discussed in Sec. 4.8. Then, as indicated in Fig. 15.23, half the lumped inductance of a loading coil is added in series at each end, thus producing a correct T equivalent for the section. The terminal relations of the section can be computed from this equivalent circuit. The line will, of course, give rise to distortion at frequencies near cutoff. The number of loading coils per mile, n, is selected large

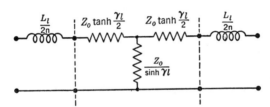

FIG. 15.23. The equivalent T network for one loading section.

enough to make the cutoff frequency occur somewhat above, or at least not lower than, the highest essential frequency to be transmitted.

PROBLEMS

1. Design a constant-k low-pass section with a cutoff frequency of 2,000 cps. The characteristic impedance at low frequencies is to be 1,000 ohms. Sketch the T and π configurations, and show values of the elements. Compute the image phase constant at 1,000 cps, and the image attenuation constant at 4,000 cps.

2. Show that, for a constant-k low-pass section, the quantity $d\beta/d\omega$ approaches the value $2/\omega_c$ at low frequencies. Observe that this is the reciprocal of the phase and group velocities (in sections per second) for signals that are well below cutoff, and that the time lag for these signals is therefore $2/\omega_c$ sec per section [see Eqs. (2.57) and (2.60)].

3. Design a constant-k high-pass filter with a cutoff frequency of 1,000 cps and a characteristic impedance that approaches 500 ohms at high frequencies. Sketch the T and π configurations and show values of the elements. Compute the image attenuation constant at 500 cps and the image phase constant at 2,000 cps.

4. Design a constant-k band-pass section with a lower cutoff frequency of 1,000 cps and an upper cutoff frequency of 2,000 cps. Use $R = 500$ ohms. Sketch the

T and π configurations and show values of the elements. At what frequency is $\beta = 0$?

5. Show that, for a constant-k filter, a pass band is obtained when $|Z_1| < 2R$.

6. Show that a constant-k low-pass section has $Z_1 = j2R(\omega/\omega_c)$ and $Z_2 = -j(R/2)(\omega_c/\omega)$.

7. A single low-pass filter section is connected as shown in Fig. 15.10. The cutoff frequency is 1,000 cps, and E_g equals 10 volts. Plot the magnitude of the receiving-end voltage as a function of frequency over the range $0 < f < 3,000$ cps.

8. A low-pass constant-k filter is composed of two T sections and is terminated in a resistance $R = \sqrt{L/C}$. It is driven by a generator that has an internal resistance equal to R.

a. Compute Z_T at $f = \sqrt{2} f_c$.

b. Use Eq. (13.11) to compute the insertion ratio of the filter at the frequency $f = \sqrt{2} f_c$. Find the insertion loss in decibels and the insertion phase shift. Observe that the quantity θ in Eq. (13.11) is the transfer constant for the whole network, not just for one section.

c. Repeat part *b* for a four-section filter.

9. Design an m-type series-derived low-pass section with a cutoff frequency of 10,000 cps. The frequency of infinite attenuation is to be 11,500 cps. Use $R = 700$ ohms.

10. Design a composite low-pass filter which consists of a prototype T section and two terminating half sections. The cutoff frequency is to be 7,000 cps. Use $R = 500$ ohms. Sketch the terminal image impedance *vs.* frequency over the pass band. Sketch the total image attenuation of the filter in decibels over the range $0 < f < 14,000$ cps.

11. Repeat Prob. 10, but use a prototype π section and two terminating half sections.

12. Design a composite high-pass filter which consists of a prototype T section and two terminating half sections. The cutoff frequency is to be 5,000 cps. Use $R = 700$ ohms.

13. Design a composite band-pass filter which consists of a prototype T section and two terminating half sections. The cutoff frequencies are to be 8,000 and 12,000 cps. Use $R = 500$ ohms.

14. Show that, in the attenuation band, the attenuation constant of an m-derived low-pass section is given by

$$\cosh \alpha = \left| \frac{(1 + m^2)(f/f_c)^2 - 1}{(1 - m^2)(f/f_c)^2 - 1} \right|$$

15. A certain telephone cable has $L = 0.001$ henry/mile and $C = 0.062 \times 10^{-6}$ farad/mile. Loading coils are added at intervals of 1.14 miles. Each coil has an inductance of 0.044 henry, and one coil is placed in series with each wire at the loading points. Determine the cutoff frequency of the loaded cable.

16. A certain telephone cable has $L = 0.0007$ henry/mile and $C = 0.062 \times 10^{-6}$ farad/mile. The inductance is to be increased to 0.0400 henry/mile by means of loading coils. If the cutoff frequency is to be 11,000 cps, what should be the spacing of the coils?

INDEX

Coupling, coefficient of, 265
Coupling network, 253
Crank diagram, 112–115
 use of chart as, 139
Crosstalk, 217–222
Crystal as detector, 182–184
Current ratio of network, 260–261
Current source, 245–246
Cutoff frequency, 306
 of loaded cable, 352–353
Cylindrical conductor, a-c resistance of, 78
 electric field of, 81–82
 high-frequency resistance of, 70
 internal impedance of, 77
 internal inductance of, 78
 magnetic field of, 80–81
 skin effect in, 71–79

D

Decibel, 43–45
Decibels per octave, 341
Decoupling sleeve, 159
Delta-wye transformation, 249–252
Depth of penetration, 58, 65
Determinant, 243
Differential equations, 8–9, 25, 32
Directional coupler, 188–190
Distance to fault, 123
Distortionless line, 48–50
 traveling wave on, 25
Distributed constants, 6
Double m-derived filter, 352
Double-stub tuner, 199–201
Driving-point impedance, 243–244

E

Earth return, 92
Energy on lossless line, 14
Envelope of standing-wave pattern, 124
Equalizing network, 325
Equivalent circuit, for sending end, 97
 of transformer, 264–268
Equivalent equilateral spacing, 227
Equivalent networks, 117–120, 255–257
Equivalent sources, 245–246
Equivalent T of network, 259
Equivalent Y and Δ, 249–252

Exponential line, 201–208
Exponential solutions, 93–101

F

Faraday's law, 62
Fault location, 123
Filter, 302
 band-elimination, 338
 band-pass, 335–337
 composite, 345, 351
 constant-k, 329–341
 high-pass, 334–335
 ladder, 303–312, 329–353
 lattice, 303–304, 321–325
 low-pass, 331–334, 345–349
 m-derived, 341–352
 pass bands of, 304–309, 330
 single-section, 339–341
Flat line, 42, 146
Flat network, 240
Foster's reactance theorem, 315–320
Four-terminal network, 253
 sending-end impedance of, 293

G

Geometric mean radius, 227–228
Group velocity, 52–55

H

H network, 254
Half-power points, 173
Half section, 277
Half-wavelength lines, 153
High-pass filter, 334–335
Higher modes, 5
 in coaxial cable, 89
Hybrid coil, 214–217
Hyperbolic functions, 101–103
Hyperbolic solution, 104–107

I

Image impedance, 271–275
 of band-pass filter, 337
 of high-pass filter, 335
 of L section, 276
 of ladder filter, 309–312